OLIGOPOLY AND THE THEORY OF GAMES

ADVANCED TEXTBOOKS IN ECONOMICS

VOLUME 8

NORTH-HOLLAND PUBLISHING COMPANY
AMSTERDAM · NEW YORK · OXFORD

OLIGOPOLY AND THE THEORY OF GAMES

JAMES W. FRIEDMAN

University of Rochester

1977

NORTH-HOLLAND PUBLISHING COMPANY
AMSTERDAM · NEW YORK · OXFORD

Library of Congress Catalog Card Number 76-16801
ISBN North-Holland for this series 0 7204 3600 1
ISBN North-Holland for this volume 0 7204 0505 X

Publishers

NORTH-HOLLAND PUBLISHING COMPANY
AMSTERDAM · NEW YORK · OXFORD

Sole distributors for the U.S.A. and Canada

ELSEVIER/NORTH-HOLLAND, INC.
52 VANDERBILT AVENUE
NEW YORK, N.Y. 10017

Library of Congress Cataloging in Publication Data

Friedman, James W.
Oligopoly and the theory of games.

(Advanced textbooks in economics; v. 8)
Bibliography: p. 301.
Includes index.
1. Oligopolies – Mathematical models. 2. Game theory. I. Title.
HD2731.F75 338.5′23 76-16801
ISBN 0-7204-0505-X

PRINTED IN THE NETHERLANDS

TO
Marcia

INTRODUCTION TO THE SERIES

The aim of the series is to cover topics in economics, mathematical economics and econometrics, at a level suitable for graduate students or final year undergraduates specializing in economics. There is at any time much material that has become well established in journal papers and discussion series which still awaits a clear, self-contained treatment that can easily be mastered by students without considerable preparation or extra reading. Leading specialists will be invited to contribute volumes to fill such gaps. Primary emphasis will be placed on clarity, comprehensive coverage of sensibly defined areas, and insight into fundamentals, but original ideas will not be excluded. Certain volumes will therefore add to existing knowledge, while others will serve as a means of communicating both known and new ideas in a way that will inspire and attract students not already familiar with the subject matter concerned.

The Editors

PREFACE

The purpose of this book is to provide a basic coverage of oligopoly theory and n-person nonzero sum game theory. While there is no pretense at being exhaustive, I have included what seem to me the main lines of work in these areas. One notable omission must be mentioned – differential games. I have passed by this topic because its coverage demands of the reader considerable mathematical tools used nowhere else in the book. Furthermore, though this area has great promise, there are not yet many results.

This book is aimed at the bright advanced undergraduate, the graduate student or economist who has some mathematics background. A knowledge of analysis at the level of a book like Bartle (1964) should be sufficient for the understanding of everything; however, a more modest background should suffice to allow the reading of most of the book. The level of difficulty does vary from place to place due to the inherent difficulty of the material.

Some new material, previously unpublished, may be found. Ch. 6 is entirely new. Though the results of chs. 10 and 11 are not new, the existence proofs are newly worked out. They follow familiar lines, however.

It is a pleasure to acknowledge the help of many friends, colleagues and students. William Fellner and Martin Shubik helped to start and to encourage my interest in the areas dealt with here. Over the years, conversations with Michael Farrell, John Harsanyi, Austin Hoggatt and Lionel McKenzie have influenced me in various ways. Comments on an earlier draft from Daniel Christiansen, Dipankar Dasgupta, Ikushi Egawa, Koji Okuguchi and Nasser Saidi have been helpful. The comments of Reinhard Selten and Akira Yamazaki have been especially helpful in revision. I greatly appreciate the patience and accuracy with which Marjorie Adams has typed and retyped the manuscript, and the aid of P. Srinagesh in proof-reading and making the index. All of these people have my thanks; however, I remain solely responsible for the errors which remain.

J. W. Friedman
Rochester, New York
18 March, 1976

CONTENTS

CHAPTER 1

INTRODUCTION AND OVERVIEW

1. Definition of a market

Oligopoly, hence this book, is concerned with the study of certain kinds of markets; and a market is usually defined with reference to some specific good or collection of goods. Examples are the treasury bill market, the bond market, the Atlanta housing market, the Gloucester fish market and the municipal produce market. These examples vary in two interesting ways. First, they vary with respect to the homogeneity of the goods traded within the given market. Compare the treasury bill market with the bond market or even the financial market. Second, some markets have a well defined and known physical location, while others do not. The Atlanta housing market is surely in Atlanta, but at what address? Perhaps at all the addresses of all the real estate brokers who deal in that market, plus the classified advertising departments of the Atlanta newspapers. By contrast, a local wholesale produce market has a single particular physical location where buyers and sellers know they must go to participate.

In most markets, a particular economic agent (that is, an individual or a firm) is always a buyer or always a seller. For example, individuals nearly always are sellers of labor and buyers of a long list of consumer goods. They do not generally sell meat or shoes or steam turbines unless that happens to be their normal line of work. A grocer will be a buyer in the local wholesale produce market and a seller in the retail produce market. He buys, but does not sell, electricity. There are markets and individuals which are exceptions. The most prominent are securities markets and those who buy and sell in them. It is normal for an individual to be a buyer of, say, common shares of a company on one day and a seller on another. Each person who trades in such a market may easily have in mind two prices for a given security: a price above which he will sell from his holdings and a price below which he will buy to add to them.

The markets in which an individual agent is always a buyer, or always a seller, tend to be markets in which the traded item is produced by the sellers and is consumed or used as an input to further production by the buyers. The markets in which a given agent may sometimes be a seller and sometimes a buyer are typically asset markets. Throughout this book, the appropriate sort of market to keep in mind is the first, in which sellers are producers and buyers are either consumers or firms which use the traded good as an input.

Another important difference among markets concerns the number and relative importance of participants. Markets in which there are a very large number of both buyers and sellers, and in which each participant is a party to only a negligible fraction of the transactions, are called *competitive.* The key characteristic of a competitive market is that no single participant can, by his own actions, have a noticeable effect on the prices at which transactions take place. So many agents are engaging in so many transactions that, at any given time, there is a *market price* at which trades take place. No seller need sell for less and no buyer need pay more. With participants divided into buyers and sellers, it is meaningful to address separately whether the buyers' side or the sellers' side of the market is competitive. When both sides are competitive, the market is said, without qualification, to be competitive. If the buyers' side were competitive, but the sellers' side consisted of only one firm, the market would be a monopoly. The reverse situation, one buyer and a competitive collection of sellers, is called monopsony.

Where the buyers' side of the market is competitive, the study of market equilibrium requires from the buyers only one form of information – the market demand function. The demand function relates quantity demanded to market price. Thus, for any given market price, the demand function indicates exactly how many units buyers wish to purchase. Similarly, where the sellers' side is competitive, there is a supply function, indicating for any market price the number of units which sellers are willing to provide. Market equilibrium in a competitive market is found by determining a price at which the market will clear. That is a price at which the amount the sellers wish to supply equals the amount the buyers wish to purchase. Equilibrium is characterized by the market clearing price and the amount traded at that price.

For a monopolistic market, it is necessary to know the seller's cost function. Then equilibrium is found by locating the price–quantity combination on the demand function which maximizes the seller's profits. It is usually assumed that the seller names a price and sells as many units as are demanded at that price. The price he names is that which maximizes his profits. A monopsonistic market may be described in a parallel fashion.

Thus far markets have been considered in which at least one side is competitive, and if one side is not competitive, it has only one agent. These markets have in common an absence of *strategic elements*. Finding the best decision for an agent involves straightforward maximization, with no need to be puzzled about how others might choose to act. For all competitive agents, prices are given and any amount can be bought or sold by the individual at those prices. The monopolist or monopsonist can set a price in his market and know precisely how many transactions will result. Compare to this a market in which the buyers' side is competitive and the sellers' side consists of two firms. For example, say they each make one model of refrigerator with the two models being exactly the same in overall size, but being different in their allocation of space between the refrigerating and freezing sections. Consumers' tastes differ. Some would be willing to pay more for seller A's product than B's; some, the reverse. It is possible to talk of the demand for A's refrigerators, but that demand will depend on B's price as well as A's (similarly for B). Neither can choose a profit maximizing price without knowing what the other chooses.

In a nutshell, this *strategic interdependence* is the source of what is called the oligopoly problem. The problem is not even solved by assuming the two firms collude. Even if they collude, they must settle on a mutually agreeable plan; and whatever pair of prices they choose, another pair can be found which gives more profit to firm A (at the expense of B), or vice versa. Then, too, perhaps it is not a feasible alternative for them to collude, which would be the case if each had an incentive to cheat on any agreement they might make and there were no way for them to enforce their agreements.

If the buyers' side of the market can have one, several or many agents and the sellers' can also; there are nine possible combinations. All except competition, monopoly and monopsony exhibit the strategic interdependence mentioned above. Of the rest, oligopoly and bilateral monopoly (one seller, one buyer) have received the most attention in the literature and, between them, exemplify the gamut of strategic problems which may arise. While bilateral monopoly is not explicitly discussed in this book, some of the models of game theory which are developed in chs. 7–12 are useful in understanding it. With respect to models of economic markets, only oligopoly is covered in any depth.

2. The market and the completeness of market models

It has already been noted that the buyers' side of an oligopolistic market can be characterized by a demand function, or perhaps n demand functions, one for each seller. The sellers' side can be characterized by n cost functions, one for each firm. These relationships could be derived from more fundamental sources: the demand functions from the preferences of the consumers plus their incomes and other prices, and the cost functions from firms' production functions and the prices of inputs. Such derivations are standard exercises in micro-economic theory and are not done here. The reader is referred to standard textbooks such as Malinvaud (1972).

A more serious source of incompleteness in the models studied in this volume stems from their partial equilibrium nature. The principal lesson from Walras (1954), dating back more than a century, is that the demand for a good is a function of the prices of all goods in the economy, and that the equilibrium prices of all goods are mutually determined by all the economic relationships in the economy. Similarly, because input prices would be determined by the general equilibrium of an economy, the cost functions of a firm depend on all prices in the economy, which, in turn, are interconnected with the behavior of the firms in any one market. Triffin (1940) argues from the Walrasian point of view on the validity of partial equilibrium market models.

Should partial equilibrium analysis be abandoned? Certainly not! Other things equal, it would be better to get one's oligopoly results in a general equilibrium setting; but other things are far from equal. Indeed, to get all the presently existing results in oligopoly theory within the context of fully disaggregated general equilibrium models implies a considerable generalization of both oligopoly theory and general equilibrium theory. A start on such a program is provided by Negishi (1961), Gabszewicz and Vial (1972), Nikaido (1975), Farrell (1970), Shitovitz (1973) and Marschak and Selten (1974). Within the confines of partial equilibrium, oligopoly can be studied with far greater breadth and depth than has proved possible otherwise – at least until now. The results so obtained are interesting and insightful, and worth the sacrifice of the narrower setting.

Basically, any theoretical structure is subject to the same complaint of narrowness. It is the very narrowness which gives the possibility of substantive results. It may be justly argued that the economic theory of consumer choice is greatly weakened because it takes no psychological or sociological phenomena into account. That is true in the sense that if it could take account of such things and also be developed to its present level of completeness and sophistication, it would be a much better theory than it is at present. This sort of argument may be extended indefinitely.

Thus the boundaries of oligopoly theory are drawn somewhat arbitrarily, reflecting the judgment of those who have worked in the area about the ways in which useful and interesting results may be obtained. Everyone must judge for himself whether the choices made are justified, and, hopefully, some will be encouraged to break new and important ground.

3. Some distinguishing characteristics of oligopolistic markets

In comparison with competition and monopoly, oligopoly is distinguished by *fewness.* As Fellner (1949, p. xi) writes: "Fewness is an important characteristic of the contemporary economic scene. Many prices and wage rates are determined under conditions which are neither atomistic nor monopolistic. They are determined under conditions of fewness: a few decision-making units shape their policies in view of how they mutually react to each other's moves." For each of the few, of the several firms making up a particular oligopoly, the decisions of any one have an observable effect on each of the others. Compare this to a competitive market, where the decisions of one economic agent have no observable effect on any firm or consumer. Or compare it to monopoly, where the decisions of the monopolist have a noticeable impact on the consumers, but no single agent's decisions have any effect on the monopolist. The mutual effects are absent.

Competitors and monopolists alike can choose their actions so as to maximize their objective functions (their profits, or utility, constrained by a budget set) knowing all the pertinent consequences of the actions of others. But for the oligopolist the optimal action depends on what his rivals choose to do, and what his rivals do depends on what they think he will do. This is seemingly an endless vicious circle, though, as will be seen later, it need not be.

There is an important distinction to be made between structural and behavioral interdependence. Sometimes models are set up in which the interdependencies sketched above are present, but the firms are assumed to ignore them. Ignored or not, they are present in the economic structure of an oligopoly and are what is meant by *structural* interdependence. In the refrigerator example used in §1 it is a fact of the model that when firm

A raises its price, the number of refrigerators demanded of firm B rises. That fact is not less true if firm B is somehow unaware of it. If firms in an oligopoly actually ignored other firms totally in their choice of decisions, behavioral interdependence would be absent. However, in oligopoly models it is usually assumed that both structural and behavioral interdependence are present. Generally, where structural interdependence is present, a lack of behavioral interdependence assaults common sense. An exception of this is where the number of firms is very large, their sizes fairly uniform and the interconnections between firms very small ("negligible"). The conditions of competition are approximately met, and it might be thought that the competitive model is an excellent approximation. It is intuitively appealing to imagine that as the number of firms increases, an oligopoly approaches a competitive model in the limit. Were this so, the structural connections would remain, but would get arbitrarily small. Eventually they would be small enough to be ignored for most purposes. As to the behavioral connections, it is plausible they would disappear after the industry reached some crucial size; however, it is an (interesting) empirical question if and when this takes place.

4. Games and their relationship to oligopoly

Just as the study of probability has some of its beginnings in the study of gambling, the theory of games is, to some degree, suggested by parlor games such as chess, bridge and tic-tac-toe. Parlor games have certain elements in common. When a game ends, a particular player has won, lost or tied. If a game is played for money, the total amount won by all winners combined equals the total amount lost by all losers. That is, in terms of money, the outcome is zero sum. The second distinguishing feature of parlor games is that they are games of strategy. Compare chess to solitaire. Solitaire is a one person game or *game against nature.* The player must act on a combination of knowledge and hunches about how cards are arranged; but he does not presume that there is a second conscious decision-maker whose actions depend on what he does. In chess, clearly one player, in choosing what to do, tries to guess how his opponent will react to the various moves he might make. Which of a player's choices looks best to him depends on how he thinks his opponent will play in response to each.

The simplest games are two person, zero sum. In such games, one player's loss is the other's gain and there is no way for two or more players to take advantage of others through joint action. Preserving the zero sum condition, with more than two players it is possible that *coalitions* (cooperating groups) may form and act jointly in such a way as to guarantee greater gain to themselves and greater loss to the others than they could ensure independently. To take a somewhat artificial example, imagine a game in which five people pay \$12 each and then divide the \$60 in any way which a majority agrees upon. Three players can decide to divide the money with \$20 going to each of the three and zero to the other two. The payoffs are then +8, +8, +8, −12, −12. A situation which could profitably be regarded as a two person nonzero sum game is a labor–management negotiation. The size of the pie to be divided is not fixed. The most obvious way to alter it is through a strike, which causes a time of low income to both sides simultaneously. Cooperation is necessary to avoid low payoffs (incomes) on both sides if the absence of a contract implies a strike.

For a monopolist or a competitor, including a buyer in a market which is competitive on the buyers' side even though it may be an oligopoly, the circumstances are analogous to those of the solitaire player. The individual has either no effect on others (a competitor) or an effect which is thoroughly predictable (a monopolist). The latter is meant to include the possibility of randomness, such as would obtain if a monopolist faced a demand function which has a stochastic component. The essential element is that there are no economic agents whose behavior would noticeably affect the monopolist and who also try to second guess what the monopolist will do before choosing their own actions.

An oligopoly is clearly interpretable as a game in which the players are the sellers. The mutual interdependence of their payoffs means they are in an n-person variable sum game. As with the labor–management example, the total profit summed over all the firms is sensitive to their actions; however, it is not necessary that the firms have any agreements with one another. Indeed, it may be impossible for them to do so.

The historical connections between game theory and economics are of some interest. The first important published result in game theory is the minimax theorem, which gives conditions for existence of an equilibrium for two-person zero sum games.[1] That was published by von Neumann (1928), and was followed by the influential treatise of von Neumann and Morgenstern (1944) in which discussion of economic applications is very prominent.

[1]The development of game theory was anticipated by Émile Borel who did not develop any significant results. See Borel (1953a,b,c) and Fréchet (1953a,b).

After being introduced to game theory, even in the very limited two-person zero sum form, one becomes sensitized to the *game theoretic point of view*, which is characterized by an awareness of strategic considerations and a desire to consider the action appropriate for an individual by first trying to see all the conceivable actions he might take, and second by looking for equilibria in which all players are behaving in ways which appear mutually consistent.

Before the birth of game theory two important precursors can be found in the literature of economics. The earliest is Cournot (1838) whose treatment of oligopoly is precisely in the modern manner of a non-cooperative n-person variable sum game.[2] This is seen clearly in chs. 2 and 7. The second is Edgeworth (1881), where he discusses the general equilibrium of a trading model and introduces the Edgeworth box, indifference curves, and the contract curve. Recall that he first considers equilibrium for a trading model with two goods and two traders and concludes that any trade could be an equilibrium if it satisfied two criteria: each trader should find his equilibrium consumption bundle to be at least as desirable as his initial endowment, and there should be no other trade which would make at least one trader better off than at the proposed equilibrium and make no one worse off. What is remarkable here is the complete absence of any reference to prices and the explicit recognition that the traders can make any trades they wish, constrained only by their endowments. Edgeworth then proceeds to enlarge the model by introducing more traders who are identical to the two with which he started. Letting the number of traders of each of the two types grow together, he shows that the number of possible equilibrium trades diminishes, and, in the limit as the number of traders goes to infinity, the only equilibrium trades which remain are those which are also competitive equilibria relative to some system of prices.

The prospects opened up by von Neumann and Morgenstern seemed immense, and for a few years many thought game theory would shortly revolutionize economic theory. In the nineteen fifties no revolution had yet taken place and it even appeared to some that the theory of games had contributed exceedingly little to economics. The pessimism was ill-founded and it remains to be seen whether the extreme optimism was overdone. It is now over three decades since the publication of the *Theory of games and economic behavior*, and it is apparent that the contributions of game theory to economics are major and continuing. With respect to

[2]References to Cournot are to the English translation, Cournot (1960).

oligopoly, the reader may form some judgment for himself from this volume. Another area in which large contributions have come is the theory of general equilibrium where the investigations of Edgeworth have been greatly expanded upon. The sort of progress, which is deep and makes lasting contributions, takes considerable time to develop and mature.

5. Plan of the book

Beyond this chapter, the book is divided into two parts which are, in turn, divided into several chapters each. The first part deals with traditional models of oligopoly treated in traditional ways, including some recent extensions; and the second part is devoted to game theory.

In part I, ch. 2 covers single-period models of the Cournot type. These are models in which all firms produce a perfectly homogeneous good and each firm's decision variable is its output level. Such models are the subject of ch. VII of Cournot (1960). Brief attention is also given in ch. 2 to price models of the Bertrand–Edgeworth type where the firms, as in Cournot, produce a homogeneous good; hence, consumers prefer to buy from the firm charging the lowest price. Chapter 3 is concerned with price models in which the firms make differentiated products. These models, which stem from Chamberlin (1956), are far more interesting than the Bertrand–Edgeworth models because the demand assumptions are more sensible. They are also more reasonable than the Cournot models, because most products appear to be differentiated and because price, rather than quantity, appears to be a firm's strategic variable.

Chapters 4 and 5 deal with extensions to dynamic models using *reaction functions*. A *reaction function* is a decision rule for a firm which gives its current period choice as a function of the observed choices of all the firms in the market in the preceding time period. These make an embryonic appearance in Cournot (1960) and there is further development in the nineteen twenties and thirties from Bowley and Stackelberg. Chapter 4 covers the pre-1960 developments and ch. 5 goes on to more recent work. In ch. 6 a model of a monopolistic firm is developed in which investment is explicit and time is continuous. Introducing adjustment costs when the firm changes the size of its plant, it is seen that the model naturally reduces to one in which time is discrete, with time periods of uneven length. The model is then expanded to a model of oligopoly.

Part II, comprising chs. 7–12, is taken up with game theory. The famous theorem of Nash (1951) for *noncooperative games* is presented, along with a number of useful, relatively straightforward variants of it. Chapter 7 is a game theoretic counterpart of chs. 2 and 3 in the sense that the game models are interpretable as single period. Chapters 8–10 all use the Nash type theorems of ch. 7 as a basic tool in building models of multi-period games. In ch. 8, as in chs. 4 and 5, the multi-period models are constructed by assuming temporal repetition of the single-period models with the same participants making decisions in all periods. In ch. 9 the model of ch. 8 is generalized to permit the sort of intertemporal connection between time periods which is typified by a market in which the demand for a product in the current time period is dependent upon past as well as current prices. Chapter 10, on stochastic games, generalizes single-period models in the manner of ch. 8, except that the sequence of games encountered over time by the players is affected by random elements whose distributions may depend on the actions of the players.

In chs. 11 and 12 various cooperative game models are examined. Ch. 11 deals with *value* models such as Nash (1950), (1953), Harsanyi (1963) and Harsanyi and Selten (1972). In this work, the approach is to determine a *value* to measure the contribution which each player makes to each coalition of which he is a member and to base his own payoff on these contributions. Core theory, which is presented in ch. 12 and based on Scarf (1967, 1971), is essentially a generalization of Edgeworth's contract curve.

6. *On the aims of this book*

Knowledge in economics accumulates and progresses in many different ways. Even knowledge within economic theory advances as a result of activity of many sorts, including formal model building, at one extreme, and informal (and informed) speculation at another. Almost exclusively, this book is in the tradition of formal model building, eschewing any serious direct concern with the empirical world or with questions of policy. The things which are avoided are not avoided from a belief that they are any less important than what is included. I am quite content to remain agnostic on the relative importance of these pursuits, though all are vital to the progress of the discipline. In my view, the material covered in the following chapters is important, interesting and very much worth communicating. Though I personally find it fascinating, it is possible to find it so without believing that it is the center of knowledge.

There are certain threads which run through the book: (1) the adherence to formal models, (2) the scope for noncollusive behavior, (3) the need to develop multi-period models, and (4) the importance of game theory as a tool for the theory of oligopoly. These are briefly addressed in turn.

(1) It is already admitted that formal models are not the only way we can "know" something of the nature of economic forces; however, such models are at the core of our general knowledge and inform our judgments in carrying out empirical research and making policy prescriptions. Often we think certain "results" are true of most formal models when, in fact, they are not. Only by the development, refinement and exposition of the formal structure of economics can our knowledge along these lines be furthered.

(2) Though this is a caricature, it is regrettably close to the truth that very many economists seem to believe that oligopolists behave like (single-period) Cournot oligopolists or they collude. On the one hand, collusion is difficult, sometimes impossible, to carry out; and, on the other hand, noncollusive behavior has a much wider scope than straight Cournot behavior. Indeed, as is shown in ch. 8, it is possible for firms engaging in reasonable noncollusive behavior to attain profits every bit as large as they could receive if they colluded.

(3) Oligopoly is certainly an area in which intuition suggests that optimal behavior in models which are explicitly multi-period is importantly different from optimal behavior in single-period models. To see whether, and how, this is true, there is no substitute for developing multi-period models and analyzing them. This is done in part I and part II. Both parts begin with basic, static one-period models which are later used to develop multi-period models. In part I, chs. 2 and 3 develop single-period models which are of interest both because they are typical of the single-period approach to oligopoly and because they form the building blocks for the multi-period models of ch. 4 and, more importantly, chs. 5 and 6. In part II, ch. 7 lays a foundation of models of noncooperative games which are reasonably interpreted as being single-period. These models form the building blocks of the multi-period, noncooperative games (*supergames*) which are taken up in chs. 8–10.

(4) Game theory is, of itself, a fascinating field. Its inherent interest is sufficient reason for devoting considerable space to it; however, there are

other reasons. The intimate connection between basic ideas in the theory of games and in economics, which is mentioned in §4, is no accident. Game theory influences the theory of oligopoly and one's understanding of oligopoly in two ways which are similar to the two ways that economic theory influences applied economics. These are through direct application and through the approach taken to the area. Some models of game theory may be directly applied to oligopoly models, yielding formal results for oligopoly. Some results may not, as yet, be applied; however, after studying game theory, one undergoes a permanent change in the way he perceives and contemplates game-like situations (such as oligopoly). This has a profound effect on how models are formulated, what is taken to be "reasonable" behavior, etc. In particular, one becomes very much more thoughtful about the *strategic element* in human behavior and quite impatient with superficial and *ad hoc* inventions to describe these strategic elements. It should not be overlooked that game theory is widely applied outside of oligopoly theory and the material presented in part II has considerable interest even for those who are not much concerned with the theory of oligopoly.

7. *A reader's guide*

The contents of the remaining chapters may be approximately characterized according to logical structure. Parts I and II may be read separately, though in reading one there is some gain from having read the other. Within part I, quantity models are covered in ch. 2 §§2–5, ch. 4 §2, §3 and ch. 5 §3, which stand together as a unit. The following may be read independently of one another: (a) ch. 4 §5, (b) ch. 5 §1, §2, §4 and (c) ch. 6; however, each uses the model developed in ch. 3 §3, §5 as a basic tool. In part II, ch. 7 §1 provides an introduction to everything which follows. Chapters 11 and 12 (cooperative games) may be read independently of one another, with no other background. Also, chs. 8, 9 and 10 (noncooperative games) may be read independently of one another; however, each uses ch. 7 §2, §3 as background. Within any chapter that is partly read, it is often helpful to read the opening and concluding sections. Also, in each of the chapters on noncooperative games (chs. 7–10), there is a section with comments on applications to oligopoly models.

I Traditional oligopoly models and their extensions

CHAPTER 2

SIMPLE QUANTITY MODELS

1. Outline of the problem

Throughout this book, the central focus is on equilibrium in oligopoly or game theoretic models. Among the ways models studied here can be divided is into *single-period* and *multi-period* models. In either context, the questions arise of how the firms would find the equilibrium and whether they might be expected to choose it. Such questions do challenge the practical validity of the equilibria. Surely if there is no way for the firms to find a particular equilibrium or there are incentives to keep them from going to it, it is cold comfort to know that the firms, should they ever find themselves at the equilibrium, would never seek to deviate from it.

There are two very good reasons for the study of equilibria as a beginning point. Intuition strongly suggests that equilibria exist and market processes, when out of equilibrium, tend toward one. Indeed, if these intuitions were generally false, it would be impossible to develop theory in economics. If one's aim were to develop a theory of how a market moved when out of equilibrium and he believed it likely that disequilibrium processes tended toward equilibrium, it would be most helpful to know toward what the disequilibrium processes were tending. A second reason for the study of equilibrium is that it is easier to pursue than the study of disequilibrium processes.

Single-period models have a quite different place in oligopoly theory than they have in value theory generally. In a competitive model, it is reasonable to follow Walras (1954) and consider static decision-making for consumers and firms within a time period as if there were no future. Even with capital formation in the model, the individual consumer's decisions may well be the same at a given time whether he takes the future into account or not. Oligopoly models differ due to the strategic interactions of the firms; because a natural source of information on how a firm is likely to behave today comes from how it has behaved in the recent past. Meanwhile, the expected behavior of other firms affects the current choices of a firm. Further, a firm is affected in its current choices by the way it believes its current choices alter the behavior of its rivals in the future. Considerations of this sort have been at the heart of discussions of oligopoly since the subject got its start with Cournot (1960).

Meanwhile, this chapter and the one following are devoted almost exclusively to single-period models, despite the arguments that such models are inadequate. There are both historical and practical reasons for this. One practical reason is that they form the basis for multi-period models. They also provide a simpler framework than the multi-period models for the introduction of important concepts, such as the non-cooperative equilibrium, which are important in the study of behavior in multi-period models. Historically, single-period models were the only sort dealt with in a moderately clear and consistent way until quite recently. Even from the beginning multi-period models were discussed in an *ad hoc* way; but such discussion would always have as a basis and starting point a much more clearly formulated single-period model. To understand the *ad hoc* attempts at dynamic (i.e. multi-period) models as well as completely formulated dynamic models, it is essential to understand the single-period models.

The class of models studied in the present chapter is, with variations of detail, the class studied by Cournot. It is characterized by a homogeneous product and pure competition on the buyers' side of the market. By *homogeneous product* is meant that each firm produces exactly the same good as any other within the industry. *Competition on the buyers' side of the market* means that a demand function for the good contains all the relevant information on buyers' behavior, if buyers are faced with one price at any time. The model allows for a fixed number of firms, n, with no entry or exit from the industry taking place, nor any capital formation. Through most of the chapter, until the discussion of Bertrand's (1883) critique of Cournot, the firms are all assumed to choose output levels with the market price being that at which buyers want to purchase the total quantity produced by the firms.

The structure of the model is completed with the cost functions of the firms. Questions of investment policy, growth of the firm and choice of production techniques either naturally do not arise or are put to one side. In a single-period model, investment and growth do not naturally come up. Choice of techniques may; however, it is assumed the firm always produces in a least-cost way and has calculated its cost function from its production function and the relevant market price information for its inputs. Thus the model of a firm consists of the market demand function and the firm's cost function.

In this and all succeeding chapters firms are assumed to have full information about themselves, the market and their rivals.[1] That is, a firm knows the demand function for its market, its own cost function and the cost functions of its rivals. If there are no stochastic elements affecting demand, a firm knows there are none. This is a condition of full information concerning structure. A firm cannot know that another firm will take a given action at some particular time in the future, though it may have good reason to suppose a given action and may plan its own behavior accordingly. Throughout the present chapter and all succeeding chapters, except the chapter on stochastic games, no randomness enters the structure of the models studied.

Equilibria may be divided into *cooperative* and *noncooperative*. For the moment these are provisionally defined, then these concepts are more fully discussed in §4, after some specific equilibrium concepts have been presented. To see the meaning of *cooperative*, imagine the firms in a market may, if they wish, get together with one another before making their output decisions and discuss the possibility of joint action – of coming to a mutually satisfactory agreement about what the production levels of all the firms should be. Further, assume that may even write a contract to which all are party which specifies the output level of each and every firm and that such a contract is enforceable. That is, if all of them agree to a set of output levels and all sign the contract, then there is no way for a firm to avoid doing what it agreed to do.

A market is called *cooperative* if it is possible for the firms in it to make binding, unbreakable agreements. Note that *cooperative market* is a structural concept. It depends on whether or not certain forms of behavior are technically possible and not at all on whether they are likely to be chosen. An equilibrium is called cooperative if it is an equilibrium which results from a binding contractual agreement.

At the other extreme from cooperative markets are markets in which the firms not only cannot make binding agreements, they cannot even make explicit informal agreements. This would hold if they had none of the usual means of communication open to them such as conversation, letters, etc. These markets are called *noncooperative* because coordination among the firms is ruled out from the start by the structure of the market itself. A *noncooperative equilibrium* is an equilibrium at which, in

[1]Unless the contrary is explicitly stated.

some clearly specified sense, the firms are doing as well for themselves individually as they can and each firm makes its decisions independently of every other. To say *independently* does not mean that a firm ignores the interactions which everyone knows are present in the market. It means there is no collusion among the firms. The question of what ought to be meant by *cooperative, noncooperative, collusion*, etc., recurs often. It is very important to an understanding of oligopoly theory, both to what people have written in the past and what they ought to have meant. After several equilibrium concepts have been introduced in §3, these questions are taken up again in §4.

The basic market model used in this chapter is developed in §2. §5 goes into the early forms of the reaction function which are found in Cournot, §6 is concerned with the criticisms of Cournot by Bertrand and followed up by Edgeworth, §7 with a few historical observations on Cournot's achievements and §8 contains summary remarks.

2. The market model

In the models of oligopoly considered in this chapter, each firm has a cost function, the industry has a demand function, it is impossible for any new firm to enter or old firm to leave the industry, and the only decision the firm is able to make is the choice of its output level. These characteristics are all structural. Behavioral considerations center about the decisions the firms make. Even in a single-period Cournot oligopoly, behavioral features are relatively simple and straightforward. It is only a question of how much to produce. But it is seen that even this question could conceivably have more than one answer. This is in contrast to a competitive trade model in which, as part of the structure, an individual always faces a known vector of market prices. In the face of these, he must decide what trades to make in the market. He wishes to end, after trade, with a most preferred (i.e. *utility maximizing*) consumption bundle. Even this condition on what is to be maximized is structural. The behavioral matter is how to achieve the best bundle. Exactly what should be traded for what? As prices are unaffected by the individual actions of one consumer, and the consumer knows prices, his own endowment and preferences – and he is assumed to have all the ability to calculate that he might need – it is a simple matter to determine the best attainable consumption bundle and buy it. Though the best bundle may not be unique, it makes no difference to him which of the best bundles is chosen.

The oligopolists considered in this chapter and in the rest of the book are assumed to be profit maximizers; however, the most profitable output level for one firm to choose depends on the output levels which others choose. What others choose depends on what they think the first firm will choose. What the first firm thinks the others will do depends on what the first firm thinks the others think it will do, etc. This situation is not so behaviorally simple as the competitive pure trade model. Here, optimal behavior depends on information which is not available and what one does, or is thought to do, can affect the behavior of others in ways which affect oneself.

In this section only the structural side of the model is described. The market is conceived as having n sellers, where n is finite and fixed. On the buyers' side, the numbers are *large*, with no single buyer important enough to have a noticeable effect on the market. The buyers cannot organize or collude. In other words, pure competition prevails on the buyers' side of the market. The output of one firm is impossible to distinguish from that of another, so price apart, buyers are indifferent from whom they buy. Their preferences for the output of the industry can be summarized by a demand function: $Q = F(p)$, where Q is the total amount demanded and p is the market price. This is a market demand function in the classic sense – if the price were p, then $Q = F(p)$ is the number of units which would be demanded, and which, if available, would also be bought.

It is assumed that the demand function is monotone, strictly decreasing and that it cuts both axes. This is illustrated in fig. 2.1. Because the firms are quantity choosers, it is more convenient to represent the demand function in inverse form: $p = F^{-1}(Q) = f(Q)$. Where references are made below to the *demand function*, what is meant is the inverse demand function, $f(Q)$. Formally, the conditions on $f(Q)$ are:

A1. *The demand function, $f(Q)$, is defined and continuous for all $Q \geq 0$. There is $\bar{Q} > 0$, such that $f(Q) = 0$ for $Q \geq \bar{Q}$ and $f(Q) > 0$ for $Q < \bar{Q}$. Furthermore, $f(0) = \bar{p} < \infty$, and, for $0 < Q < \bar{Q}$, f has a continuous second derivative and $f'(Q) < 0$.*

The internal aspects of the firms are modeled in the simplest possible way. It is assumed that the production function of the firm is fixed and input costs are given. These matters do not even enter explicitly because all relevant information they contain is summarized in the firms' cost functions. The minimal assumptions on them are that they imply

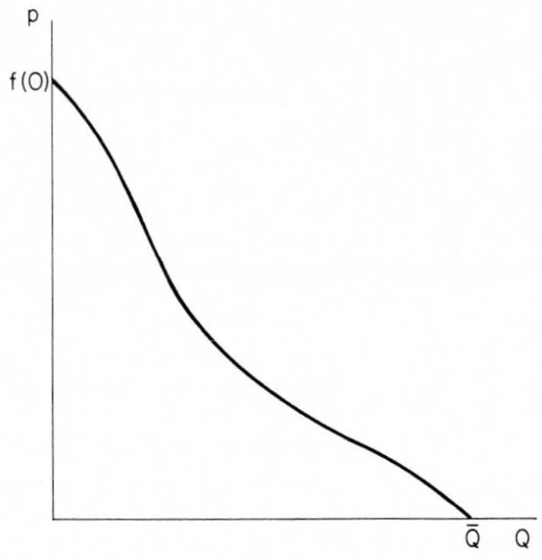

Fig. 2.1

nonnegative fixed cost and positive marginal cost. These conditions are stated formally below:

A2. *The cost function of the ith firm is denoted $C_i(q_i)$, where q_i is the output level of the ith firm. C_i is defined and continuous for all output levels $0 \leq q_i$. $C_i(0) \geq 0$. C_i has a continuous second derivative for $0 < q_i$ and $C_i'(q_i) > 0$.*

Recalling that $Q = \Sigma_{i=1}^{n} q_i$, it is now possible to write a profit function for each firm:

$$\pi_i(q_1, \ldots, q_n) = q_i f(Q) - C_i(q_i), \qquad i = 1, \ldots, n. \tag{2.1}$$

The operation of the market is this: each firm simultaneously chooses an output level. These together determine a total quantity of the (homogeneous) industry output on the market. This output sells at the market clearing price, which is determined by the demand function.

As this model has been formulated, a single firm never considers output levels less than zero or above $\bar{Q}$. The lower limit applies because negative output has no economic meaning in the model and the higher limit is effective because market price is guaranteed to be zero if industry output, much less the output of a single firm, is that high. At the other end, as industry output shrinks to zero, market price goes to a finite upper bound of $\bar{p}$. Thus the profit of a firm is bounded above and below. It cannot exceed, indeed cannot even reach, $\bar{p}\bar{Q}$ and it cannot fall below $-C_i(\bar{Q})$. It is reasonable to regard q_i as being restricted to the closed interval $[0, \bar{Q}]$. Then the output vector $\boldsymbol{q} = (q_1, q_2, \ldots, q_n)$ is restricted to the product of n such intervals $[0, \bar{Q}] \times [0, \bar{Q}] \times \cdots \times [0, \bar{Q}] = \mathscr{D}$, which form a connected and compact set on which the profit functions are continuous. $\mathscr{D}$ is a subset of $\mathscr{R}^n$, the *n-dimensional Euclidean vector space.* It is well known that under these conditions the set of all attainable profit levels for the *i*th firm is compact and connected. Letting $\boldsymbol{\pi}(\boldsymbol{q}) = (\pi_1(\boldsymbol{q}), \pi_2(\boldsymbol{q}), \ldots, \pi_n(\boldsymbol{q}))$, a vector of profits, the set of all attainable profit vectors associated with output levels in $\mathscr{D}$ is $\boldsymbol{\pi}(\mathscr{D}) = \{\boldsymbol{\pi}(\boldsymbol{q}) | \boldsymbol{q} \in \mathscr{D}\}$.[2] This may be summarized:

Theorem 2.1. *If $\mathscr{D} \subset \mathscr{R}^n$ is compact and connected, and $\boldsymbol{\pi}(\boldsymbol{q})$ is continuous for all $\boldsymbol{q} \in \mathscr{D}$, then $\boldsymbol{\pi}(\mathscr{D})$ is compact and connected.*

A *payoff space* or *attainable profit set,* $\boldsymbol{\pi}(\mathscr{D})$ is illustrated in fig. 2.2. It is drawn only to conform to the conditions in the theorem. It is possible that there are "holes" in the set, as well as the set having a peculiar shape. The set of points in $\boldsymbol{\pi}(\mathscr{D})$ which comprise the boundary from *a* to *b*, running through the positive quadrant, where the boundary has heavy dots has a special significance. These are the *Pareto*

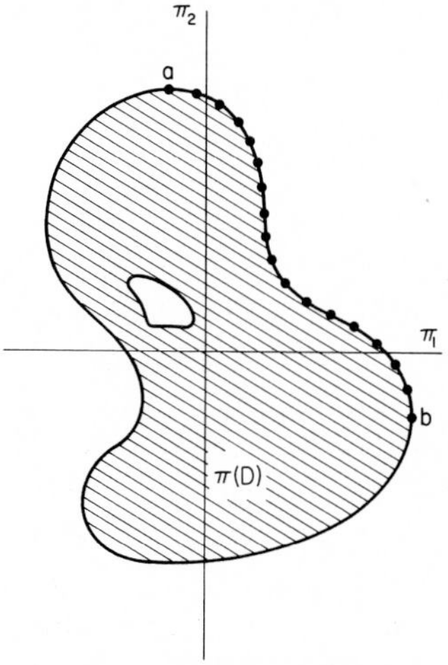

Fig. 2.2

[2]For the definitions and theorems used here, see a standard text on analysis such as Bartle (1964).

optimal points of the attainable profit set. In this book a weak, somewhat overly inclusive definition of Pareto optimality is used. A profit vector in $\boldsymbol{\pi}(\mathcal{D})$ is Pareto optimal if there is no profit vector in $\boldsymbol{\pi}(\mathcal{D})$ at which all firms have strictly greater profit. Before giving a formal definition of the Pareto optimal set, notation for vector inequalities is introduced. Let $\boldsymbol{v}$ and $\boldsymbol{u}$ be elements of $\mathcal{R}^n$. $\boldsymbol{v} \leq \boldsymbol{u}$ means $v_i \leq u_i$ $(i = 1, \ldots, n)$. $\boldsymbol{v}$ is not larger than $\boldsymbol{u}$, and, in particular, is not larger in any component. It is possible that $\boldsymbol{v} = \boldsymbol{u}$. $\boldsymbol{v} < \boldsymbol{u}$ means $\boldsymbol{v} \leq \boldsymbol{u}$ and $\boldsymbol{v} \neq \boldsymbol{u}$. $\boldsymbol{v}$ is not larger than $\boldsymbol{u}$, but is also not the same. It is strictly smaller in at least one component. It could be said that $\boldsymbol{v}$ is *weakly* smaller. $\boldsymbol{v} \ll \boldsymbol{u}$ means $v_i < u_i$ $(i = 1, \ldots, n)$. $\boldsymbol{v}$ is *strongly* smaller than $\boldsymbol{u}$; it is smaller in each component.

The Pareto optimal set is denoted $\mathcal{H}$ and is defined as $\mathcal{H} = \{\boldsymbol{\pi}(\boldsymbol{q}) | \boldsymbol{q} \in \mathcal{D}$ and for any $\boldsymbol{q}' \in \mathcal{D}$, $\boldsymbol{\pi}(\boldsymbol{q}') \not\gg \boldsymbol{\pi}(\boldsymbol{q})\}$. This definition allows the possibility that two vectors are in $\mathcal{H}$, say $\boldsymbol{\pi}(\boldsymbol{q}')$ and $\boldsymbol{\pi}(\boldsymbol{q}'')$ with $\boldsymbol{\pi}(\boldsymbol{q}') > \boldsymbol{\pi}(\boldsymbol{q}'')$. Thus $\boldsymbol{\pi}(\boldsymbol{q}')$ could give higher profits to some firms than $\boldsymbol{\pi}(\boldsymbol{q}'')$, yet still $\boldsymbol{\pi}(\boldsymbol{q}'')$ is defined to be Pareto optimal because $\boldsymbol{\pi}(\boldsymbol{q}')$ does not give higher profits to all firms. The reason for this definition lies in "technical" convenience. In later chapters some results are proved about Pareto optimality which depend on this definition. The reader should keep in mind that these results, like any other results, are limited by the conditions under which they are proven.[3]

3. *Equilibrium concepts for oligopoly*

There are four equilibrium concepts to be reviewed in this section. Only two have a history in the pregame theory literature of oligopoly. These are the Cournot equilibrium and the joint profit maximum. Another, the von Neumann–Morgenstern solution, comes from Mayberry *et al.* (1953) and Shubik (1959) and is based upon von Neumann and Morgenstern (1944), but it fits in well here because it only uses notions developed to this point and does not require any of the special machinery of game theory. The last of the equilibria, the efficient point, is found in Shubik (1959) and Mayberry *et al.* (1953) and is not an equilibrium in the usual sense that one might reasonably expect to find firms choosing efficient point output levels. Its interest

[3]Exit from the industry is not discussed anywhere in this volume; thus it is implicitly assumed that profits at any equilibrium point are sufficiently high that no firm would rather quit the market.

derives from comparisons between it and other equilibria, for it is the equilibrium which would result if the firms acted as if they were perfect competitors – that is, acted as price takers and equated price to marginal cost.

Prior to examing equilibria, particularly the Cournot equilibrium, it is essential to state clearly what is meant by a *single-period model*. It so often happens that in written or verbal discussions of oligopoly a single-period model seems to be under discussion at one moment, then a multi-period model the next; and the reader has no hint when the basis for discussion has changed.

There are two ways in which single-period models might be characterized. First, the firms are imagined to be in a market for one time period only. The market has no past history and no future prospects. The firms' profits are only those earned in the current (and only) period; hence, there is no need to consider the effects of a firm's present actions on the future choices of the others. Second, the firm may be imagined to be in the market for any number of periods; however, it must choose an output level at the beginning which it can never alter. The artificiality of having to make an output choice at one point in time and stay with it forever is so great that it may be just as well to use the explicit single-period assumption.

The very first expansion on single-period models to many periods is accomplished by assuming firms are in a market which will function over an extended time horizon and that they maintain a single-period mentality. The single-period mentality resides in the firm's belief that the current time period is the last or, what amounts to the same, the firm seeks to choose its current period output level so as to maximize its current period profits. It puts no weight on the future. It is aware of the past, knowing all past period choices of all firms, and may use this information to guide its current decision. This framework characterizes the first reaction function models which appear in Cournot's discussion.

3.1. The Cournot equilibrium

Revert now to the strict single-period model. There is no past, no future; hence, the output choices of a firm cannot be a reaction to or function of past actions of any firm. Communication and agreements among firms are ruled out, and the firms all choose their output levels simultaneously. The essence of the *Cournot equilibrium* is that it is characterized by a vector of output levels upon which no single firm can unilaterally improve. That is, if the output levels $\boldsymbol{q}^c = (q_1^c, q_2^c, \ldots q_n^c)$ are the output levels corresponding to a Cournot equilibrium, then it is impossible for firm 1 to increase its profit by changing from q_1^c, given the already chosen levels of the others, it is impossible for firm 2 to increase its profits by changing from q_2^c, given the output levels of the others, etc. Here and throughout the later chapters it is often convenient to write a vector of outputs, prices or strategies in a compact way which separates the choice of one firm from the others. This is done by writing $\boldsymbol{q} = (q_i, \bar{\boldsymbol{q}}_i)$ for output vectors and by an analogous notation for other variables.

The Cournot equilibrium, formally stated is: $\boldsymbol{q}^c$ is the output vector associated with a Cournot equilibrium if $\boldsymbol{q}^c \geq 0$ and $\pi_i(\boldsymbol{q}^c) = \max_{q_i} \pi_i(q_i, \bar{\boldsymbol{q}}_i^c)$, $(i = 1, \ldots, n)$. Under the assumptions made thus far (A1 and A2), the Cournot equilibrium need not exist, and if it exists, it need not be unique. Existence and uniqueness are taken up in ch. 4, where conditions for both are given.

In coming to an understanding of the Cournot equilibrium, bear in mind that there is no room for the "if I do this, he'll do that." kind of reasoning because each firm makes one and only one decision and does so at the same time as all the others. As a result, the choice actually made by one cannot affect the choices of others. Thus if firm 1 had specific hunches about the output levels firms $2, \ldots, n$ would choose, it could do no better than to choose that output level for itself which maximizes its profit, given the output levels it guesses for the others. Keep in mind, too, that the market structure is taken to be noncooperative; so the possibility of making joint plans is totally ruled out.

Now consider a hypothetical thought process for, say, firm 1. Imagine firm 1 to be considering whether firms $2, \ldots, n$ might be inclined to choose $q_2^e, \ldots, q_n^e$, respectively. First, firm 1 finds that his profit maximizing output is q_1^0 when the others use $q_2^e, \ldots, q_n^e$. Firm 1 might then check whether firm 2 would want to choose q_2^3 if he (i.e., firm 2) thought the others were choosing $q_1^0, q_3^e, q_4^e, \ldots, q_n^e$. The same check may be made for all the others – to see whether q_i^e is the profit maximizing output level for the ith firm when the others choose $q_1^0, q_2^e, \ldots, q_{i-1}^e, q_{i+1}^e, \ldots, q_n^e$. Assume there is at least one firm, i, for which q_i^e is not the profit maximizing output level. That firm would not choose q_i^e if its expectations of the others were $q_1^0, q_2^e, \ldots, q_{i-1}^e, q_{i+1}^e, \ldots, q_n^e$. Thus, either the ith firm cannot be assumed to choose q_i^e or it cannot have the same expectations concerning the other firms as firm 1 has. If firm 1 insists that its expectations should be consistent with those of every other firm and that each firm should, subject to its expectations, be maximizing its profit, then firm 1 cannot expect firm i to choose q_i^e. As a result, $(q_1^0, q_2^e, \ldots, q_n^e)$ cannot be an equilibrium.

Now consider the second possibility. For each firm, 2 through n, q_i^e is the profit maximizing output level when the others choose q_1^0, $q_2^e, \ldots, q_{i-1}^e, q_{i+1}^e, \ldots, q_n^e$. In this case, if all firms have the same expectations of one another, none has any incentive to choose a different output level from that which is expected of it. Then

$$(q_1^0, q_2^e, \ldots, q_n^e) = (q_1^c, q_2^c, \ldots, q_n^c) = \boldsymbol{q}^c,$$

the Cournot equilibrium output vector.

There is another way to view the Cournot equilibrium output vector which is useful subsequently. Imagine any arbitrary vector of outputs, $\boldsymbol{q}^0$. For each firm, i, there is a *best reply* to $\boldsymbol{q}^0$ which is the output level for that firm which maximizes its profits, given $\bar{\boldsymbol{q}}_i^0$ as the outputs of the others. Call it q_i^1. If firm i knew the others were to choose $\bar{\boldsymbol{q}}_i^0$, then it would choose q_i^1. For the n firms, taken as a whole, $\boldsymbol{q}^1$ could be called the best reply against $\boldsymbol{q}^0$. If $\boldsymbol{q}^0$ were a Cournot equilibrium, then $\boldsymbol{q}^0$ would be its own best reply – i.e., $\boldsymbol{q}^0 = \boldsymbol{q}^1$.

It is a characteristic of the Cournot equilibrium that the profits associated with it are not usually on the profit possibility, or Pareto optimal, frontier. With respect to the model formulated here, a more precise statement may be made:[4]

Theorem 2.2. *Let $\boldsymbol{q}^c$ be a Cournot equilibrium output vector for a market satisfying A1 and A2. If $\boldsymbol{q}^c \gg 0$ then the vector of profits $\pi(\boldsymbol{q}^c) = (\pi_1(\boldsymbol{q}^c), \ldots, \pi_n(\boldsymbol{q}^c))$ is not Pareto optimal.*

□[5] It is first shown that $Q^c = \Sigma_{i=1}^n q_i^c < \bar{Q}$, which implies that the conventional first-order conditions for the maximum of a function must hold. If $Q^c > \bar{Q}$ it is possible to reduce slightly the output level of a firm whose output is positive, which would reduce costs without lowering the firm's (already zero) revenue, thereby increasing profits. Should $Q^c = \bar{Q}$, such a firm could still lower its output slightly and increase its profits

[4]A similar result for symmetric (i.e. identical cost functions for the firms) duopoly may be seen in Roemer (1970).

[5]All proofs begin and end with the symbol □.

because its marginal revenue, $f(Q^c)+q_i^c f'(Q^c)$ is negative.[6] Therefore if $q_i^c>0$, the Cournot equilibrium is interior, implying that the conventional first-order condition is satisfied:[7]

$$\partial\pi_i/\partial q_i = \pi_i^i(q) = f(Q) + q_i f'(Q) - C_i'(q_i) = 0. \tag{2.2}$$

A simultaneous reduction in q_1 and q_2 would increase the profits of firms 1 and 2 and any others for which eq. (2.2) holds. This may be seen by noting that $\pi_i^j = q_i f'(Q) < 0$ (for $i \neq j$). Therefore it is possible to find a vector of output levels which yields strictly larger profits to each firm than it would receive at the Cournot equilibrium. □

In many of the succeeding chapters, for example chs. 3–5 and 7–10, the Cournot equilibrium is generalized in various ways. One of the generalizations, found in ch. 8, is an equilibrium in the spirit of Cournot which yields Pareto optimal profits. This proves possible because the *best reply* of a firm (player) is not a continuous function of the decisions (strategies) of the others.

3.2. *Two cooperative equilibria*

A comparison of the joint profit maximum as a cooperative equilibrium and the von Neumann–Morgenstern solution gives some insight into how oligopoly has benefited from game theory apart from specific substantive contributions, just by way of encouraging a more intelligent approach to the area. The joint maximum is, as the name implies, that output vector and associated profits at which the sum of the profits of the firms is maximized. Fig. 2.3 illustrates the joint maximum. It is on the profit frontier, marked *JM*, at the point of tangency with the broken line. The Cournot equilibrium is marked at π^c.

While it is true that at the joint maximum, the firms wring from the market the largest possible aggregate profit, it may easily be that to obtain this profit, the output levels chosen result in a peculiar distribution of

[6] At $Q^c = \bar{Q}$ derivatives of $f(Q)$ are not defined. Directional derivatives do exist, and, where derivatives are indicated, these are always directional derivatives. Context should make explicit which direction is meant.

[7] Throughout the book derivatives of functions of several variables are denoted by superscripts, and of functions of one variable, by primes. For example, $\partial\pi_i/\partial q_j = \pi_i^j$, $\partial^2\pi_i/\partial q_j\partial q_k = \pi_i^{jk}$, $\partial C_i/\partial q_i = C_i'$, etc.

profits among the firms. Note, for example, fig. 2.3 in which the lion's share of joint maximum profit goes to firm 1. Perhaps this is unacceptable to firm 2. It is also possible that the relative positions of the firms is such that the joint maximum depicted in fig. 2.3 gives unacceptably little to firm 1. The lesson to be learned here is that there is no reason why the joint maximum should obey any conditions of reasonableness or acceptability other than the one condition which defines it. Furthermore, it may actually be very difficult for several firms to agree on any particular Pareto optimal point, even though they all know it is in their mutual interest to agree on something.

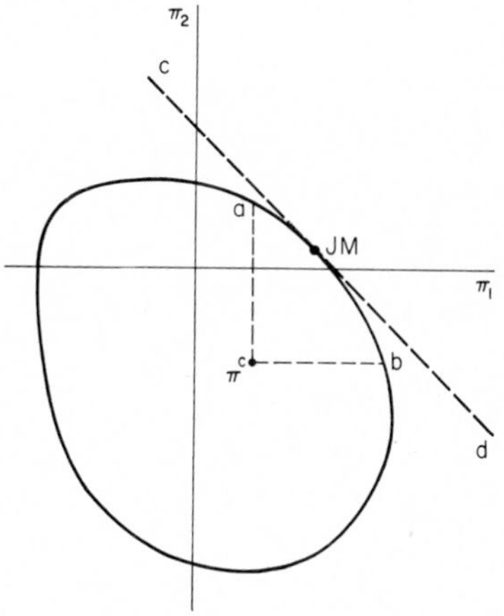

Fig. 2.3

There are two notable exceptions to the preceding remarks. The first is that if the firms have the ability to distribute a fixed total profit any way they see fit, then there is an argument in favor of producing at the joint maximum, followed by a redistribution of profits in a mutually agreed upon manner. The line *cd* in fig. 2.3 represents total profits equal to those at the joint maximum, so if redistributions were possible, they could be anywhere along *cd*. In an abstract single-period model there is no objection to making *side payments* which redistribute profits; however, in a fuller model with a long time horizon and perhaps some kind of consumer loyalty or explicit capital stock, an agreement which has some firms producing little or nothing may put them in an atrophied position over time. Then it later becomes possible for the others to reduce the subsidy to this group. In principle, account may be taken of this in advance.

The second exception is a symmetric market in which the several firms are absolutely identical. In such a case, the joint maximum gives equal profits to each firm and a natural sense of fairness makes this seem the only acceptable outcome. Unfortunately it is impossible to understate the importance of this case. The doubting reader is invited to think of any cases of near perfect symmetry he can among existing industries.

The *von Neumann–Morgenstern solution* consists of all the Pareto optimal points which cannot clearly be thrown out.[8] There is a problem deciding upon what rules to disallow Pareto optimal points. One, which I find unreasonable, is to look at the highest profit a firm could guarantee itself, irrespective of what output levels the others might choose and disallow any Pareto optimal points which give less than that to the firm. Do this for all firms, and the points which remain are all regarded as possible equilibria. This procedure makes the set of equilibria too large in the sense that the profit a firm can guarantee itself if others are seeking to minimize its profit, is very low; and assumes unreasonably malevolent behavior on the part of the others. This is especially true considering that each firm is simultaneously taking the same paranoid attitude; while all any firm really wants is the largest profit for itself that can be attained.

Another possibility is to include as equilibria those Pareto optimal points at which no firm gets less profit than it gets at the Cournot equilibrium. The justification for this is that, in the absence of a joint agreement, the firms can be expected to choose Cournot output levels; hence, any Pareto optimal point which does at least as well for every firm cannot be ruled out. In fig. 2.3, the equilibria under this rule would be all profit frontier points from a to b.

It might be objected that the method of the preceding paragraph is a suitable beginning but does not take account of the ability of subgroups of firms, less than n, to make agreements among themselves. Such subgroups (coalitions) may be able to obtain profits greater than they get at the Cournot equilibrium if they cooperate with one another and, vis-a-vis the other firms, behave noncooperatively. Further discussion may be found in Shubik (1959) and von Neumann and Morgenstern (1944).

[8]This name is given by Shubik (1959, p. 67); however, it is really the core rather than the von Neumann–Morgenstern solution found in von Neumann and Morgenstern (1944). See ch. 12 §1.

3.3. *The efficient point*

It is a commonplace of the theory of general equilibrium under competition, in the absence of externalities, that price should equal marginal cost in all markets as part of the conditions for efficiency. In an oligopolistic market it is to be expected that price is not equal to marginal cost. This is illustrated by the equilibrium condition, eq. (2.2) for the Cournot equilibrium where marginal cost is equated with $f(Q)+q_if'(Q)$, which is marginal revenue for the firm and is less than price ($f(Q)$). While there is no reason to expect firms to act as if they are competitors, taking price as given when it actually is influenced perceptibly by the actions of each, it may yet be interesting to have an idea of the extent to which a market deviates in observed behavior from what would be seen if price were equated to marginal cost by all firms.

The *efficient point* is characterized by the output vector for which price equals marginal cost for all firms simultaneously:

$$f(Q)=C_i'(q_i), \qquad i=1,\ldots,n. \tag{2.3}$$

It is possible that, as the number of firms in the market gets larger, the Cournot equilibrium converges to the efficient point. This question is examined in §3.4. Were one to have a cooperative equilibrium of interest, defined for any n, it would be possible to measure the relative nearness of the Cournot equilibrium to the cooperative equilibrium as compared with the efficient point as a means of indexing the relative inefficiency of the Cournot equilibrium. One such measure is d_{ec}/d_{ev}, where d_{ev} is the distance from the efficient point to the cooperative equilibrium, say the nearest von Neumann–Morgenstern solution point, and d_{ec} is the distance from the Cournot equilibrium to the efficient point.[9] This measure may be sensitive to the parameters of a specific model as well as to n, the number of firms. It could be computed in either profit or output space. The efficient point, as an equilibrium for oligopoly is, to the best of my knowledge, introduced by Mayberry *et al.* (1953). It is also discussed in Shubik (1959). As a final note on the topic, it is conceivable that for a sufficiently large, finite n, firms which had hitherto chosen Cournot output levels may choose efficient point levels because the cross effects among the firms may be small enough that they ignore them.

[9]As a distance measure, Euclidean distance is probably most appropriate here.

3.4. Equilibrium in a Cournot market as the number of firms increases

Intuition suggests that a Cournot oligopoly converges to a competitive market as the number of firms in the market increases without limit. Such convergence has two aspects: on the one hand, the Cournot equilibrium would be expected to converge to a competitive equilibrium (i.e. to the efficient point equilibrium), and, on the other, it would be expected that the total output in the industry would increase with the number of firms. The latter comes from a widely held belief that under oligopoly output is restricted as compared with what it would be under competition. As with many such issues in economics, the correct answer is "it all depends . . . ." For an interesting range of models, the results do hold, but not for all models of interest.

An example will illuminate some of the important points. Let $f(Q) = a - bQ$, where a and b are positive constants, and let C_i be the same for all firms: $C_i = cq_i$. The ith firm's profit is

$$\pi_i = q_i(a - bQ) - cq_i, \tag{2.4}$$

and the first-order conditions for the Cournot equilibrium are

$$\pi_i^i = a - bQ - bq_i - c = 0, \qquad i = 1, \ldots, n. \tag{2.5}$$

Taking advantage of the symmetry of the model to note that in equilibrium all firms have identical output levels, eq. (2.5) can be easily solved for the Cournot output levels: $q_i^c = (a - c)/(n + 1)b$ and $Q^c = n(a - c)/(n + 1)b$. The efficient point output for the market is $Q^e = (a - c)/b$. In this example the Cournot equilibrium does converge to a competitive equilibrium, with aggregate output rising, as the number of firms increases.

Even this simple example need be altered very little to destroy the results. Change the cost functions to $C_i = d + cq_i$, where d, the fixed cost, is positive. It is now necessary to add to the equilibrium conditions explicit recognition that a viable equilibrium requires nonnegative profits for all firms. The marginal conditions for equilibrium are unaltered by the addition of a fixed cost. Profits per firm in an n-firm Cournot equilibrium are $\pi_i^c = (a - c)^2/(n + 1)^2 b - d$; and the requirement that they be nonnegative is equivalent to $n \leq (a - c)/\sqrt{(bd)} - 1$. The larger is d, the smaller the maximum number of firms that can have nonnegative profits at a Cournot equilibrium.

Inquiring into the nature of fixed costs, presumably they are related to the size of the firm's *plant* or, more generally, its capital stock. If so, it is natural to suppose that capital per firm would be less, the more firms there were in a market. At the same time, the firm's variable costs should also be related to the level of fixed cost. Making c, the firm's constant marginal cost, a function of d, rising as d falls, makes it no longer obvious that aggregate market output at the Cournot equilibrium rises as the number of firms increases. Undoubtedly, with freedom to choose the exact relationship between c and d, either result could be obtained.

An important lesson to learn here is that it is important to be very careful about what is meant by allowing the number of firms to grow. Exactly how is the market modeled? How do the firms change when their numbers change? Results are very sensitive to the precise assumptions made. Shubik (1959), Shapley and Shubik (1969), Shubik (1970) and Ruffin (1971) discuss the effects on market equilibrium resulting from increasing numbers. Ruffin gives a number of additional references and his analysis, based on a symmetric market, is extended to an asymmetric market by Okuguchi (1973).

4. *On the meaning of cooperation, collusion and noncooperative behavior*

In §1 cooperative and noncooperative markets are defined with reference to two polar cases. Concerning the cooperative, it is noted that the firms could make contractual agreements with one another which are legally binding in the sense that once made, they literally could not be broken. An equilibrium is said to be cooperative if it results from a binding contractual agreement. A noncooperative market is defined as one in which the firms cannot make binding agreements, and, in fact, cannot make agreements in any explicit fashion because they cannot communicate with one another by talking, writing, etc. Between these two sets of conditions lies a very large territory to which attention is turned soon.

Making an agreement means giving up some freedom of action. This would only be done when others are also restricting their choices and the prospects in the market for each firm are improved as a result of all the restrictions taken together.

Between the extreme of complete communication and binding contracts on one side and the total lack of both on the other lies (a) partial communication and (b) contracts which are not binding, though they alter the conditions under which firms act.[10] Literally there is probably no such

[10]On these issues, see also Telser (1972, p. 179).

thing as a 100 percent binding agreement; however, for practical purposes a binding agreement may be thought of as one which changes payoffs in such a way that it is necessarily more profitable to keep than break the agreement. Imagine, for example a one year property lease at a rent of $100 per month with the additional condition that the renter posts $1500 with a bank. The bank is instructed to pay to the owner at the end of the lease period any money, plus interest, which the renter failed to pay. The leftover funds revert to the renter when the contract is satisfied. The renter can refrain from paying some of his rent, but, should he do so, the bank will pay for him with his own money. A more usual contract, less strongly binding than the preceding, is an ordinary apartment lease under which the renter pays at the beginning a deposit of an amount equal to one or two month's rent, and then is to pay rent monthly over the term of the lease. Should the renter stop paying, the owner must go to some expense and trouble to evict him; however, as a practical matter, it is impossible for the owner to get more money from the renter than he has already paid. Imagine a renter who decides to move out of his apartment half a year before the lease is over and who intends to cease paying rent. The owner can no doubt keep the deposit and he may be able to rent the apartment sufficiently soon that he loses no income; however, he probably cannot get more money than the deposit plus previously paid rent from the first renter.

Applied to firms in a market, if the firms can make agreements which are not literally enforceable, but which do provide some penalties for noncompliance, then firms can be expected to honor their commitments if keeping them is more profitable than breaking them. Remember the present context is that of a one-period market. In a broader context, two sources of loss from breaking agreements are (1) that one may not be much trusted in the future, which could cause some cost or inconvenience, and (2) that one may value honoring contracts as part of one's code of behavior.[11] Neither of these considerations would preclude the breaking of contracts; they merely raise the cost of doing so.

In some countries contracts between firms in the same market covering commercial policy are illegal. That deprives firms of the two enforcement mechanisms, legal action through the courts and use of a third party such as a bank. In the United States it is even illegal for the representatives of two or more firms to discuss the possibility of agreements, formal or informal.

[11]See Selten (1975), where this incentive to keeping agreements is discussed.

Imagine a market in which agreements have no legal standing, but where firms can freely communicate with one another. It is often argued that the opportunity to communicate is enough to put firms in a position to make agreements which raise their profits at the expense of the general public. That view goes back at least two hundred years to Adam Smith (1937) who wrote (p. 128):

> People of the same trade seldom meet together, even for merriment and diversion, but the conversation ends in a conspiracy against the public, or in some contrivance to raise prices. It is impossible indeed to prevent such meetings, by any law which either could be executed, or would be consistent with liberty and justice. But though the law cannot hinder people of the same trade from sometimes assembling together, it ought to do nothing to facilitate such assemblies; much less to render them necessary.

That Smith thought unenforceable agreements may be made to work is implicit in his decision to comment on the subject. That view is, I believe, widely shared and, in the way it is held, mistaken. The issue can be only partially dealt with here. The remainder must wait for ch. 8. Staying strictly within the single-period framework, what makes a firm carry out the action to which it agreed? Say, for example, that the firms decide on a vector of output levels which lead to a mutually acceptable point on the profit frontier. The mutually agreed upon outputs are $q_1^0, \ldots, q_n^0$. After agreeing, each firm, in isolation from all the others, actually makes its output decision. The first, if it really believes the others will do as they said, has every incentive to choose q_1 so as to maximize π_1 on the assumption that the other output levels are $\bar{\boldsymbol{q}}_1^0$. But firm 1 may realize that the others are likely to think as he thinks; hence, they cannot be expected to choose $q_i^0 (i = 2, \ldots, n)$. Thus, unless $\boldsymbol{q}^0$ is a Cournot equilibrium output vector, there is no reason to suppose that an agreement to choose it will be honored. On the other hand, were it a Cournot equilibrium, if unique, there would be no reason to make agreement. Should there be several Cournot equilibria, an agreement on one of them would be honored because each firm would have every reason to suppose the others would do as they promised and the firm itself would, under the circumstances, do as it promised.

Anticipating somewhat the discussion to come in ch. 8, imagine a market which operates for more than one time period. There is now a possibility for threats to be made and carried out; whereas, in a single-period market, after a firm learns others did not do as they promised, it is too late to do anything about it. If, for example, the market ran for two periods, with a fresh output decision made in each, firms could make an agreement for the two periods at the outset and if, after the first period results are in, some have not done as they said, there is time for reprisals from the others. In the Cournot model, any firm can threaten to produce $\bar{Q}$, forcing the market price to zero and preventing every firm from making positive profits. The threat, then, is to adhere to the agreement in period 1 or else the threatening firm will produce $\bar{Q}$ in period 2.

Alas! The threat is not to be believed. Consider the position of the threatener if one of the others violates the agreement. He has promised to produce $\bar{Q}$. If the others believe him, their best choice is to produce zero; because that minimizes their costs, while their revenue would, in any case, be zero. If the threatener believed the others would produce zero, he really should not produce $\bar{Q}$. They would not find out what he actually chose until after they were committed; hence, the threatener would be free to choose his output in the way which would maximize his second-period profit, given that the other firms all have zero output.

But when the other firms think through the position of the threatener, they see how he reasons and that he has no incentive to produce $\bar{Q}$. So they would not find it best to produce zero. No matter who makes what threats, when the second period comes, it is no different from a one-period market in isolation, which means the only credible threat is to produce the output level associated with the Cournot equilibrium.

In a single-period market when firms can only make agreements which are not binding and have no contractual costs associated with violation, cooperative equilibria appear to be ruled out. The only viable equilibria are the Cournot equilibria. Should there be more than one of these, there can be agreement on which to choose. In a two-period market, the second period is identical to a single-period model; hence, the only possible play is the Cournot equilibrium for the second period. But settling the second-period choices in a way which is independent of what is done in the first period means that the first period also stands alone as if there were no second period. This leaves the Cournot equilibrium prevailing for both periods. It is clear that the same reasoning would hold for a market like that considered in this chapter which functioned for any finite number of periods.[12]

[12]Evidence from experiments suggests that this *backward induction* argument does not account very well for actual behavior. See, for example, Selten (1974), Friedman (1963) and, particularly, Rapoport and Chammah (1965). The latter two references report experiments in which the subjects' time horizon is finite, yet apparently cooperative behavior is observed for a large fraction of the time that games are in progress.

5. The beginnings of analysis with reaction functions

In a quantity model a reaction function for a firm is a function which associates a current output choice for the firm with output levels chosen at some time in the past. One common form which will be considered in this section associates the current period output of the firm with the previous period output levels of all the other firms. Following Cournot (1960) and later writers such as Bowley (1924) and Stackelberg (1934), the dynamics are implicit rather than explicit. That is, the variables are time subscripted.

There is a particular reaction function for a firm associated with Cournot. It is found by solving the *i*th firm's first-order condition, eq. (2.2), for the Cournot equilibrium for q_i as a function of $\bar{\boldsymbol{q}}_i$.[13] This function, $w_i(\bar{\boldsymbol{q}}_i)$, is called the *Cournot reaction function.* w_i is the best response of the *i*th firm to any given $\bar{\boldsymbol{q}}_i$ of the others. Let $\boldsymbol{w}(\boldsymbol{q}) = (w_1(\bar{\boldsymbol{q}}_1), \ldots, w_n(\bar{\boldsymbol{q}}_n))$. Then for any output vector, $\boldsymbol{q}^1$, $\boldsymbol{w}(\boldsymbol{q}^1) = \boldsymbol{q}^2$ is the vector of best responses to it. These functions can easily be used to characterize a Cournot equilibrium: $\boldsymbol{q}^c$ is the Cournot equilibrium output vector if it is the best reply to itself, i.e. if $\boldsymbol{q}^c = \boldsymbol{w}(\boldsymbol{q}^c)$. Cournot (1960) describes the reaction functions by means of a diagram, fig. 2.4

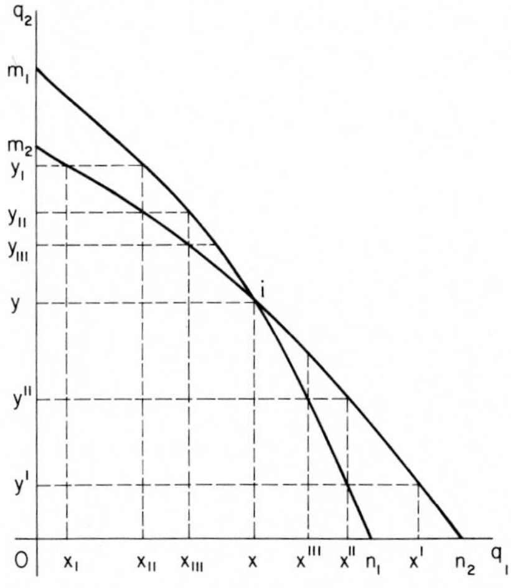

Fig. 2.4

[13]If the conditions of the implicit function theorem are met, which are the second-order conditions for a maximum (i.e. that $\pi_i^{ii} < 0$), then it is possible to solve for w_i. For the time being it is assumed the conditions hold. On the implicit function theorem see Bartle (1964) or Dieudonné (1960).

which illustrates a duopoly model for which the reaction functions are

$$q_1 = w_1(q_2), \tag{2.6}$$

$$q_2 = w_2(q_1). \tag{2.7}$$

On page 81, he writes:[14]

> Let us suppose the curve m_1n_1 (fig. 2.4) to be the plot of eq. (2.6), and the curve m_2n_2 that of eq. (2.7), the variables q_1 and q_2 being represented by rectangular coordinates. If proprietor 1 should adopt for q_1 a value represented by ox_1, proprietor 2 would adopt for q_2 the value oy_1, which, for the supposed value of q_1, would give him the greatest profit. But then, for the same reason, producer 1 ought to adopt for q_1 the value ox_{11}, which gives the maximum profit when q_2 has the value oy_1. This would bring producer 2 to the value oy_{11} for q_2, and so forth; from which it is evident that an equilibrium can only be established where the coordinates ox and oy of the point of intersection i represent the values of q_1 and q_2.
>
> The state of equilibrium corresponding to the system of values ox and oy is therefore *stable*; i.e. if either of the producers, misled as to his true interest, leaves it temporarily, he will be brought back to it by a series of reactions, constantly declining in amplitude, and of which the dotted lines of the figure give a representation by their arrangement in steps.

Clearly there is the notion of each firm choosing its output as a response, or reaction, to the output chosen by the other, which means that at this point Cournot is discussing a multi-period process. The analysis is neither carried far nor made entirely explicit. For example it is not evident whether Cournot has in mind that the firms make their decisions (1) simultaneously or (2) alternately. Both alternatives lead to difficulties. It can be argued that alternate period decision is not viable and is also nonsensical. The former is argued in ch. 6, where a continuous time model of the firm is presented in which the firm chooses to make new decisions only at intervals rather than continuously varying

[14]In this passage the symbols are altered to conform to those used in the chapter. In quoting the works of others, notation is generally changed to conform to the notation of this volume.

the values of its decision variables. When the model is expanded to an oligopoly, it is found that in equilibrium the firms still choose to make decisions at intervals, and they all choose the same identical points in time to make their new decisions. The alternate period decision version also begins to appear strange when the model is generalized from two to n firms. Then the ith firm can change its output in periods i, $n+i$, $2n+i$, etc. The artificiality of this round robin decision-making is hard to defend.

Presuming simultaneous decision leads to other difficulties. Notice that the behavior choice of the firm is no longer its output level. It is the rule by which the output level is to be chosen. The rule which Cournot assumes for firm 1 is w_1, in particular that $q_{1t} = w_1(q_{2,t-1})$, where t denotes time period. Similarly for firm 2, $q_{2t} = w_2(q_{1,t-1})$ is the supposed behavior rule. The shortcoming is that w_1 is not a best reply to w_2 and, conversely, w_2 is not a best reply to w_1. That robs an equilibrium based upon these two reaction functions of an internal consistency without which the equilibrium is much less interesting.

Consider the market from the point of view of firm 1. Assume that w_2 does describe the behavior of firm 2. Firm 1 knows that firm 2 behaves according to w_2; for if firm 2 should do otherwise, firm 1 would observe it from the market information he receives. It is immediate that, except under very special circumstances, w_1 cannot be optimal for firm 1. Choosing $q_{1t} = w_1(q_{2,t-1})$ maximizes the tth period profits of firm 1 on the condition that $q_{2t} = q_{2,t-1}$. Except at the Cournot equilibrium, this condition will not hold. In fact, the decision rule which would be optimal for firm 1, given that firm 2 uses w_2, is $q_{1t} = w_1(w_2(q_{1,t-1}))$.

An *ad hoc* defense of the Cournot reaction functions, w_1 and w_2, may be constructed by assuming that they would lead very quickly to output levels which are negligibly far from the Cournot equilibrium; hence, the firms would not have enough time to perceive that their behavior was nonoptimal and change it. After only a few periods, the firms appear to be keeping their output levels constant, so, as a working assumption, constancy is not bad.

This defense has as a weakness that it is imprecise and too much contrived. There remains another objection which can be raised as soon as time explicitly enters the model. Why does the firm ignore the future? Should future profits, discounted perhaps, be of interest? This issue is addressed in chs. 4 and 5, which are devoted to reaction function models.

6. Bertrand's objections to the Cournot model

Joseph Bertrand, a noted French mathematician, reviewed Cournot's *Researches*... in 1883, some 45 years after it was published. His remarks on the oligopoly chapter are reproduced in full below:[15]

> The results [of Cournot's book] seem of minor importance; sometimes they appear unacceptable.
> Such is the study, made in chapter VII, of the fight between two entrepreneurs who own identical mineral springs and have no fear of other competition. Their interest would be to merge or at least to fix a common price which would maximize their joint profits; but this solution is rejected. Cournot assumes that one of the competitors will lower his price to attract all the customers to himself, then the other, to get them back, lowers his even more. They will stop this procedure only when each of them, even if his competitor would give up the fight, would not gain anything by lowering his price further. One decisive objection presents itself: with this hypothesis, the lowering of prices would have no limit. What common price would be adopted if, when one of the competitors lowers his, he attracts to himself, neglecting unimportant exceptions, all of the sales and he doubles his profits if his competitor does nothing. If the formulas of Cournot mask this obvious result, it is because, inadvertently treating the quantities sold by the two competitors as independent variables, he assumed that if one entrepreneur changes his, the other will keep his constant. The opposite is indicated by the evidence.

Bertrand criticizes Cournot's oligopoly model on two grounds. The first is that the firms would collude, rather than act noncooperatively. The second is that, by some accident, Cournot used output levels rather than prices as the independent variables of the firms. The main body of the passage is taken up with a description of the consequences of noncooperative behavior if prices are assumed to be the independent variables. Bertrand's comment that the firms should collude may be dismissed out of hand, for there is no evidence that collusion is so easy as his remark suggests. Indeed, the empirical record suggests that collusion is quite hard to achieve when binding agreements cannot be made. See, for example, Herfindahl (1959).

[15] Translation mine.

That price is the decision variable Cournot intended is suggested by Cournot's use of price as the independent variable in the two chapters on monopoly, chs. V and VI; however, on p. 80 he writes: "Instead of adopting $Q = F(p)$ as before, in this case it will be convenient to adopt the inverse notation $p = f(Q)$;" indicating a deliberate choice of quantity. Probably the change of variables was made in order to avoid having to deal with the model Bertrand outlines in his review. Bertrand, like Cournot, clearly makes an assumption that the firms produce a perfectly homogeneous good; hence any consumer prefers to buy from the firm charging the lowest price. Very likely Cournot thought along the same lines and decided the model with prices as independent variables is less interesting or less tractable than the model he chose to work with.

There is an important objection to quantities as the decision variables of oligopolists, with a market price being determined by the total quantity produced. By what mechanism is the market price established? It is no problem imagining firms choosing prices and either producing output to order or producing for inventory. But what institutional arrangements accomplish price determination if the firms do not choose prices? There is, also, an objection to the Bertrand formulation. In the world, one is hard put to find markets in which all consumers want to buy from the firm charging the lowest price. To the contrary, examples are endless of apparently stable markets in which different firms charge different prices and all have positive sales. Virtually any consumer good is an example. In these markets it seems that small price changes by a firm lead to small changes in its sales and in the sales of its rivals. Of course, Bertrand is merely taking up the homogeneous goods assumption made by Cournot and tracing through, correctly, the logical consequences of price choice. Given that differentiated products had not come into the theory of oligopoly yet, one is faced with a choice between Cournot's version in which firms use the "wrong" variable and the model behaves reasonably, and Bertrand's version in which firms use the "correct" variable and the model behaves absurdly. I opt for the Cournot version because it does give a much better account of how an oligopoly might function. In ch. 3, where differentiated products price models are taken up, it is seen that the insight gained from such models is much more nearly approached by the Cournot than the Bertrand version. Bertrand's analysis was expanded upon by Edgeworth (1925) and has been further developed by Shubik (1959).

7. Historical comments on Cournot

Cournot's book, which is very modest in size as well as scope, is concerned with the partial equilibrium analysis of markets in a static setting. He looks at monopoly, oligopoly and competition – in that order. In all cases, the buyers' side of the market is represented by a demand function. These demand functions are assumed, not derived; thus Cournot is not directly concerned with utility and consumer preferences. From the sellers' side, the most fundamental data are cost functions for the firms. As with the demand functions, the cost functions are taken as given, not derived from anything more basic.

Having thus limited his scope, he formulates explicit mathematical models which differ in a very important way from any mathematical modeling which preceded his own. He does not specify particular functional forms. He uses general functions which may be subject to some restrictions, but exact forms are avoided. For example, a demand function is not, say, $Q = a - bp$, linear with positive coefficients a and b; rather, it is $Q = F(p)$, where F is differentiable, with a negative first derivative, etc. This practice is commonplace now; however, Cournot was the first to do it, and it was certainly not suggested by other sciences. The most prominent source from which an example might be set is physics, where specific functional forms abound to explain and measure physical phenomena. That works for physics, but it does not work for economics. Cournot was the first to perceive this and eventually his example prevailed.

His handling of simple monopoly is essentially the same as is found today in many intermediate level price theory textbooks. The same holds for his oligopoly model. Monopoly and competition were known and discussed in his time, though not with anything like his clarity and rigor, but oligopoly does not appear in the earlier literature. It seems that the topic was launched by Cournot, probably because it appeared as an intermediate step between monopoly and competition. That discovery and the progress he made with it must rank as one of the outstanding achievements in the development of economics.

In passing, it is worth noting that Cournot is often regarded as having been neglected. This is apparently because of the 45 year lapse between publication of the *Researches* in 1838 and Bertrand's review in 1883. However, Cournot was a friend of Walras' father and a great influence on Walras. In this connection, see Walras (1954) and Schumpeter (1954). Marshall apparently read Cournot about the year 1868 and gives Cournot and von Thünen credit for important influence on him. See Marshall (1975, v. 1, p. 40) and the preface to the first edition of the *Principles* [Marshall (1961, p. x)]. Effects of this sort, through the middle of the nineteenth century, are a peculiar form of neglect.

The quantity model used throughout this chapter is said to be a *Cournot model*. The assumptions used are not exactly Cournot's, so, to see more precisely what he accomplished, it is useful to review his own assumptions. The demand function is introduced (p. 49), assumed continuous, monotone decreasing (p. 50) and differentiable (p. 53). Cournot points out here that the revenue function $pF(p)$, is at first increasing, then decreasing (i.e. *quasi-concave*), and that its maximum is characterized by the condition $F(p)+pF'(p)=0$.[16] He has also assumed that as p goes to zero, $F(p)$ goes to some finite upper limit, which assures a maximum to the revenue function.

His approach to oligopoly is seen well in the first part of ch. VII (pp. 79–80):[17]

> To make the abstract idea of monopoly comprehensible, we imagined one spring and one proprietor. Let us now imagine two proprietors and two springs of which the qualities are identical, and which, on account of their similar positions, supply the same market in competition. In this case the price is necessarily the same for each proprietor. If p is the price, $Q = F(p)$ the total sales, q_1 the sales from the spring (1) and q_2 the sales from the spring (2), then $q_1 + q_2 = Q$. If, to begin with, we neglect the cost of production, the respective incomes of the proprietors will be pq_1 and pq_2; and *each of them independently* will seek to make this income as large as possible.
> We say *each independently*, and this restriction is very essential, as will soon appear; for if they should come to an agreement so as to obtain for each the greatest possible income, the results would be entirely different, and would not differ, so far as consumers are concerned, from those obtained in treating of a monopoly.
> Instead of adopting $Q = F(p)$ as before, in this case it will be convenient to adopt the inverse notation $p = f(Q)$; and then the profits of proprietors (1) and (2) will be respectively expressed by $q_1f(q_1+q_2)$, and $q_2f(q_1+q_2)$, i.e. by functions into each of which enters two variables, q_1 and q_2.

[16]Let $y = f(x)$. $f(x)$ is said to be *quasi-concave* if, for any x^0 in the domain $\mathscr{D}$ of f, the set $\{x | x \in \mathscr{D}, f(x) \geq f(x^0)\}$ is convex.

[17]The italics are in the original.

The "spring" is, of course, the famous mineral spring of the zero cost duopoly model for which Cournot is most remembered. Cournot's model readily generalizes to n firms and nonzero costs with each firm having its own cost function, rather than all having identical costs. The generalization is done by him later in the chapter. It is true that the essentials of the model and of what he has to contribute can be expressed adequately in terms of symmetric (i.e. identical cost) duopoly.

It is clear from the passage quoted above that Cournot seeks explicitly to avoid analysis of cooperative behavior. He points out that the intent to maximize profits on the part of the firms, and their acting independently is (pp. 80–81) "... equivalent to saying that q_1 will be determined in terms of q_2 by the condition

$$d[q_1 f(q_1+q_2)]/dq_1 = 0,$$

and that q_2 will be determined in terms of q_1 by the analogous condition

$$d[q_2 f(q_1+q_2)]/dq_2 = 0,$$

whence it follows that the final values of q_1 and q_2, and consequently of Q and of p, will be determined by the system of equations

$$f(q_1+q_2)+q_1 f'(q_1+q_2)=0, \qquad (1)$$

$$f(q_1+q_2)+q_2 f'(q_1+q_2)=0." \qquad (2)$$

While the derivatives are implicitly assumed to exist, as well as the solution to Cournot's eqs. (1) and (2), his conditions are nearly a subset of the conditions given in §3.1 for a Cournot equilibrium.[18] Cournot then goes into a discussion, quoted and discussed above in §5, whose meaning is not entirely clear. It may be intended only as a stability argument or as an attempt to inject dynamic elements into the model, introducing reaction functions.

It should be clear from what has been reviewed that the presentation of the Cournot model in this chapter does not break ground beyond Cournot (though some topics are discussed which he did not get into) and is probably not significantly superior for clarity or rigor.

[18]The only point of exception is that Cournot allows $C_i'=0$; whereas, in A2, it is required that $C_i'>0$.

8. Summary comments

This chapter has dealt mainly with the single-period Cournot quantity model, partly for historical and partly for practical reasons. The model is still very much taught and written about; hence, there is considerable historical interest in seeing just what the model is and what results Cournot wrung from it. It is also an excellent vehicle to use to introduce the nature of oligopolistic interdependence and of equilibria which use the *best reply* principle, such as the Cournot equilibrium. The *best reply* principle is seen over and over again. It characterizes all equilibria which are noncooperative, and even plays a role in some cooperative equilibria.

Cournot's market model is broadly similar to the differentiated products single period model to be studied in ch. 3, and serves as an introduction to and standard of comparison for it. Both the model of this chapter and that of the next may be used as the basis of a multi-period model, as will be seen in chs. 4 and 5. Even some of the equilibria of the single-period models may carry over; though, in a multi-period model, more equilibria become possible.

CHAPTER 3

SINGLE-PERIOD PRICE MODELS WITH DIFFERENTIATED PRODUCTS

1. *Product differentiation*

A firm's product is *differentiated* if it is very much like that of certain other firms, but exactly like that of no other. Generally one thinks of a differentiated product as being obviously of a particular commodity classification. A suitable example is tennis shoes, a reasonably clearly defined item, made by many firms. The tennis shoes made by one firm are not quite the same as those made by another. Meanwhile, other commodities are not, in general, so easily substituted for tennis shoes as one brand of them is for another. Virtually all consumer goods are differentiated products.

Informal discussions of product differentiation sometimes are concerned with whether certain goods are "truly" differentiated. These goods often fall under two headings, and, from an economic standpoint, both are differentiated. The first are goods which are, in a sense, identical, but which are sold in combination with other goods and services and cannot be separated from them. The second are goods which may be identical from any *objective* viewpoint one wishes to use, but which are not regarded as identical by those who buy them.

An example of the first category is gasoline. Assume for the moment that no matter where it is bought, gasoline is chemically identical. Still the product is differentiated, if only because of the location of the sellers. In addition to location, in the process of buying, the consumer has dealings with employees of the seller who may be more or less polite and pleasant, and who may choose to render fewer or more peripheral services in the course of selling the gasoline. All the circumstances of the purchase form part of the product (or bundle of products) which are bought. It is possible that these fringe matters vary so little from one firm to another for certain products that they may be left out of account. Nonetheless, there are commodities for which they are quite important.

Now consider a product available in profusion from any single seller – aspirin. It is claimed by some that the product is chemically identical from one supplier to another. Others claim this is not so, and in the stores widely varying prices can be observed as between aspirin sold under well-known brand names and aspirin sold under the store's own label or a brand hardly anyone has heard of. Is aspirin differentiated, even if all bottles carry chemically identical products? Clearly, from an economic standpoint, it is a differentiated product. Many consumers think the tablets in the jar with the well-known label are substantively different from the tablets in other jars. Insofar as one wishes to model the world more or less as it is, it is unnecessary to worry about the chemical differences or have hair-splitting arguments about what differentiation "really is." Aspirin is an example of economic differentiation, whatever the chemical considerations. If a campaign of education changes consumer attitudes to the point where every consumer is indifferent among all brands of aspirin, except on grounds of price, then aspirin becomes an undifferentiated product. Indeed, it would be economically undifferentiated, whether or not there were chemical differences.

It is Chamberlin (1956) who made a permanent place for differentiated products within the body of economic theory. That place is a peculiar one in a way, as, within models of general equilibrium, products are usually assumed homogeneous. Differentiated products usually are found in models of oligopoly and, of course, monopolistic competition, the type of market with which Chamberlin's name is particularly associated.

It is argued in ch. 2 §6 that price is a more natural decision variable for the firm than output, where models are to be restricted to have only one decision variable per firm. A clear justification of price over quantity may be seen by imagining a multi-period model in which firms choose both and can carry inventories from one period to another. If there were no costs to carrying inventory and marginal cost were constant, the firm would have an incentive to carry sufficient inventory to provide a cushion against the unpredictable element in its demand stemming from changes in the prices of the other firms. Meanwhile, its price would be the only *strategic* variable – that is, the only variable which affects the other firms – as it, not quantity, would enter the demand functions of the others. Were the model to allow for inventory carrying charges and marginal costs which increased with the level of the firm's output, still the output level would remain essentially a variable relevant to the internal efficiency of the firm and would not be a strategic variable. Though a firm's production policy might influence, or be mutually determined with, its price policy, the effect of a firm's output decisions on the behavior of another firm would be at most a second-order effect, through the minor influence it might have on the firm's own price.

A more difficult problem arises in differentiated products models with the notion of *industry* or *market*. In a competitive model of general equilibrium with m goods, there are m markets. If there are M non-competitive producers of differentiated products, one product per firm, how many markets are there? It all depends on the criteria by which they are to be sorted out into industries. Our intuitive vision is that each product belongs naturally to a particular group. Tennis shoes in group 1, gasoline in group 2, aspirin in group 3, and so on. It is then supposed that the cross effects in the demand functions between any two members of the same group are *large* compared with the cross effects between two firms from different groups.

To see the problems with these groups, consider radios as a group. Surely the cross effects between a small, inexpensive AM radio and an expensive AM/FM short wave radio are too small to measure. The cross effects between the inexpensive radio and a phonograph record may be larger, those between the expensive radio and a camera may be large. Should the one be grouped with phonograph records and the other with cameras? Consider also that between the two radios is an array of many which are intermediate, between any two of which the differences are rather small. Any two *adjacent* radios should be in the same group; however, if that rule is followed, the two extreme radios with which the example began are in the same group, linked by a series of steps pairing close substitutes.

These problems about the nature and definition of the market are not raised because they are solved in this chapter. They are raised because some thought about the nature of differentiated products brings them naturally to mind, and, even if they cannot be dealt with, their existence should be recognized. The method of handling them in this and later chapters is to assume them away. The market is defined by the demand functions. There is one function for each of the n firms in the market and the prices of all n enter into the demand function of each.

The idea of product differentiation predates Chamberlin. Schneider

(1962, pp. 284–299) discusses differentiated products models and credits Launhardt (1885, pp. 161–163) with the first treatment of this type of market. The earliest formal model in English is in Hotelling (1929), which is discussed in §2. In §3 the formal demand conditions are set out and discussed for the price model of main concern; §4 contains a discussion of other variables which might be decision variables of the firm; §5 examines equilibria for the price model and §6 contains concluding comments.

2. Hotelling's model of spatial duopoly

After Launhardt, the seeds of monopolistic competition and differentiated products can also be seen in a remark by Marshall (1922) about a firm's "own peculiar market," (p. 458) apparently meaning that even a competitive firm is somewhat in the position of a monopolist with respect to a small market in its immediate vicinity. This, at least, is the reading of Sraffa (1926) who elaborated on Marshall's remark and apparently influenced Hotelling. In setting his problem Hotelling writes (pp. 467–8):[1]

> After the work of the late Professor F. Y. Edgeworth one may doubt that anything further can be said on the theory of competition among a small number of entrepreneurs. However, one important feature of actual business seems until recently to have escaped scrutiny. This is the fact that of all the purchasers of a commodity, some buy from one seller, some from another, in spite of moderate differences of price. If the purveyor of an article gradually increases his price while his rivals keep theirs fixed, the diminution in volume of his sales will in general take place continuously rather than in the abrupt way which has tacitly been assumed.
>
> A profound difference in the nature of the stability of a competitive situation results from this fact. We shall examine it with the help of some simple mathematics....
>
> Piero Sraffa has discussed the neglected fact that a market is commonly subdivided into regions within each of which one seller is in a quasi-monopolistic position. The consequences of this phenomenon are here considered further.

[1]Page references are to Stigler and Boulding (1952) where the article is reprinted.

And on pp. 470–471:

> The feature of actual business to which, like Professor Sraffa, we draw attention, and which does not seem to have been generally taken account of in economic theory, is the existence with reference to each seller of groups of buyers who will deal with him instead of with his competitors in spite of a difference in price. If a seller increases his price too far he will gradually lose business to his rivals, but he does not lose all his trade instantly when he raises his price only a trifle. Many customers will still prefer to trade with him because they live nearer to his store than to the others, or because they have less freight to pay from his warehouse to their own, or because his mode of doing business is more to their liking, or because he sells other articles which they desire, or because he is a relative or a fellow Elk or Baptist, or on account of some difference in service or quality, or for a combination of reasons. Such circles of customers may be said to make every entrepreneur a monopolist within a limited class and region – and there is no monopoly which is not confined to a limited class and region. The difference between the Standard Oil Company in its prime and the little corner grocery is quantitative rather than qualitative. Between the perfect competition and monopoly of theory lie the actual cases.

To illustrate, Hotelling constructs a model in which the buyers of a commodity are assumed to be distributed uniformly along a line of length l. As shown in fig. 3.1, firm A is located at a distance a from the left end of the line and firm B at a distance b from the right. Production costs are zero and total demand is perfectly inelastic; one unit of the commodity being consumed per unit time for each unit length of the line. A consumer buys from the firm from which price plus transportation cost is lower. The cost of transportation is c per unit of distance for each unit of commodity.

Assuming the difference between the two prices is less than the cost of transport between the two locations (i.e. less than $c(l-a-b)$), there is a point on the line between the two firms which divides the customers between them. In fig. 3.1 it is at a distance x from A and y from B, where $x+y=l-a-b$. A consumer located at that point is

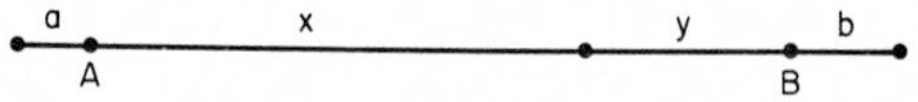

Fig. 3.1

indifferent from whom he buys. Thus, denoting A's price by p_1 and B's by p_2,

$$p_1 + cx = p_2 + cy. \tag{3.1}$$

Therefore

$$x = \tfrac{1}{2}(l - a - b + (p_2 - p_1)/c), \tag{3.2}$$

$$y = \tfrac{1}{2}(l - a - b + (p_1 - p_2)/c), \tag{3.3}$$

and the firm's profits are

$$\pi_1 = p_1 q_1 = p_1(a + x) = \tfrac{1}{2}(l + a - b)p_1 - p_1^2/(2c) + p_1 p_2/(2c), \tag{3.4}$$

$$\pi_2 = p_2 q_2 = p_2(b + y) = \tfrac{1}{2}(l - a + b)p_2 - p_2^2/(2c) + p_1 p_2/(2c). \tag{3.5}$$

Hotelling then looks for a Cournot style equilibrium at which each firm's price is a best reply to the price of the other (p. 473):

> Each competitor adjusts his price so that, with the existing value of the other price, his own profit will be a maximum. This gives the equations
>
> $$\partial \pi_1/\partial p_1 = \tfrac{1}{2}(l + a - b) - p_1/c + p_2/(2c) = 0, \tag{3.6}$$
>
> $$\partial \pi_2/\partial p_2 = \tfrac{1}{2}(l - a + b) + p_1/(2c) - p_2/c = 0, \tag{3.7}$$
>
> from which the equilibrium prices are
>
> $$p_1 = c[l + (a - b)/3], \tag{3.8}$$
>
> $$p_2 = c[l - (a - b)/3], \tag{3.9}$$
>
> and equilibrium output levels are
>
> $$q_1 = a + x = \tfrac{1}{2}[l + (a - b)/3], \tag{3.10}$$
>
> $$q_2 = b + y = \tfrac{1}{2}[l - (a - b)/3]. \tag{3.11}$$

This model is very simple and is intended by Hotelling merely to provide an illustration of how little it takes to remove the discontinuity of the Bertrand–Edgeworth model. Recall in that model the consumers all want to buy from the firm offering the good at the lowest price. The price paid by the consumer equals the price received by the seller; hence, should seller A have the lowest price for one consumer, he has the lowest price for all. In Hotelling's model, the price paid by the consumer and that received by the seller are not the same. They differ by the transport cost of the consumer, which varies from one consumer to the next. Demand for one firm is a continuous function of both prices, as can be seen from eqs. (3.2) and (3.3). Hotelling takes the view that the simple spatial differentiation illustrated in his model provides a rough analogy for other kinds of differentiation.

The formal spatial model does not generalize readily as a spatial model of economic markets; however, it has been made the basis of some very interesting work in the theory of political science. Hotelling makes a political analogy (p. 482), but does not carry it very far in the context of the model. His model is used by Downs (1957) to explain competition among political candidates. A simple version of the model is this: the line in fig. 3.1 along which people are located is an axis measuring positions in a one issue political campaign. For example, it might be the size of the federal budget or the level of income which should be guaranteed as a minimum to poor families. To say a voter is located at a point on this line is to say that the location is his most preferred position on the issue. There is a known distribution of voters across preferred positions, and there are two candidates who must decide where to locate on the line – that is, decide what position to enunciate. A voter will vote for the candidate whose position is nearest his own. The interested reader is referred to Riker and Ordeshook (1973) where the model is exposited and many references are given.

3. Conditions defining the demand functions for the price model, and Chamberlin's monopolistic competition

The demand functions here take Hotelling's demand functions as a clear point of departure.[2] First, there is a separate demand function for each firm, with all the firms' prices entering as arguments of each function. Second, a firm's sales fall when its own price rises and rises when the price of a rival firm rises. The ways in which the demand conditions used below depart from Hotelling's is that the demand functions are much more general than his linear demand and the market demand is not, in the aggregate, inelastic. His model is also exp-

[2]My first introduction to the general sort of modeling used in this chapter comes from Martin Shubik rather than from Hotelling, whom I read, or at least appreciated, only later.

licitly formulated as a spatial model with two firms selling identical products at different locations, while the model to follow deals with n firms and an unspecified source of differentiation.

After setting out the demand conditions in §3.1, §3.2 contains a discussion of Chamberlin's contribution, from which the present formulation also stems. It is much easier to understand and discuss Chamberlin's contribution after both Hotelling's model and that used below are understood.

3.1. Modeling demand under product differentiation

It is with some consequences of product differentiation that this section deals. The fundamental consequence is that the amount demanded of one firm is a continuous function of all prices charged in the market. There is no special significance, for example, to all firms charging the same price. Indeed, for them to do so would be a rare event. Recall an oligopoly model in which products are homogeneous and firms are price choosers. As we have seen, if one firm has a lower price than all the others, every consumer wishes to buy from it. Product differentiation completely alters the situation. Imagine, for example, the market for *radios*. As a matter of common sense, no one expects that two radios, chosen at random, will sell for the same price. They may differ in several dimensions: portability, tone quality, sensitivity to weak signals, sturdiness of the case, general attractiveness, etc. Given a price of p^* for radio #1 and a set of consumers labelled by the index h $(=1,\ldots,H)$, each consumer who would buy a radio #1 for p^* (given his income, prices of other goods, and assuming that radio #2 is priced so high that it would not be bought) has a "switching price" p_h at which he is indifferent between #1 at p^* and #2 at p_h. In general the switching prices vary, perhaps quite widely, from one consumer to another due to differences in individual taste. One person wants to listen to far away stations, another does not care, some are more sensitive to tonal quality or appearance than others, etc. Unless one radio dominates (i.e. is superior in every way to) the other, some of the p_h may exceed p^* and some may be less.

Throughout this and the next three chapters, the product of each firm is taken as given. It is true that in actual markets, decisions about which products to make and how to design them are among the most important that firms face; however, it is also true that economic theory has relatively little to contribute on how to model the nature of a product as a decision variable. In §4.3 product design is briefly discussed.

The assumptions made directly on demand are A3–A5. They stipulate that each firm faces its own demand function, which gives the amount demanded of the firm as a (twice continuously) differentiable function of the prices of all the firms in the market. The amount demanded of a firm is assumed to be a decreasing function of the firm's own price and an increasing function of each of the other prices, making the products of the firms gross substitutes. It is possible for the price of a firm to be so high that it sells nothing, in which case, the firm could be said to be *out of the market*. If a firm's price is that high, a further increase should have no effect whatever on the demand facing other firms. The various demand conditions are divided into several assumptions, the first of which is:

A3. *The demand of the ith firm is given by* $q_i = F_i(p_1, \ldots, p_n) = F_i(\mathbf{p}) \geq 0$. *$F_i$ is defined, continuous and bounded for all* $\mathbf{p} \geq \mathbf{0}$. *For* $p_j^0 > p_j^1$ *and* $\bar{\mathbf{p}}_j^0 \geq 0$, $F_i(p_j^1, \bar{\mathbf{p}}_j^0) \leq F_i(p_j^0, \bar{\mathbf{p}}_j^0)$, $(i \neq j)$ *and* $F_j(p_j^1, \bar{\mathbf{p}}_j^0) \geq F_j(p_j^0, \bar{\mathbf{p}}_j^0)$.

In addition to specifying that the amount demanded of a firm is defined, and no less than zero, for any vector of nonnegative prices, A3 stipulates that as a firm increases its own price, the amount demanded of it cannot rise; and as another firm raises its price, the amount demanded of the original firm cannot fall. For a firm, the price space $\mathcal{R}_+^n$, consisting of price vectors with nonnegative components, is naturally divided into two broad regions. In the first of these its demand is greater than zero and in the second, its demand is zero. For the *i*th firm, the former set is denoted $\mathring{\mathcal{A}}_i$ $(= \{\mathbf{p} | \mathbf{p} \geq \mathbf{0},\ F_i(\mathbf{p}) > 0\})$. In fig. 3.2 the set $\mathring{\mathcal{A}}_1$ consists of all price vectors to the left of the solid curve labelled "$F_1 = 0$," and $\mathring{\mathcal{A}}_2$ consists of all price vectors below the dotted curve marked "$F_2 = 0$." In fig. 3.3 the shaded area, not including the broken line boundary, is $\mathring{\mathcal{A}}_1$.

The effect of changes in p_1 is different outside $\mathring{\mathcal{A}}_1$ from what it is inside. If $\mathbf{p} \in \mathring{\mathcal{A}}_1$, an increase in p_1 causes q_1 to fall, and if $\mathbf{p} \in \mathring{\mathcal{A}}_i$ $(i \neq 1)$, q_i rises; but if $\mathbf{p} \notin \mathring{\mathcal{A}}_i$, an increase in p_1 has no effect on q_1 or q_i. These effects are illustrated in fig. 3.2, where a broken line is drawn in to represent the price vectors for which $q_2 = q_2^0$, a constant. Inside $\mathring{\mathcal{A}}_1$ an increase in p_1, because it causes an increase in q_2, must be accompanied by an increase in p_2 to force q_2 to remain constant. Outside of $\mathring{\mathcal{A}}_1$ further increases in p_1 have no effect on q_2, so increases in p_1 must be accompanied by no change in p_2 if q_2 is to be unchanged.

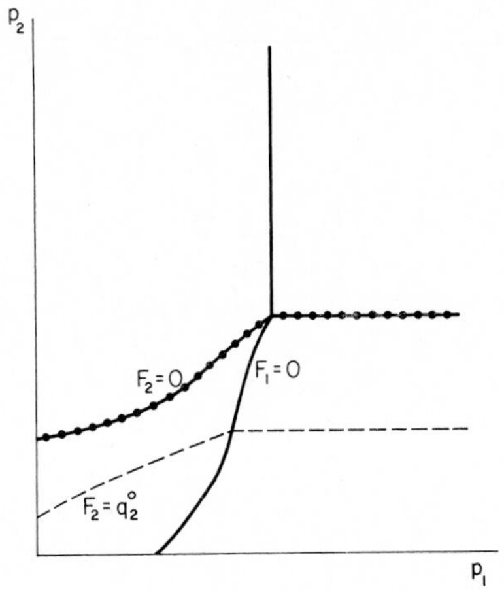

Fig. 3.2

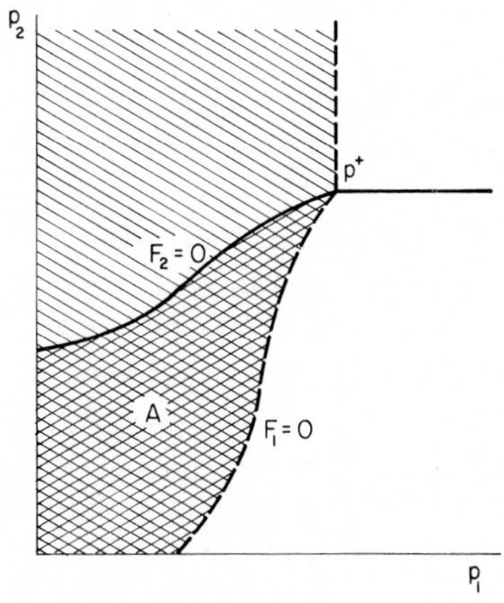

Fig. 3.3

The boundary of $\mathring{\mathscr{A}}_i$ where q_i just becomes zero is of particular interest. This set is denoted $\mathscr{a}_i$. If $\mathscr{A}_i$ is defined to be the closure of $\mathring{\mathscr{A}}_i$, the set $\mathring{\mathscr{A}}_i$ plus all points on the boundary of $\mathring{\mathscr{A}}_i$, then $\mathscr{a}_i$ consists of all price vectors in $\mathscr{A}_i$ but not in $\mathring{\mathscr{A}}_i$. Formally, $\mathscr{a}_i = \{\boldsymbol{p} | \boldsymbol{p} \in \mathscr{A}_i \text{ and } \boldsymbol{p} \notin \mathring{\mathscr{A}}_i\}$. In specifying that the F_i have derivatives, problems arise with price vectors which fall into one or more of the sets $\mathscr{a}_i$. Imagine $\boldsymbol{p}^0 \in \mathscr{a}_1$ and $\boldsymbol{p}^0 \in \mathring{\mathscr{A}}_2$. Then $F_2(\boldsymbol{p}^0) > 0$ and $F_1(\boldsymbol{p}^0) = 0$; however, the price vector is borderline for firm

1. Were it to increase its price, no firm's demand would change, but were it to lower its price, its sales would rise and those of firm 2 would fall. Derivatives of F_1 with respect to any price need not be defined at $\mathbf{p}^0$ and derivatives of F_2 with respect to p_1 need not be defined. Imagine a sequence of price vectors, $\mathbf{p}^k$ $(k = 1, \ldots)$, which are all in $\mathring{\mathcal{A}}_1 \cap \mathring{\mathcal{A}}_2$ and which converge to $\mathbf{p}^0 \in \mathscr{a}_1 \cap \mathring{\mathcal{A}}_2$. Perhaps one might wish to stipulate that $\lim_{k\to\infty} F_2^1(\mathbf{p}^k) = 0$, which means that the effect of a change in p_1 and q_2 goes to zero as the demand of firm 1 (q_1) goes to zero; however, one might well be reluctant to force the same restriction onto $\lim_{k\to\infty} F_2^{11}(\mathbf{p}^k)$, $\lim_{k\to\infty} F_2^{12}(\mathbf{p}^k)$ or $\lim_{k\to\infty} F_1^1(\mathbf{p}^k)$, etc. But because F_1 is zero outside of $\mathring{\mathcal{A}}_1$, all these sequences of derivatives would converge to zero if the price sequence were outside $\mathring{\mathcal{A}}_1$, though it still converges to $\mathbf{p}^0$. Taking account of these boundary considerations serves to make the statement of A4 somewhat unwieldy.

A4. *$F_i(\mathbf{p})$ is twice continuously differentiable for all $\mathbf{p} \gg \mathbf{0}$, except for $\mathbf{p} \in \bigcup_j \mathscr{a}_j$. If $\mathbf{p} \in \mathscr{a}_j$ for $j = i_1, \ldots, i_k$ and $\mathbf{p} \notin \mathscr{a}_j$ for $j = i_{k+1}, \ldots, i_n$, then continuous second partial derivatives with respect to $p_{i_{k+1}}, \ldots, p_{i_n}$ exist at $\mathbf{p}$. All derivatives are bounded, and, if $F_i(\mathbf{p}) = 0$, $\mathbf{p} \notin \mathcal{A}_i$, all derivatives of F_j $(j = 1, \ldots, n)$ with respect to p_i are zero. For $\mathbf{p} \in \mathring{\mathcal{A}}_i \cap \mathring{\mathcal{A}}_j$ $(j \neq i)$, $F_i^j(\mathbf{p}) > 0$, and, for $\mathbf{p} \in \mathring{\mathcal{A}}_i \cap \mathring{\mathcal{A}}_{i_1} \cap \cdots \cap \mathring{\mathcal{A}}_{i_k}$, $F_i^i(\mathbf{p}) + \Sigma_{j=1}^k F_i^{i_j}(\mathbf{p}) < 0$.*

All the conditions in A4 have been discussed except the final one which stipulates that $F_i^i(\mathbf{p}) + \Sigma_{j=1}^k F_i^{i_j}(\mathbf{p}) < 0$, when $F_i(\mathbf{p}) > 0$ and the sum of first partial derivatives is taken over all firms having positive demand. In addition to assuring that F_i^i is negative, it guarantees that total demand in the market will, in some sense, fall as all prices rise. The exact condition is that if all firms raise their prices by equal amounts, all firms having positive demand encounter a decrease in demand.

The final assumption on demand limits the size of the region in which all firms, simultaneously, have positive demand. Let $\mathcal{A} = \bigcap_{i=1}^n \mathcal{A}_i$. $\mathcal{A}$ consists of the set of all price vectors for which every firm has positive demand, plus the boundary of that set. For the two-firm case, the set $\mathcal{A}$ is the cross hatched area in fig. 3.3 (including boundary).

A5. *$\mathcal{A}$ is bounded.*

As fig. 3.3 suggests, the set $\mathcal{A}$ has a unique maximal element. That is there is a particular price vector in $\mathcal{A}$, call it $\mathbf{p}^+$, which is at least as large in every component as any other member of $\mathcal{A}$. $\mathbf{p}^+$ is also the only price vector in $\mathcal{A}$ for which all firms have zero demand.

Theorem 3.1. *Under assumptions A3–A5, the set $\mathscr{A}$ has a unique maximal element, $\mathbf{p}^+$, such that for any $\mathbf{p} \in \mathscr{A}$, if $\mathbf{p} \neq \mathbf{p}^+$ then $\mathbf{p} < \mathbf{p}^+$. In addition, $F_i(\mathbf{p}^+) = 0$, $i = 1, \ldots, n$; but for $\mathbf{p} \neq \mathbf{p}^+$, $\mathbf{p} \in \mathscr{A}$, $F_i(\mathbf{p}) > 0$ for at least one firm, i.*

□ The method of proof is to first find for each firm a least upper bound price, p_i^*, such that if it charges p_i^* or more its demand is zero irrespective of the prices of the others, and if it charges less than p_i^*, there are prices of the others for which it would have positive demand. Then it is shown that $\mathbf{p}^* \in \mathscr{A}$, that $\mathbf{p}^* = \mathbf{p}^+$, the unique maximal element, and that no other element of A is associated with zero demand for all firms.

Define $\mathbf{p}^* = (p_1^*, \ldots, p_n^*)$ as follows: $p_i^* = \sup_{\mathbf{p} \in \mathscr{A}_i} p_i$. p_i^* is the least upper bound on prices of the ith firm to be found in the set $\mathscr{A}$. Therefore, $\mathbf{p}^*$ is the *upper right corner* of the smallest (n-dimensional) rectangle which can contain $\mathscr{A}$. That is, for any $\bar{\mathbf{p}}_i \geq 0$, $F_i(p_i^*, \bar{\mathbf{p}}_i) = 0$ $(i = 1, \ldots, n)$; therefore, for any $\mathbf{p} \in \mathscr{A}$, $\mathbf{p} \leq \mathbf{p}^*$. Existence of $\mathbf{p}^*$ follows directly from A5, and its definition implies that for $p_i < p_i^*$, $F_i(p_i, \bar{\mathbf{p}}_i^*) > 0$. Therefore, $\mathbf{p}^* \in \mathscr{A}_i$ for all i; hence $\mathbf{p}^* \in \mathscr{A}$, and $\mathbf{p}^* = \mathbf{p}^+$ and is the unique maximal element of $\mathscr{A}$. It remains to show that no other member of $\mathscr{A}$ yields zero demand to all firms. Let $\mathbf{p}$ be any other element of $\mathscr{A}$, and choose firm k by the criterion that $p_k^* - p_k \geq p_i^* - p_i$ $(i = 1, \ldots, n)$. By A4 $F_k(\mathbf{p}) > 0$. □

Retaining the cost assumption, A2, profit functions for the firms may be written:

$$\pi_i(\mathbf{p}) = p_i F_i(\mathbf{p}) - C_i(F_i(\mathbf{p})). \tag{3.12}$$

3.2. *Chamberlin's monopolistic competition and modeling of demand*

Chamberlin's (1956) principal concern is with markets having many of the characteristics of competitive markets, particularly those in which the number of firms is very large and the sizes of the firms does not vary greatly. If, in such a market, each firm supplies a product somewhat different from that of every other firm, then each firm could be thought to be in a quasi-monopolistic position with respect to part of the overall market. With many firms, none of which is large in relation to the overall market, it appeared natural to assume the cross effects between any two firms to be zero. The analytic key to the whole construction is the way in which demand is modeled. Roughly speaking, it is done along the lines of §3.1; however, there are some peculiarities in important details. Fig. 3.4 is a reproduction of Chamberlin's fig. 14 [Chamberlin (1956, p. 91)]. To get the flavor of his construction, his own words serve best (pp. 90–92):

> The curve *DD'*, as heretofore drawn, describes the market for the "product" of any one seller, *all* "products" and *all other* prices being given. It shows the increase in sales which he could realize by cutting his price, *provided* others did not also cut theirs: and conversely, it shows the falling off in sales which would attend an increase in price, *provided* other prices did not also increase.

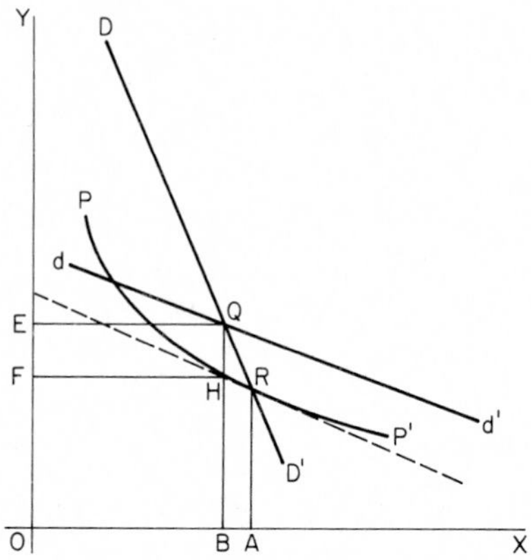

Fig. 3.4

> Another curve may now be drawn which shows the demand for the product of any one seller at various prices on the assumption that his competitors' prices are always identical with his. Evidently this latter curve will be much less elastic than the former, since the concurrent movement of all prices eliminates incursions by one seller, through a price cut, upon the markets of others. Such a curve will, in fact, be a fractional part of the demand curve for the general class of product, and will be of the same elasticity. If there were 100 sellers, it will show a demand at each price which will be exactly 1/100 of the total demand at that price (since we have assumed all markets to be of equal size). Let *DD'* in fig. 3.4 be such a curve, and let the price asked by all producers be, for the moment, *BQ*. The sales of each

> are *OB*, and the profits of each (in excess of the minimum contained in the cost curve) are *FHQE*. Now let *dd'* be drawn through *Q*, showing the increased sales which any one producer may enjoy by lowering his price, provided the others hold theirs fast at *BQ*. Evidently, profits may be increased for any individual seller by moving to the right along *dd'*; and he may do this without fear of ultimately reducing his gains through forcing others to follow him because his competitors are so numerous that the market of each of them is inappreciably affected by his move. (Each loses only 1/99 of the total gained by the one who cuts his price.) The same incentive of larger profits which prompts one seller to reduce his price leads the others to do likewise. The curve *dd'*, then, explains why each seller is led to reduce his price; the curve *DD'* shows his actual sales as the *general* downward movement takes place. The former curve "slides" downwards along the latter as prices are lowered, and the movement comes to a stop at the price of *AR*. Evidently it will pay no one to cut beyond that point, for his costs of producing the larger output would exceed the price at which it could be sold.

In the modeling of demand Chamberlin has some unnecessary, partially incorrect baggage which helps obscure his contribution. One is the assumption that all firms are symmetric with one another and, tied to that, the other is the *DD'* curve which is called "... a fractional part of the demand curve for the general class of product...". With respect to the symmetry of firms, the assumption has no logical inconsistency; however, from an economic standpoint it is peculiar. He is attempting to model a market for a heterogeneous grouping of goods which are meant to be very similar but not identical. A lack of symmetry among them would seem at the heart of what it is desirable to capture. For example, it is of interest that the model cover goods which differ in overall quality. Where such differences occur, it is not expected that when all firms charge the same price, they all sell equal quantities.

The *DD'* curve has an interesting and useful interpretation without any need to refer to an aggregated demand curve. The notion of the "demand curve for the general class of product" is without meaning unless many special assumptions are made which vitiate the possibility of dealing in an interesting way with differentiated products. His *DD'* curve as an aggregated demand curve is valid when he assumes symmetry and assumes that all firms charge the same price; all of which gets too close to throwing the baby out with the bath water.

Reasonable meaning may be put into the dd' and DD' curves by defining them with reference to the demand function, F_i, developed in §3.1. dd' is merely a graph of p_i against $q_i = F_i(p_i, \bar{\boldsymbol{p}}_i)$ for some fixed value of $\bar{\boldsymbol{p}}_i$. A curve similar to DD' may be obtained by assuming that, as p_i varies up or down, some or all of the other prices vary in the same direction. It could be supposed that all other prices change by the same amount as p_i changes from some specific value. That is, let $\bar{\boldsymbol{\delta}}_i(p_i)$ be a vector having $n-1$ components, all of which are equal to $p_i - p_i^0$; and suppose a DD' curve to be defined by $q_i = F_i(p_i, \bar{\boldsymbol{p}}_i^0 + \bar{\boldsymbol{\delta}}_i(p_i))$. Let dd' be given by $q_i = F_i(p_i, \bar{\boldsymbol{p}}_i^0)$. These two curves are illustrated in fig. 3.5. They intersect where $p_i = p_i^0$ and, because under A4, $F_i^j > 0$ $(i \neq j)$ and $\Sigma_{j=1}^n F_i^j < 0$, both curves are negatively sloped with the DD' being steeper.

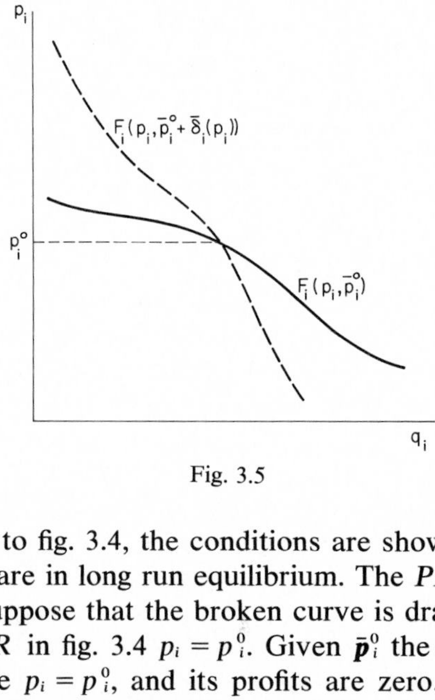

Fig. 3.5

Referring again to fig. 3.4, the conditions are shown under which the firm and industry are in long run equilibrium. The PP' curve is average total cost. Now suppose that the broken curve is drawn for $\bar{\boldsymbol{p}}_i = \bar{\boldsymbol{p}}_i^0$ and that at the point R in fig. 3.4 $p_i = p_i^0$. Given $\bar{\boldsymbol{p}}_i^0$ the ith firm maximizes profits at R where $p_i = p_i^0$, and its profits are zero. If the situation is the same for all firms at $\boldsymbol{p}^0$ then Chamberlin's equilibrium holds. It is a Cournot type, or best reply, equilibrium in the sense that each firm is maximizing its own profits with respect to its own price and taking other prices as fixed. As compared with Cournot there are two notable differences. First, there are no (or *negligible*) cross effects among firms, so firms do not worry about how their own actions or speculations about their actions will affect other firms. Second, a zero profit condition is incorporated, which implies no incentive for firms to leave or enter the industry.

There is a loose end to tie up in connecting more precisely the model of §3.1 with Chamberlin's monopolistic competition. A3–A5 are a means of modelng an n-firm industry with differentiated products; however the cross effects among the firms are explicitly nonzero (i.e. the F_i^j), and to say that sometimes they might be *negligible* begs the question. Imagine specifying a family of models satisfying A3–A5 and subject to the additional condition that $F_i^j(\boldsymbol{p}) \leq k|F_i^i(\boldsymbol{p})|/n$ for some fixed, positive k. Then as $n \to \infty$, F_i^j goes to zero. That is, as the number of firms increases, the model converges to a model of monopolistic competition in which the cross effects between pairs of firms is zero.

There is one final observation to be made on Chamberlin's monopolistic competition. There may not be a market in the world which approximately meets the conditions of monopolistic competition. Think about examples of markets in which the product is differentiated and the number of firms very large. Gasoline stations, dry cleaners, grocery stores in the days before the large food chains – all have something in common. They are industries in which the product is differentiated and the number of firms large, and perhaps the firms tend to be of approximately the same size; however, the cross effects among firms are not in general small. With respect to any one firm, certain other firms are "close" and have large cross effects, while others are "farther" and the cross effects are less. With respect to most other firms, the cross effects might well be negligible; but with a few they are not. Furthermore, the set of firms with which one firm is in close competition is not going to be precisely the same as the set for any other. To have a picture of the situation, imagine a market like Hotelling's, except that the line, or road, on which the firms are located stretches indefinitely far to right and left, with firms located evenly at one inch intervals. A firm may be identified by its position $(\ldots -5, -4, \ldots, 0, +1, +2, \ldots)$. Firm j is in closest competition with firms $j-1$ and $j+1$; however, firm $j+1$ is in closest competition with firms j and $j+2$, etc. Chamberlin (1956) actually suggests this type of market (p. 103): "Retail establishments scattered throughout an urban area are an instance of what might be called a "chain" linking of markets. Gasoline filling stations are another. In either of these cases the market of each seller is most closely linked (having regard only to the spatial factor) to the one nearest him, and the degree of connection lessens quickly with distance until it becomes zero. Under such circumstances subgroups cannot be distinguished. Were an area to be marked off arbitrarily, stores at its border would compete with those on the border of the adjoining area more than with those in other portions of the area in which they were placed. Classes of custom are often indistinct, and shade into each other in a similar way."

Chamberlin's chain market appears to me as the sensible way to model monopolistic competition because it does correspond to something recognizable in actual markets – quite unlike the customary version of monopolistic competition. Unfortunately, though the idea has been noticed, for example by Henderson (1954), it has not been developed.

4. Other decision variables

This section is devoted to a brief look at the possibilities for going beyond models having one decision variable per firm. The most obvious first step in this direction is a model in which both price and quantity are chosen by the firm. This is examined in §4.1. Especially in the light of product differentiation, advertising is a natural variable to wish to incorporate. It is discussed in §4.2. Product design, itself, is another variable of obvious empirical importance, which is taken up in §4.3 and §4.4 deals with miscellaneous other variables. It must be admitted from the outset that there is regrettably little substance on these topics.

4.1. Models with price and quantity as variables

When both the firm's price and its production level are chosen, there is the possibility that sales and demand do not coincide, and that demand and production do not coincide. It is much more natural to have both price and quantity produced as decision variables in a multi-period model than in a single-period model precisely because the foregoing discrepancies suggest allowing the firm to carry inventories, hence to have an inventory policy which governs production. Such considerations are meaningless in a one-period framework; therefore, the following discussion is mainly carried on in the context of an incomplete multi-period framework.

Consider the model of §3 adapted to a multi-period context in the following manner: demand for the firm is given by F_i, as before, and the costs of production by C_i. The firm chooses both price and output; hence, demand can be larger or smaller than output, with the firm's inventory supplying or absorbing the additional units, as needed. Requiring that inventory cannot be negative means that sales can never exceed the amount produced during a time period, plus the inventory on hand when the time period began.[3] It might be desirable to add to the model a cost of holding inventory, making the firm's profits in a time period be its revenue (p_iF_i) minus its production cost (C_i) minus its inventory carrying cost, an increasing function of the level of inventory.

Now note that in an equilibrium the firm has zero inventory, production equals demand and the variable which it is interesting to study is price because it is price which enters into the demand functions of all firms. Production is merely adjusted to the right level, dictated by the equilibrium prices. An equilibrium is a steady state in which all firms continually repeat the same decisions, so there is no need to carry inventories, either as a cushion against uncertainties or to engage in production smoothing to reduce costs.

If the preceding model were changed to have stochastic elements, say in the demand functions, inventory policy would acquire some interest. Also, if disequilibrium adjustment processes were being studied, even within the simple framework described above, again inventory policy would come to life. But in both instances, though inventory policy would have some effect on the firms' price policies – to the extent that it is the study of the way the firms interact that is of main interest – inventory policy would not matter very much. That is, it is doubtful that there would be any notable new insights from such models.[4] There is an interesting and rich literature on the theory of the firm in which inventory policy plays an important role. See, for example, Zabel (1972) where other references may also be found.

[3]It might be natural to assume, when a firm cannot supply the total demanded of it in a time period, that demand for it, and perhaps all firms in the industry, is greater in the subsequent period due to the held-over demand of the unsatisfied customers.

[4]The reader should keep in mind that this conjecture is only a conjecture; hence, it may turn out to be incorrect.

4.2. Advertising as a decision variable

Intuitively, advertising seems the obvious candidate for a second important strategic decision variable in an oligopoly model. Firms would appear to compete with advertising – to use advertising to gain new customers for themselves, often, though not invariably, at the expense of rival firms. Textbook treatments of the theory of consumer choice leave no place for advertising because consumers are assumed to know their own preferences perfectly well and to have complete information on the characteristics and prices of all commodities.

Information may be incomplete with respect to preferences, prices or characteristics of commodities. Concerning preferences, a consumer may in fact be very uncertain what it would be like for him to have, say, a tennis court in his back yard or to live on a farm. With inexpensive items, especially those which are short lived or perishable, it is easy and cheap for the consumer to gather information by buying and trying many things. For a trivial amount of money, a person can find out whether he likes a particular type of cheese or if a certain sort of shirt will be comfortable and practical. Even with these inexpensive items which are easily sampled, the consumer must be made aware of their existence; and with expensive, long lived items, especially those which have large transactions costs associated with their purchase, like the tennis court and farm, merely getting the requisite information on one's own preferences may be difficult, even impossible, except at the large cost associated with the purchase of the good. In between fall a host of commodities which are longer lived and more expensive than food and clothing, and much less expensive than farms and back-yard tennis courts, about which the consumer may know some things but not nearly all that would be useful to him.

In the preceding paragraph it is supposed that the consumer is fully informed about the technical characteristics of all commodities, but that he does not know precisely how they fit into his scheme of preferences. In many instances, the consumer does not know all the relevant technical data about goods he might buy. Thus, even if his knowledge concerning preferences were complete, he could be uncertain where to classify many commodities.

It is a commonplace that advertising carries information about the characteristics of products and how they might be used. Even advertising which only makes the consumer aware of the existence of an item may be said to convey useful information. It is in the advertiser's interest to convey information because doing so lowers the cost to the consumer of obtaining information, probably increasing the likelihood of purchase, on balance; and the advertiser can choose which information is conveyed as well as the manner in which it is presented. The advertiser is not merely trying to sell, say, refrigerators, he is trying to sell his own particular models. Whether it is in the advertiser's interest to convey information which is completely accurate, or, to some extent false is another matter. It is no problem to imagine circumstances in which an advertiser aids himself by being dishonest or by failing to provide voluntarily information which consumers would greatly benefit from having, but which they do not realize they would find useful. Nonetheless, in the end, advertising is a matter of information, be it false or true. A good discussion of the general nature of advertising may be found in Braithwaite (1928).

For our present purposes, all that matters about advertising is that it affects the sales of the firm; hence, it should be an argument of the demand function. Beyond noting that when a firm advertises, its own sales should rise as a result, and that diminishing returns are likely to set in, what more might be said at a relatively abstract level? It does seem plausible that advertising has both cooperative and competitive sides.[5] To explain what is meant, it is clearest to first explain purely cooperative advertising and purely competitive advertising separately and then to imagine a mixed model. Assume that the firm makes no choices about the nature of advertising; it only decides on an amount of money to spend. Purely cooperative advertising is characterized by two conditions: (1) every firm has its demand rise when only one firm advertises, and (2) the demand increases for any one firm are independent of which firms undertake to advertise. Where advertising is close to being purely cooperative, it is not much in the interest of the individual firm to engage in it unless the firm dominates the market. Otherwise, a trade association to which all firms contribute would be more likely to advertise. Advertising campaigns exhorting the public to "drink milk" are probably one of the few examples of quite purely cooperative advertising. Milk may be a sufficiently standardized product that advertising by one dairy would help milk sales in general without giving especial help to the dairy itself.

Purely competitive advertising has the characteristic that, on balance, it does not raise sales in the aggregate even though a firm which

[5]This explicit recognition of cooperative and competitive aspects to advertising was first brought to my attention by Martin Shubik.

advertises finds its own sales increased. Two conditions would characterize purely competitive advertising: (1) when one firm advertises, its sales rise and those of other firms in the industry fall, and (2) there is no way that all firms can advertise simultaneously with all firms having an increase in demand as a result. If some firms have increases, others have decreases. When advertising is purely competitive, the gains of one come completely at the expense of others in the market. Cigarettes may be an example of a product for which advertising is close to purely competitive. Were that so, it would mean that the combined advertising of all firms has almost no effect on bringing in new customers or in increasing the amounts consumed by present buyers; but that only the distribution of customers among the various brands is affected. In such a case, it would be possible for all firms to simultaneously reduce advertising expenditures in a way which would keep sales unchanged. The result would be a reduction in costs for all firms. Such a reduction would be hard to accomplish for the same reasons that cooperative equilibria are hard to attain when binding contracts cannot be made. While it would be in the interest of all firms to agree to a reduction, it would also be in the interest of an individual firm to continue to advertise heavily if all the others were to cut back.

Most advertising undoubtedly falls into a middle ground of having both competitive and cooperative elements. Thus, if all firms advertise, it is possible that all have higher demand than if one advertised; and the demand of one rises more when it advertises than when another firm advertises in the same amount. Indeed, when both cooperative and competitive elements are present, the amount demanded of a firm may either rise or fall in response to advertising undertaken by rival firms. The presence of a competitive element means only that a given expenditure on advertising increases the sales of a firm more when it does the spending itself than when another firm does it. The presence of a cooperative element means that it is possible to find amounts for each firm to spend on advertising such that all firms experience an increase in amounts demanded (prices being fixed) simultaneously.

Turning from single-period to multi-period models allows a new facet of advertising to be recognized, namely that the effects of advertising in a given time period may extend beyond that period. Arrow and Nerlove (1962) model advertising in an analogous way to investment. A dollar spent on advertising adds one unit to the firm's stock of "goodwill" which is subject to exponential depreciation. The stock of goodwill enters the demand function of the firm, with the derivative of quantity with respect to goodwill being positive. Treating advertising like investment goes back to Hoos (1959) if not farther. In §5, where equilibria for the single-period price model is discussed, some comments are made concerning equilibria with additional variables.

4.3. *Product design as a decision variable*

Little work has been done on this tantalizing problem, probably because it is uncommonly hard to come to grips with it theoretically in a way which yields insights and results. Brems (1951) made an attempt in which the underlying notion is that a product has a number of measurable characteristics which can be used to define it, and values of which are chosen by the firm. Lancaster (1975) complements this approach and fills it out by taking a novel, insightful view of the nature of a good. He starts from the premise that there are, say, m consumption characteristics. An appropriate analogy is nutrients in food. A particular good is a bundle of the m characteristics in fixed proportions and the products of the firms in a market differ according to the chosen proportions for the characteristics.

4.4. *Other variables*

It is easy to name other variables which it would be desirable to take account of in formal models. Examples are selling effort, explicit location variables, research and development effort, ability of management, form of corporate organization and the relationship of ownership to control. Working them in is another matter, for we generally lack a good sense of how to relate a given variable to other variables of a model or, in some instances, we do not even know how to specify what is to be meant by a variable. As an instance, how should *form of corporate organization* be specified? An easier, though difficult, problem is how to measure research and development effort. To merely count expenditures begs the question of what is to be meant.

5. *Equilibria in the single-period price model*

The equilibria introduced in ch. 2 can be readily defined for the price model. An equilibrium in the manner of Cournot is defined by the following: $\boldsymbol{p}^c$ is the price vector corresponding to a Cournot type equilibrium if (a) $\mathbf{0} \leq \boldsymbol{p}^c \leq \boldsymbol{p}^+$ and (b) $\pi_i(\boldsymbol{p}^c) = \max_{0 \leq p_i \leq p_i^+} \pi_i(p_i, \bar{\boldsymbol{p}}_i^c)$, $(i = 1, \ldots, n)$. As with the Cournot equilibrium in the quantity model, this equilibrium is characterized by a price vector such that no firm can, by itself, change its price and obtain higher profit. It is strictly correct to call the equilibrium associated with Cournot in ch. 2 the Cournot equilibrium. My usage is to refer also to the price model equilibrium above as a Cournot equilibrium. This equilibrium is sufficiently close to and in the spirit of Cournot's quantity model equilibrium that the usage is justified. Questions of existence, uniqueness and whether all firms have positive output are taken up in ch. 7.

Among the three single-period models encountered thus far, Cournot, Bertrand–Edgeworth and Chamberlin, the Chamberlinian differentiated products price model is by far the most satisfactory. All three are at approximately the same level of abstraction, but the Chamberlinian model more nearly mirrors reality. In comparing it to the Cournot and Bertrand models, it is clear the Cournot model is superior to the Bertrand because it lacks the unbelievable discontinuities of the latter.

Theorem 2.2, stating conditions under which the Cournot equilibrium does not yield Pareto optimal profits, carries over to the price model.

Theorem 3.2. *Let $\boldsymbol{p}^c$ be a Cournot equilibrium price vector for a market satisfying A2–A5. If $\mathbf{0} \ll \boldsymbol{p}^c \ll \boldsymbol{p}^+$, then the vector of profits $\boldsymbol{\pi}(\boldsymbol{p}^c) = (\pi_1(\boldsymbol{p}^c), \ldots, \pi_n(\boldsymbol{p}^c))$ is not Pareto optimal.*

□ Proof is almost immediate from noting that the condition that the price vector is interior to the price space implies the equilibrium is characterized by interior first-order conditions:

$$F_i(\boldsymbol{p}) + (p_i - C_i')F_i^i(\boldsymbol{p}) = 0, \qquad i = 1, \ldots, n. \tag{3.13}$$

Because F_i is positive and F_i^i is negative, $p_i - C_i'$ must be positive. Now look at the derivative of π_i with respect to some price other than p_i:

$$\partial \pi_i / \partial p_j = (p_i - C_i')F_i^j > 0, \qquad i \neq j, \quad i, j = 1, \ldots, n. \tag{3.14}$$

Therefore if all prices were increased marginally above $\boldsymbol{p}^c$ by, say ϵ, all profit levels would rise. □

The other equilibria, joint maximum, von Neumann–Morgenstern and efficient point, may all be defined in the obvious ways for the price model. This exercise is left to the reader.

Equilibria in the price model when there are additional decision variables is a trivial extension of equilibria when there is only one, at least so far as the conditions defining the equilibria are concerned. For this reason, adding variables to the model is an uninteresting enterprise, unless and until more is known about how the variables ought to be incorporated and the results cease being trivially obvious. To illustrate, imagine a model in which the firm has two decision variables, price and advertising, and consider the conditions which would hold at a Cournot equilibrium when that equilibrium is interior – that is, corresponds to positive levels of price, output and advertising for all firms. Denoting the vector of advertising expenditures by $\boldsymbol{a} = (a_1, \ldots, a_n)$ and the profit function by $\pi_i(\boldsymbol{p}, \boldsymbol{a}) = p_i F_i(\boldsymbol{p}, \boldsymbol{a}) - C_i(F_i(\boldsymbol{p}, \boldsymbol{a})) - a_i$, the first-order conditions for the equilibrium are

$$\partial \pi_i / \partial p_i = F_i + (p_i - C'_i)F^i_i = 0, \tag{3.15}$$
$$\partial \pi_i / \partial a_i = (p_i - C'_i)F^{i+n}_i - 1 = 0. \tag{3.16}$$

Eq. (3.15) is the usual sort of marginal revenue equals marginal cost condition, which may be more easily seen by rewriting it as

$$F_i / F^i_i + p_i = C'_i. \tag{3.17}$$

Eq. (3.16) states that advertising should be carried to the point where the last dollar of advertising expenditure adds a dollar to net revenue, where net revenue is the difference between price and the marginal cost of production, multiplied by $\partial q_i / \partial a_i$, the additional sales which result from the marginal advertising expenditure:

$$(p - C'_i)F^{i+n}_i = 1. \tag{3.18}$$

Having used the *marginal principle* once, we may use it twice! Unfortunately much "analysis" of new variables amounts to no more than this. While it is true they can be incorporated into the models, and the marginal conditions are correct, there are surely no new insights obtained from such an exercise.[6]

6. Summary comments

In this chapter a differentiated products price model is developed which is used as the basis for the reaction function studies of ch. 5. Among simple

[6]In Chamberlin (1956) the analysis of advertising and selling costs amounts to no more than this sort of exercise.

models, it appears to be the most realistic and satisfying. The equilibria developed in ch. 2 are seen to have their counterparts in the price model; and, the Cournot equilibrium in particular is seen to play an important role in the following two chapters, partly as a point of comparison and partly as an example from which an appropriate noncooperative equilibrium may be developed for the reaction function models.

CHAPTER 4

REACTION FUNCTION EQUILIBRIA DEVELOPED PRIOR TO 1960

1. The reaction function

The developments discussed in this chapter represent a transition from single-period to multi-period models. On the one hand, it is clear that they are concerned with the behavior of firms in a multi-period setting; but on the other, the multi-period models are not explicitly formulated.[1] From the context it is often possible to specify completely the multi-period model which a writer must have had in mind and, as a rule, these are not models which capture well the problem which they were intended to explain.

It is of interest to review these models purely for historical reasons, to better understand how results in oligopoly theory have unfolded and to be able to put into a clearer perspective the nature and worth of more recent work. The study of the models presented in this chapter serves as an excellent introduction to dynamic oligopoly models in general and reaction function models in particular. It greatly facilitates the understanding of the reaction function models of ch. 5 to have an introduction to many of the facets of such models in the simpler setting of the present chapter.

To fix an example in mind, assume a Chamberlinian differentiated products model, where the price a firm charges is its only decision variable. Assume also that time is divided into discrete time periods indexed by t and numbered $t = 1, 2, \ldots$, and that at the beginning of each time period, each of the firms chooses a price which will be in effect for the whole of the period. The demand, cost and profit functions measure rates of sales, cost and profit per unit time. A reaction function for a firm is a decision rule which determines for that firm its current period price as a function of the prices which prevailed in the market during the previous

[1]An exception is my version of the Stackelberg model, presented in §5 after Stackelberg's version.

period. The essential characteristic of the reaction function is that the period t price of, say firm i, depends on prices of some or all firms observed during one or more past periods; however, a common form is $p_{it} = \psi_i(\boldsymbol{p}_{t-1})$, where all previous period prices, and no other variables, determine the current period price.

In the remaining sections the time subscript is sometimes used and sometimes suppressed. In Cournot (1960), Bowley (1924) and Stackelberg (1934), whose work forms the basis of this chapter, time is never explicit in the models and it is not always clear how it is intended that it enter. One of the useful lessons to be learned is that results which are conjectured from incompletely specified models are often wrong, and that the models under which the results are true often require strange and unacceptable assumptions.[2]

All the reaction function models considered below and in ch. 5 are built upon single period models as a foundation. Assumptions are added from time to time as needed, and, in §2, both the quantity and price single-period models are further developed. In §3 Cournot reaction functions are formulated and discussed. They form a point of departure for Bowley's suggestion, which is taken up in §4 and are part of the underpinnings of Stackelberg's model, reviewed in §5. Also in §5 is a reformulation of Stackelberg's model in which time is explicit. This model gives additional insight into Stackelberg's work and forms an excellent introduction to the models in ch. 5.

2. *Extending the market models*

Two new assumptions are added below to those made for the quantity model and another two for the price model. In each case, the first of the pair is sufficient, along with previous assumptions, to ensure the existence of the Cournot equilibrium for the model; and the second guarantees uniqueness. These theorems will not be proved below. Their proofs are delayed until ch. 7.

2.1. *Quantity models*

Both of the assumptions which are being added are stated in terms of the profit functions; thus they involve both demand and cost functions and the relationship between them.

[2]The Stackelberg leader–follower model is a particularly good example.

A6. *For all* $\boldsymbol{q}$ *such that* $q_i > 0$ *and* $Q < \bar{Q}$, $\partial^2\pi_i(\boldsymbol{q})/\partial q_i^2 < 0, i = 1, \ldots, n$.

A7. *For all* $\boldsymbol{q}$ *such that* $\boldsymbol{q} \geqslant \boldsymbol{0}$ *and* $Q < \bar{Q}$, $\partial^2\pi_i(\boldsymbol{q})/\partial q_i^2 + \Sigma_{j\neq i}|\partial^2\pi_i(\boldsymbol{q})/\partial q_i\partial q_j| < 0$, $i = 1, \ldots, n$.

The existence condition, A6, is not unreasonable in the sense that a fairly large class of models satisfy it. Note that $\partial^2\pi_i(\boldsymbol{q})/\partial q_i^2 = 2f' + qif'' - C_i''$, which is negative if the demand function is downward sloping ($f' < 0$) and concave ($f'' < 0$) and the cost function is convex ($C_i'' > 0$). Such models are not all which are of interest, but they form a fairly broad class. Furthermore, A6 can be satisfied with $F'' > 0$ and/or $C_i'' < 0$ as long as these terms are *dominated* by the negative slope term. A6 is also the second-order condition for maximizing π_i with respect to q_i.

Condition A7 is stronger than A6 in the sense that it implies A6. It is not as readily acceptable as A6, which may be seen by writing it out in terms of the derivatives of the cost and demand functions,

$$\partial^2\pi_i(\boldsymbol{q})/\partial q_i^2 + \sum_{j\neq i}|\partial^2\pi_i(\boldsymbol{q})/\partial q_i\partial q_j| = 2f' + q_if'' - C_i'' + (n-1)|f' + q_if''|. \tag{4.1}$$

If $f'' < 0$, eq. (4.1) becomes

$$-(n-3)f' - (n-2)q_if'' - C_i''. \tag{4.2}$$

For n greater than 2, the first two terms of eq. (4.2) are positive and, as n increases, the condition that eq. (4.2) be negative requires that C_i'' be larger and larger.[3] The unreasonableness of A7 is one of the weaknesses of the quantity model.

2.2. Price models

The parallel assumptions to A6 and A7 are made for prices which fall into a subset, $\mathring{\mathscr{A}}_i^*$ (or the interior of $\mathscr{A}_i^*$), of the price space which is of special interest; the region within which the price of a firm (or all firms) is at least as great as its marginal cost. $\mathscr{A}_i^* = \{\boldsymbol{p} | \boldsymbol{p} \in \mathscr{A}_i \text{ and } p_i \geq C_i'(F_i(\boldsymbol{p}))\}$. Let $\mathring{\mathscr{A}}_i^*$ denote the interior of $\mathscr{A}_i^*$, and $\mathscr{A}^*$ the intersection of the $\mathscr{A}_i^*$.

A8. *For all* $\boldsymbol{p} \in \mathring{\mathscr{A}}_i^*$, $\partial^2\pi_i/\partial p_i^2 < 0$.

[3]A7 implies stability of the Cournot reaction functions of the quantity model. This is seen in §3.

A9. *For all* **p** *in the interior of* $\mathscr{A}^*$, $\partial^2\pi_i/\partial p_i^2 + \Sigma_{j\neq i}|\partial^2\pi_i/\partial p_i\partial p_j| < 0$.

A8 specifies a reasonably large class of models and, like A6, is the second-order condition for a maximum of π_i with respect to the firm's decision variable – in this instance, p_i. Writing out A8 in terms of derivatives of the demand and cost functions yields

$$\partial^2\pi_i/\partial p_i^2 = (2 - C_i''F_i^i)F_i^i + (p_i - C_i')F_i^{ii}. \tag{4.3}$$

A negative sign is assured by $F_i^{ii} \leq 0$ and $C_i'' \geq 0$, which are concavity of the demand function with respect to the firm's own price and nondecreasing marginal cost. The condition given in A9 is quite acceptable – unlike its counterpart, A7. This may be seen by writing out

$$\begin{aligned}\partial^2\pi_i/\partial p_i^2 + \sum_{j\neq i}|\partial^2\pi_i/\partial p_i\partial p_j| &= (2 - C_i''F_i^i)F_i^i + (p_i - C_i')F_i^{ii} \\ &\quad + \sum_{j\neq i}|(1 - C_i''F_i^i)F_i^j + (p_i - C_i')F_i^{ij}| \\ &= F_i^i + (1 - C_i''F_i^i)\sum_{j=1}^{n} F_i^j \\ &\quad + (p_i - C_i')(F_i^{ii} + \sum_{j\neq i}|F_i^{ij}|). \end{aligned} \tag{4.4}$$

A9 is implied by requiring nondecreasing marginal cost ($C_i'' \geq 0$), or, $C_i''F_i^i < 1$, along with $F_i^{ii} + \Sigma_{j\neq i}|F_i^{ij}| < 0$. These are not conditions which appear unreasonable or which become more restrictive as n increases. The demand condition states that the effect of a change in p_i on F_i^i dominates the combined effects of all other prices. A8 guarantees the existence and A9 uniqueness of equilibrium.

3. Cournot reaction functions

For the quantity model, Cournot reaction functions are introduced in ch. 2 §5, where it is pointed out they may be obtained by solving eqs. (2.2) for current output in terms of the output levels of the other firms in the preceding time period. To be precise, the ith firm assumes that all other firms choose in period t the same output which they chose in period $t-1$. Taking this assumption into account, the first-order condition for profit maximization becomes

$$\pi^i_i(q_{it}, \bar{\boldsymbol{q}}_{i,t-1}) = f(q_{it} + Q_{i,t-1}) + q_{it}f'(q_{it} + Q_{i,t-1}) - C'_i(q_{it}) = 0, \tag{4.5}$$

where $Q_{i,t-1} = Q_{t-1} - q_{i,t-1}$. By A6 and the implicit function theorem, eq. (4.5) may be written in the form $q_{it} = w_i(\bar{\boldsymbol{q}}_{i,t-1})$; thus the following theorem is proved:

Theorem 4.1. *In a model satisfying A1, A2 and A6, Cournot reaction functions exist.*

Theorem 4.1 does not imply the existence of a Cournot equilibrium; for, to make an example of duopoly, the reaction functions need not intersect in the positive quadrant. This is illustrated in fig. 4.1. As noted above, A1, A2 and A6 do imply existence, which is proved in ch. 7. Meanwhile, existence is assumed.

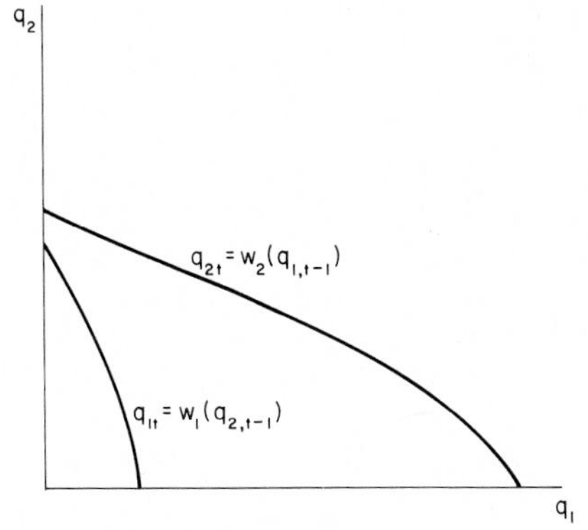

Fig. 4.1

In this model it is possible to consider whether the Cournot equilibrium, which satisfies the condition $\boldsymbol{q}^c = \boldsymbol{w}(\boldsymbol{q}^c)$, is stable. The equilibrium may be called stable if, starting from an arbitrary output vector, $\boldsymbol{q}_1$, and generating a sequence of output vectors using the Cournot reaction functions, $\boldsymbol{q}_t = \boldsymbol{w}(\boldsymbol{q}_{t-1})$, $t = 2, 3, \ldots$, it is found that $\boldsymbol{q}_t$ converges to $\boldsymbol{q}^c$ as $t \to \infty$. Adding A7 to the other three assumptions guarantees stability.

Theorem 4.2. *In a model satisfying A1, A2, A6 and A7, in which the Cournot equilibrium exists, and in which the Cournot reaction functions describe dynamic behavior, the Cournot equilibrium is stable.*

□ The Cournot reaction functions are defined for output vectors satisfying the conditions that at least two output levels are positive and the combined output of all firms sums to less than $\bar{Q}$. The functions may be extended so that they are defined for all output vectors, $\boldsymbol{q}$, which satisfy $\mathbf{0} \leq \boldsymbol{q} \leq (\bar{Q}, \ldots, \bar{Q})$. This is done by letting $w_i(\bar{\boldsymbol{q}}_{i,t-1}) = 0$ for $\bar{\boldsymbol{q}}_{i,t-1}$ such that $\bar{Q} \leq Q_{i,t-1} \leq (n-1)\bar{Q}$. Then w_i remains undefined only for $\bar{\boldsymbol{q}}_{i,t-1} = 0$, where it is now defined as $\lim_{\bar{q}_i \to 0} w_i(\bar{\boldsymbol{q}}_i)$. The extended w_i are now defined for all output vectors for which each firm produces between zero and $\bar{Q}$. They are continuous and, for any output vector for which they are defined, they give the profit maximizing response of the *i*th firm.

It remains now only to show that the system of reaction functions, $\boldsymbol{w}(\boldsymbol{q}_{t-1})$, which is a function from a subset of $\mathcal{R}^n_+$ to the same subset, is a contraction. For a differentiable function such as $\boldsymbol{w}$ to be a contraction, it is necessary and sufficient that $\Sigma^n_{j\neq i} |w^j_i(\bar{\boldsymbol{q}}_i)| < 1$. Where $Q_{i,t-1} > \bar{Q}$, the partial derivatives are all zero; hence, the condition is met. Where $Q_{i,t-1} < \bar{Q}$, the condition is derived from

$$w^j_i = -\pi^j_i / \pi^i_i. \tag{4.6}$$

Therefore,

$$\sum_{j\neq i} |w^j_i| = -\sum_{j\neq i} |\pi^j_i| / \pi^i_i, \tag{4.7}$$

which is less than one by A7. There remains the possibility that $Q_{i,t-1} = \bar{Q}$, in which case w_i does not have derivatives; however, directional derivatives exist which satisfy one or the other of the two cases which are discussed above. Hence, $\boldsymbol{w}$ is a contraction which establishes the stability of the Cournot equilibrium. □

Exactly parallel results may be obtained for the price model, in which the first-order condition for profit maximization for a firm, given that it expects the previous prices of the other firms to be repeated, is

$$\pi^i_i(p_{it}, \bar{\boldsymbol{p}}_{i,t-1}) = F_i(p_{it}, \bar{\boldsymbol{p}}_{i,t-1}) + (p_{it} - C'_i(F_i(p_{it}, \bar{\boldsymbol{p}}_{i,t-1})))F^i_i(p_{it}, \bar{\boldsymbol{p}}_{i,t-1}) = 0. \tag{4.8}$$

Corollary 4.3. *In a price model satisfying A2–A5 and A8, Cournot reaction functions exist. If, in addition, A9 is satisfied and the Cournot equilibrium exists, the Cournot equilibrium is stable.*

The proof exactly parallels the proofs of theorems 4.1 and 4.2 and need not be repeated.

There are some aspects of these models which deserve comment. First, the nature of the models needs to be explicitly pointed out. In them each firm is assumed to be a maximizer of current period profits. That is, in period t the ith firm's objective is to maximize $\pi_i(\boldsymbol{p}_t)$ with respect to p_{it}; however, with time explicitly in the model and the assumption being made that the firm is in the market for many consecutive time periods, it might be more sensible to assume the firm wishes to maximize a discounted sum of profits such as

$$\sum_{t=1}^{T} \alpha_i^{t-1} \pi_i(\boldsymbol{p}_t), \tag{4.9}$$

where T is the time horizon of the firm and $\alpha_i (0 < \alpha_i < 1)$ is the firm's *discount parameter.* The firm's discount parameter is the weight attached to period $t+1$ profits in period t. Thus, if r_i is the *discount rate* of the firm, the discount rate and the discount parameter are connected by the relation $\alpha_i = 1/(1+r_i)$. Single-period profit maximization is, in my view, hard to defend in a model such as this unless it turns out that optimal behavior is the same whether the firm has the maximization of current period profits or of an expression like eq. (4.9) as its objective.

Two related questions are whether it is reasonable to assume that each firm believes that the previous prices (or output levels) of its rivals are to be repeated, and when each firm is assumed to make its decisions. Given the rigid time structure of the model, either all firms would be assumed to make decisions in each time period or only one firm would be allowed to change its price (or output) in a given time period, with the firms alternating in a regular way. Take an example of the latter case, with three firms. Say that firm 1 makes decisions in periods 1, 4, 7, . . . , firm 2 in periods 2, 5, 8, . . . and firm 3 in periods 3, 6, 9, . . . Then it would certainly be true that in period 5, when it would be firm 2's turn to make a decision, assuming $p_{1,5} = p_{1,4}$ and $p_{3,5} = p_{3,4}$ would be entirely correct. But if the former case held, and all firms could choose new prices in each time period, it would generally be false that all firms would leave their prices unchanged. Should they behave according to Cournot reaction functions, for example, only when $\boldsymbol{p}_t = \boldsymbol{p}^c$ would prices be uniformly repeated. The usual situation would be that all firms would choose new prices in each time period, causing all their assumptions about their rivals' behavior to be wrong. Furthermore, the firms would continually receive information showing their assumptions to be wrong. Each firm is assumed to know the previous period prices of all firms; hence, a firm is in a position to know if $p_{j,t-1} = p_{j,t-2}$ as it had previously assumed.[4] Thus, in a model of simultaneous decision the Cournot reaction functions do not give an entirely acceptable equilibrium because, for each firm, its mode of behavior (i.e. its Cournot reaction function) is not a best reply to the actual behavior of the others. Instead, it is a best reply to wrong assumptions about how others behave, and the information flow in the model is naturally such that the firm is bound to see this.

It is tempting to opt for models of rotating decision in order to benefit from the analytical simplicity they provide. In ch. 5 a duopoly model of Cyert and de Groot (1970) utilizing alternate-period decision-making is discussed. They cite Cournot (1960) as a precedent, and it is correct that Cournot's stability discussion (pp. 81–82) can be interpreted as a process of alternate-period decision. On the other hand, in ch. 6 a model is developed in which firms may continuously change prices if they wish, but, due to the presence of adjustment costs associated with changes in the capital stock, they only want to alter prices at discrete intervals of time. Price changes in this model need not be evenly spaced; however, it is found that in equilibrium all firms want to choose new prices at the same points in time. These results provide a strong justification for models of simultaneous decision where discrete time periods are assumed; and in most of the discrete time models presented in this and later chapters, simultaneous decision is assumed.

The stability of Cournot equilibrium has been the subject of a large number of articles, appearing mainly in the *Review of Economic Studies* during the 1960s. Some of the references are Theocharis (1960), Fisher (1961), Hahn (1962), Okuguchi (1964), Hadar (1966) and Okuguchi (1969). The interested reader can get additional references from these sources. Both quantity and price models are analyzed, mostly assuming discrete time (with simultaneous decision). This line of analysis starts out from Theocharis' brief note which treats stability in a linear version of Cournot's quantity model. All the work is characterized by the assumption of single-period profit maximization and it does not face up to the inconsistency stemming from the firm's incorrect assumptions about rivals' behavior. These shortcomings greatly reduce the potential value of the articles.

[4]It could be assumed that the firms' memories only go back one time period. While this formally solves the problem, it involves making an assumption which is unreasonable in the present model.

4. *Bowley's conjectural variation*

Bowley (1924) mentions oligopoly in passing. Like Cournot, his model is not explicitly dynamic; however, unlike Cournot, no sense is to be made of it if it is not regarded as dynamic. He formulates a simple duopoly model for illustration, apparently taking the view that the essentials are represented and the reader can perform generalizations if he wishes. His model in the notation of this volume, is

$$\pi_1 = q_1(b_1 - b_2q_1 - b_2q_2) - b_3q_1^2, \tag{4.10}$$

$$\pi_2 = q_2(b_1 - b_2q_1 - b_2q_2) - b_3q_2^2, \tag{4.11}$$

where b_1, b_2 and b_3 are positive constants. The terms in parentheses are, of course, the demand function and the total cost functions are the terms $b_3q_i^2$. What is significant in Bowley's brief treatment is his handling of the first-order conditions for simultaneous profit maximization by the two firms. These are

$$\begin{aligned} \mathrm{d}\pi_1/\mathrm{d}q_1 &= \pi_1^1 + \pi_1^2 \partial q_2/\partial q_1 \\ &= b_1 - 2b_2q_1 - b_2q_2 - 2b_3q_1 - b_2q_1\partial q_2/\partial q_1 = 0, \end{aligned} \tag{4.12}$$

$$\begin{aligned} \mathrm{d}\pi_2/\mathrm{d}q_2 &= \pi_2^2 + \pi_2^1 \partial q_1/\partial q_2 \\ &= b_1 - 2b_2q_2 - b_2q_1 - 2b_3q_2 - b_2q_2\partial q_1/\partial q_2 = 0. \end{aligned} \tag{4.13}$$

He then writes (p. 38): "To solve these we should need to know q_2 as a function of q_1, and this depends on what each producer thinks the other is likely to do." That is, the values of the "conjectural variation" terms $\partial q_2/\partial q_1$ and $\partial q_1/\partial q_2$ are not known. Were they both zero, the first-order conditions would be the same as Cournot's.

Bowley seems on the track of setting up a model in which the two firms each have two reaction functions in mind. For firm 1

$$q_{1t} = \phi_1(q_{2,t-1}), \tag{4.12}$$

$$q_{2t} = g_1(q_{1,t-1}), \tag{4.13}$$

where the function g_1 describes the behavior which firm 1 thinks firm 2 will follow and the function ϕ_1 is optimal for firm 1 if his assumption about firm 2's behavior is correct. There would be a similar pair of reaction functions for firm 2. While this gets around the rather wooden assumption that firm 1 thinks that $q_{2t} = q_{2,t-1}$, it shares with it a fundamental difficulty. Namely if g_1 does not, in fact, describe the behavior of firm 2, then firm 1 is sure to find out. That is, firm 1 has predictions about how firm 2 behaves. If they are wrong, and indeed systematically wrong, firm 1 has every opportunity to notice and alter both his estimate and his own behavior accordingly. In other words, the chosen behavior of one firm is not a best reply to the actual, observed behavior of the other.

5. The Stackelberg model

Stackelberg (1934) carries the reaction function analysis further, and carries it out in much more completeness and detail than Bowley. I know of no one who, coming between them in time, carries the subject beyond Bowley. Though Stackelberg presents his results using several different duopoly models, only one such model, based on a Cournot type of duopoly, is reviewed here. In it, the essential contribution may be seen clearly. After presenting his results in his way, the model is reformulated in an explicitly dynamic fashion.

Stackelberg allows for two kinds of behavior, called *leader* and *follower.* A *follower* is a firm which behaves according to a Cournot reaction function, which means the firm chooses its output so that profit is maximized, given that the other firm repeats in the present time period its output from the previous period. A *leader* is a firm which operates on the assumption that its rival is a follower. The leader assumes its rival chooses output using a Cournot reaction function and seeks to maximize its own profit on that knowledge. The model is formulated with time subscripts omitted so that the equilibrium may be presented in the way Stackelberg gave it. With two firms and two types of behavior, there are three possible configurations for the model: follower–follower, leader–follower and leader–leader. Follower–follower is Cournot's version and has been fully discussed. Leader–leader is correctly dismissed by Stackelberg because it implies extreme inconsistencies. Inevitably both firms are always expecting each other to behave in ways they will not, because each assumes the other to be using a Cournot reaction function. There is not even a pair of output levels which, once attained, is repeated and expected by the firms.

The leader–follower case is the novel one, for which Stackelberg is particularly remembered. Assume firm 1 to be the follower. He chooses his output level using the Cournot reaction function $q_1 = w_1(q_2)$. This is known by the leader who takes w_1 into account; hence his profit function may be written

$$\pi_2(q_1, q_2) = q_2 f(w_1(q_2) + q_2) - C_2(q_2). \tag{4.14}$$

For firm 2 the first-order condition for a profit maximum is

$$\pi_2^2 + \pi_2^1 w_1' = f + q_2 f' - C_2' + q_2 f' w_1' = 0. \tag{4.15}$$

The leader–follower equilibrium is characterized by a pair of output levels (q_1^s, q_2^s), which satisfy the Cournot reaction function of firm 1 (i.e. $q_1^s = w_1(q_2^s)$) and eq. (4.15), the first-order condition of firm 2. The output choice of the leader is made as if he looks at each output pair on the reaction function of firm 1 and chooses for his own output the member of that pair for which his own profit is greatest. This is illustrated in fig. 4.2 where iso-profit curves are drawn for firm 2 in output space. The higher the iso-profit curve, the greater is the profit of firm 2, that is, $k_7 > k_6 > \cdots > k_1$. The leader–follower equilibrium is where an iso-profit curve of firm 2 is tangent to the reaction function of firm 1. This is the point $\boldsymbol{q}^s$ in fig. 4.2. For comparison, the light, broken curve is the Cournot reaction function of firm 2. Where the two reaction functions intersect, at $\boldsymbol{q}^c$, is the Cournot (follower–follower) equilibrium for the model.

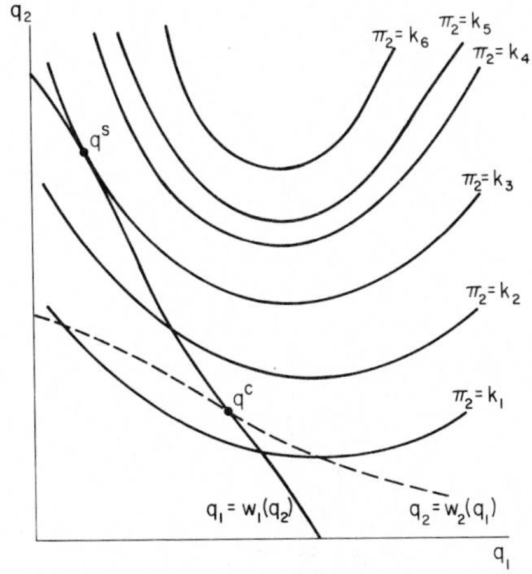

Fig. 4.2

Now consider various possible time structures for the model. First imagine a single-period model with simultaneous decision, second, a single-period model with sequential decision, third, a multi-period model with simultaneous decision and firm 2 maximizing current period profits, and fourth, a multi-period model with simultaneous decision and firm 2 maximizing a discounted stream of profits. In the first version, with simultaneous decision in a one-period model, eq. (4.15) is nonsense. That is, there is no way for firm 1 to react to firm 2, and firm 2 to take that into account. In other words, in this version, the only equilibrium which Stackelberg considers which is possible within the model is the Cournot. In the second version, with the leader choosing his output first, then the follower choosing after he is told what the leader did, the Stackelberg equilibrium is the only possible noncooperative equilibrium. Once the leader has chosen, the follower cannot possibly improve upon using w_1. Furthermore the follower has no way of influencing the leader's decision; hence, the leader can do no better than to take into account the inevitable behavior of the follower and choose his own output to satisfy eq. (4.15).

Although the preceding argument does justify the Stackelberg leader–follower equilibrium, it leaves us with a peculiar time sequence of decision-making which lacks economic justification. What is it that should make one firm go first? In a one time only model, nothing. In a multi-period model, one might argue that a particular firm could actually be a leader and take the lead in decision-making. Even there, what determines whether a firm emerges as leader, and if one does, which one? But in a multi-period model, the sequence of decisions becomes leader, follower, leader, follower, leader, ... How does this differ from a symmetric situation in which decisions are made alternately? Either not at all, or by an artificial device. The artificial device would be that in each time period, the leader announces first. An alternative is a continuous time model in which firms are free to alter rates of output continuously if they wish; but in which leader–follower behavior is optimal. Then, in effect, the leader announces at discrete intervals, and the follower announces just an instant after the leader. Such a sequence would need to be justified by being optimal for both firms, and the results of ch. 6 suggest, instead, that mutual optimality requires simultaneous announcement.

Turning now to multi-period models with simultaneous decision, the behavior of firm 1 is characterized by $q_{1t} = w_1(q_{2,t-1})$. Assuming first that firm 2 is a maximizer of current period profits, the objective in period t is to maximize $\pi_2(w_1(q_{2,t-1}), q_{2t})$ with respect to q_{2t}. At time t, $q_{2,t-1}$ is already chosen; hence, the first-order condition for firm 2 is $\pi_2^2(w_1(q_{2,t-1}), q_{2t}) = 0$, which, in conjunction with w_1, results in the Cournot equilibrium. In this version of the model, whenever the leader is to make a decision, the only choices of the follower which are relevant have already been made, and the Stackelberg leader–follower equilibrium cannot result.[5]

Consider a model just like the preceding except that the objective of the leader is to maximize a discounted stream of profits,

$$\sum_{t=1}^{T} \alpha_2^{t-1}\pi_2(w_1(q_{2,t-1}), q_{2t}). \tag{4.16}$$

Now firm 2 puts a positive weight on the profits of period $t+1$ when making its period t decision. Furthermore, its period t decision influences its period $t+1$ profits through its effect on $q_{1,t+1}$. What firm 2 needs to do is to solve for a production plan to cover its whole time horizon of T periods; to find, in other words, the values of $q_{21}, q_{22}, \ldots, q_{2T}$ which maximize eq. (4.16). It is apparent that this maximization problem can be solved recursively, starting from the last period in the firm's time horizon. Picture the firm's decision problem in period T, the last period of its horizon, when it knows $\boldsymbol{q}_t$ $(=(q_{1t}, q_{2t}))$ for $t = 1, \ldots, T-1$. In addition q_{1T} is known through w_1 and profits for all periods through $T-1$ are also completely determined. The firm has only to maximize $\pi_2(w_1(q_{2,T-1}), q_{2T})$ with respect to q_{2T}. The result of this exercise is to give the optimal q_{2T} as a function of $q_{2,T-1}$.[6] Call this function $\phi_1(q_{2,T-1})$. Stepping back one period to period $T-1$, the firm faces a two-period maximization problem for which its decision rule in the second of the two periods is already known. The two period objective function is

$$\pi_2(w_1(q_{2,T-2}), q_{2,T-1}) + \alpha_2\pi_2(w_1(q_{2,T-1}), \phi_1(q_{2,T-1})). \tag{4.17}$$

This, too, is a relatively simple maximization problem yielding the optimal value of $q_{2,T-1}$ as a function of $q_{2,T-2}$. Denote this function $\phi_2(q_{2,T-2})$. Continuing in the same manner, at period $T-k$ the firm knows the functions $\phi_1(q_{2,T-1}), \ldots, \phi_k(q_{2,T-k})$; therefore, the objective function for period $T-k$, $\sum_{l=0}^{k} \alpha_2^{k-l}\pi_{2,T-l}$, can be written as a function of only $q_{2,T-k}$ and $q_{2,T-k-1}$. Optimal behavior is given by $q_{2,T-k} = \phi_k(q_{2,T-k-1})$. The first-order condition from which ϕ_k is derived is

$$\begin{aligned}&\pi_2^2(w_1(q_{2,T-k-1}), q_{2,T-k})\\ &\qquad + \alpha_2\pi_2^1(w_1(q_{2,T-k}), \phi_{k-1}(q_{2,T-k}))w_1'(q_{2,T-k}) = 0.\end{aligned} \tag{4.18}$$

Thus the leader, firm 2, maximizes his discounted profits over a T

[5]For another Stackelberg model in which time is explicit, see Negishi and Okuguchi (1971).

[6]To simplify an unwieldy notation, π_{it} is sometimes used to denote $\pi_i(q_{1t}, q_{2t})$ and w_{it} for $w_i(q_{2t})$, etc.

period horizon by choosing output levels according to the plan $q_{2,1} = \phi_T(q_{2,0})$, $q_{2,2} = \phi_{T-1}(q_{2,1}), \ldots, q_{2T} = \phi_1(q_{2,T-1})$.[7] Note that each behavior rule is optimal independent of the actual time period, given the length of the remaining horizon. For example, ϕ_1 is the optimal rule to follow in the last period of the firm's horizon, no matter whether the last period is period T or period 6 or some other. Models of which the present one is a special case are studied intensively in ch. 5, where many results on such models are proved. For the Stackelberg model considered here, all desired results are assumed; for conditions may be found in ch. 5 under which the results hold. Meanwhile, we will see the implications of this Stackelberg model, given that results are obtainable.

In the present model, the behavior rules would be expected to be related to one another as illustrated in fig. 4.3, where it is seen that ϕ_1 lies below ϕ_2 which lies below ϕ_3, etc. ϕ_k lies below ϕ_{k+1} in the sense that

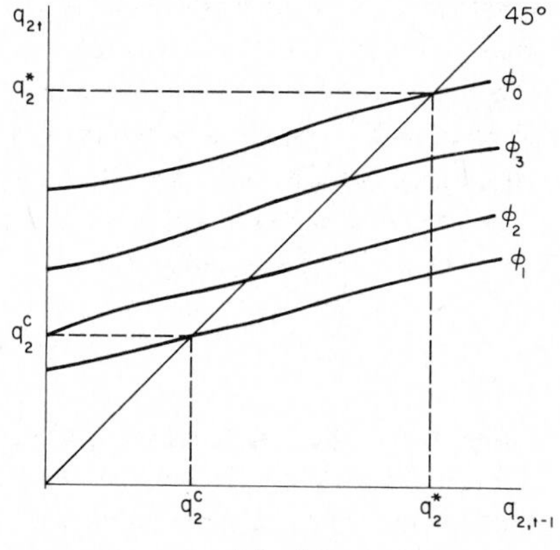

Fig. 4.3

$\phi_k(q_{2t}) < \phi_{k+1}(q_{2t})$. It is possible that as the time horizon T, goes to infinity, the behavior rule for the first time period, ϕ_T, converges to some rule, ϕ_0, and that ϕ_0 is the optimal behavior rule for the firm when its horizon is infinite. Were that so, it is likely that, with firm 1 using w_1 and firm 2 using ϕ_0, the output levels would converge to steady state values. In fig. 4.3, the steady state value of q_{2t} is shown as q_2^*. The corresponding value for firm 1 is $q_1^* = w_1(q_2^*)$. The Cournot equilibrium value of q_2, q_2^c, is

[7] $q_{2,0}$ is an initial condition which has not been introduced previously. Assume it is part of the initial information known to firm 2.

also shown (where ϕ_1 intersects the 45° line). In the limit as $T \to \infty$, with $q_{2t} = q_2^*$, the first-order condition eq. (4.18) reduces to

$$\pi_2^2(w_1(q_2^*), q_2^*) + \alpha_2 \pi_2^1(w_1(q_2^*), q_2^*) w_1'(q_2^*) = 0. \tag{4.19}$$

Compare eq. (4.19) to the Stackelberg leader first-order condition, eq. (4.15). They would be identical if $\alpha_2 = 1$. Thus, Stackelberg's leader–follower equilibrium is a limiting case of the present model.

There remains a serious flaw in the Stackelberg model, even as amended above. The follower's behavior in the leader–follower model is no more credible than that of either follower is in the follower–follower (i.e. Cournot reaction function) model. In all cases, the follower makes an assumption about how his rival behaves which is demonstrably false. The rival, be he follower or leader, does not keep his output unchanged from period to period – except if the initial output levels are the final equilibrium levels.

There is another peculiarity in the leader–follower model, a convenient asymmetry. The leader is very clever, but the follower is extremely naive and blind. One knows exactly, correctly, how his rival behaves, while the other makes the most patently false assumption about his rival. Another limitation of the leader–follower model is that it generalizes to n firms only on the basis of one leader to $n-1$ followers.

6. Concluding remarks

After stating conditions under which the single-period quantity and price models have unique Cournot equilibria, attention moved for the remainder of the chapter to the early reaction function models of Cournot, Bowley and Stackelberg. Of a model in which behavior is described by Cournot reaction function, Fellner (1949) writes (p. 63):

> It is essential to realize that, as long as the firms make the Cournot assumptions concerning their rivals' behavior, the analysis cannot be adjusted in such a way as to make the firms be right for the right reasons, instead of describing a situation in which they turn out to be right for the wrong reasons.

Being "right for the wrong reasons" refers to being, eventually, at the Cournot equilibrium where each firm's assumption that the other will repeat his previous output is correct. For them to be "right for the right reasons" requires that each firm use a reaction function which is a best reply to the reaction functions of its rivals. The Stackelberg leader–follower equilibrium in the explicitly dynamic model presents a mixed situation in which the follower is not using a best reply against the leader's reaction function; however, the leader's behavior is a best reply against the behavior of the follower. Eventually, however, the quantity choices settle at (q_1^*, q_2^*) where the follower is "right for the wrong reasons." It is worth noting that the leader's behavior rule, $\phi_0(q_{2t})$, is not of the same form as the follower's. It gives the leader's current output as a function of his own lagged output, rather than as a function of his rival's. A transformation can be used to state the leader's optimal current output as a function of the follower's previous period output: $\phi_0(q_{2t}) = \phi_0(\phi_0(q_{2,t-1})) = \phi_0(\phi_0(w_1^{-1}(q_{1t}))) = v_2(q_{1t})$. It is of fundamental importance to remember that in a single-period model, the Cournot equilibrium ($\boldsymbol{q}^c$ or $\boldsymbol{p}^c$) is characterized by the firm's being right for the right reasons. The situation of being right for the wrong reasons occurs precisely because the behavior of one or more firms is not a best reply to the behavior of the others. In the single-period models, the Cournot choice of one firm is a best reply to the Cournot choices of the others.

In ch. 5 the techniques used to find the leader's optimal behavior, ϕ_0, are used to find reaction function equilibria for quite general models. Just as the leader–follower model has a steady state output vector $\boldsymbol{q}^*$ which is different from the Cournot output vector $\boldsymbol{q}^c$, the general reaction function models which follow lead to steady state price vectors which are, in general, different from the Cournot price vector $\boldsymbol{p}^c$.

CHAPTER 5

REACTION FUNCTION EQUILIBRIA FOR EXPLICITLY DYNAMIC MODELS

1. Setting the problem

Three different models are pursued in this chapter using quite similar techniques and assumptions. The first, which is studied in greatest detail, is based upon the differentiated products models for n firms which are presented in ch. 3 and the second is a homogeneous products duopoly model of quantity choice along the lines of those of ch. 2. Both assume simultaneous decision, while the third model, utilizing an underlying market model which is a simplified version of the second, is based on alternate period decision.

Using the price model of ch. 3 as a foundation, imagine n firms in a market with objective functions which are discounted streams of profit,

$$\sum_{t=1}^{T} \alpha_i^{t-1}\pi_i(\mathbf{p}_t), \qquad i = 1, \ldots, n. \tag{5.1}$$

Note that the model possesses stationarity and lacks time dependence. By stationarity is meant that the profit function for a firm in a given time period, π_i, is in no way a function of time. If for two different times, t and t', $\mathbf{p}_t = \mathbf{p}_{t'}$, then $\pi_i(\mathbf{p}_t) = \pi_i(\mathbf{p}_{t'})$ for any firm i. *Time dependence* is present when the profit function for a given period, t, is a function of past, as well as current prices. A natural example is the market for a consumer durable in which it is to be expected that lower prices in period $t-1$ bring higher sales in period $t-1$, resulting in a larger stock of relatively new goods in the hands of consumers in period t; hence, lower sales in period t. Game theoretic models with a time dependent structure are studied in ch. 9 and applications to oligopoly are considered there.

An individual firm, like the Stackelberg leader of ch. 4 §5, must find a policy which maximizes its discounted profit stream, given its expectations concerning the decision policies of the other firms. To simplify the example slightly for the time being, assume the firm has an infinite time horizon (i.e. that T in eq. (5.1) is ∞). Now we assume further that, as in the Bowley model, the firms have beliefs about how their rivals behave; particularly, that

$$p_{it} = \psi_i(\boldsymbol{p}_{t-1}), \qquad i = 1, \ldots, n. \tag{5.2}$$

There are several characteristics of eq. (5.2) which are important. First, p_{it} depends upon all prices of period $t-1$ including $p_{i,t-1}$, the ith firm's own price. Second, ψ_i describes the expectations which all firms other than the ith have of the ith. Third, ψ_i is stationary; it is a reaction function which all other firms think the ith will use forever.

Now consider the position of firm 1, which seeks to maximize $\Sigma_{t=1}^{\infty} \alpha_1^{t-1} \pi_1(\boldsymbol{p}_t)$, subject to $\bar{\boldsymbol{p}}_{1t} = \bar{\boldsymbol{\psi}}_1(\boldsymbol{p}_{t-1})$.[1] It is intuitively quite plausible that for firm 1, discounted profits are maximized by choosing $p_{1t} = \phi_1(\boldsymbol{p}_{t-1})$ in each time period. ϕ_1 would, in that event, be a best reply to $\bar{\boldsymbol{\psi}}_1(\boldsymbol{p}_{t-1})$. Approaching the decision problem of any other firm, i, in the same way, the firm seeks to maximize $\Sigma_{t=1}^{T} \alpha_i^{t-1} \pi_i(\boldsymbol{p}_t)$ subject to $\bar{\boldsymbol{p}}_{it} = \bar{\boldsymbol{\psi}}_i(\boldsymbol{p}_{t-1})$. Say this results in a best reply for the ith firm of $p_{it} = \phi_i(\boldsymbol{p}_{t-1})$. For the firms, taken together, $\boldsymbol{\phi}(\boldsymbol{p}_t) = (\phi_1(\boldsymbol{p}_t), \ldots, \phi_n(\boldsymbol{p}_t))$ is a best reply to $\boldsymbol{\psi}(\boldsymbol{p}_t) = (\psi_1(\boldsymbol{p}_t), \ldots, \psi_n(\boldsymbol{p}_t))$.

What is the most natural way to generalize the Cournot equilibrium (for single-period models) to the present context? The distinguishing feature of the Cournot equilibrium is that the equilibrium decision of the firms $\boldsymbol{p}^c$ are, taken together, a best reply to themselves. That is, for each firm, p_i^c is a best reply to $\bar{\boldsymbol{p}}_i^c$. In the present context, the decision of a firm is its reaction function, ϕ_i; so a best reply equilibrium is one for which $\phi_i = \psi_i$ for all i. In other words, $\boldsymbol{\phi}$ should be a best reply to itself. If all firms believe the ith will use ϕ_i, and they are each using ϕ_j $(j \neq i)$, then, knowing the reaction functions of the others, the ith firm can do no better than to use ϕ_i. Should the foregoing be true for all i, then when $\phi_1, \ldots, \phi_n$ are the respective reaction functions of the firms, there is no way that one firm can, by itself, alter its behavior and increase its discounted profits.[2]

For the model under discussion it does not prove possible to show the existence of a best reply equilibrium; however, an approximation to it can be defined which does exist. These questions are pursued for the price

[1]The notation distinguished by the overbar, to set apart all firms except the ith, is used for vectors of functions as well as for vectors of scalars. Thus $\bar{\boldsymbol{\psi}}_i(\boldsymbol{p}) = (\psi_1(\boldsymbol{p}), \ldots, \psi_{i-1}(\boldsymbol{p}), \psi_{i+1}(\boldsymbol{p}), \ldots, \psi_n(\boldsymbol{p}))$.

[2]In addition to generalizing the Cournot equilibrium, the equilibrium sketched in this paragraph is a noncooperative equilibrium in the sense of Nash (1951). This Nash equilibrium is studied in ch. 7 and put to much use in chs. 8–10.

model with simultaneous decision in §2. In §3 a quantity model, essentially a Cournot duopoly with simultaneous decision, is analyzed along lines sketched above. §3 also takes up a very simple (linear demand, constant marginal cost) Cournot duopoly in which decision is sequential, with firm 1 able to change its output level in odd numbered periods and firm 2 in even. For this model, a best reply equilibrium can be shown to exist. §4 is concerned with an evaluation of the reaction function results.

2. *Price models with simultaneous decision*

The model used here assumes A2–A5, A8 and A9, and several additional assumptions which are added at various points in the present section. §2.1–2.3 are wholly taken up with the study of optimal behavior for one firm, alone, when its rivals are known to behave according to particular reaction functions. The first step, taken in §2.1 is to consider the problem when the firm has a finite time horizon. In this case, the ith firm has an optimal policy which is described by a sequence of reaction functions, $p_{it} = \phi_{i1}(\mathbf{p}_{t-1}), \ldots, p_{it} = \phi_{iT}(\mathbf{p}_{t-1})$. The function ϕ_{i1} is the optimal behavior rule for the firm for the last period of its time horizon. This is true irrespective of which time period is last or of the length of the horizon. The function ϕ_{i2} is the optimal behavior rule for the second to last period of its planning horizon, given that it behaves optimally in the last period by using ϕ_{i1} at that time. In general, ϕ_{is} is optimal for the firm when the remaining horizon is s periods, given that $\phi_{i,s-1}, \ldots, \phi_{i1}$ is followed in the succeeding periods.

The sequence of reaction functions, $\phi_{i1}, \ldots, \phi_{iT}$, constitutes a best reply to $\bar{\boldsymbol{\psi}}_i(\mathbf{p}_t)$; however, that best reply is contingent on the horizon length, T. Intuitively, it seems likely that the change from one reaction function to another would be less and less as the horizon increases (that is, that ϕ_{iT} and $\phi_{i,T-1}$ would be more nearly identical as T increases). That question is investigated in §2.2, where it is found that ϕ_{iT} converges as $T \to \infty$. The next question which naturally presents itself is whether the reaction function $\phi_i = \lim_{T\to\infty} \phi_{iT}$ is optimal for the firm when its time horizon is infinite. In §2.3 it is found that ϕ_i is optimal for an infinite horizon; hence, ϕ_i may be said to be the best reply of the ith firm to $\bar{\boldsymbol{\psi}}_i$.

Among the results sketched above is that a stationary reaction function for the firm is optimal when the other firms use stationary reaction functions and the firm's time horizon is infinite. What of market equilibrium? Is it possible to show the existence of reaction functions, $\boldsymbol{\psi} =$

$(\psi_1, \ldots, \psi_n)$ which are best replies to themselves? Not in a satisfactory general way. There are two results, one of which is as obvious as it is uninteresting. Imagine the degenerate reaction functions, $p_{it} = p_i^c$, $i = 1, \ldots, n$. Under these, each firm chooses the price which is its Cournot equilibrium price, irrespective of what any other firm does. Such behavior is a best reply equilibrium for the model.[3] Given the choices of the others, the ith firm cannot improve upon choosing p_i^c in every time period. It is useful to know that such behavior is a best reply equilibrium for the model; however, the question of the existence of a best reply equilibrium based on non-degenerate reaction functions remains open and is of great interest. What is shown in §2.4 is the existence of an *approximate best reply equilibrium.*

Much discussion of oligopolistic behavior presumes that firms either collude or follow Cournot behavior. It is then reasoned that Cournot behavior is not followed because cooperation can give each firm greater profits. Such discussion does not grapple with either of two important issues: (1) Under which *rules* or institutional arrangements can collusion occur? (2) Are other noncooperative equilibria possible than the Cournot equilibrium? It is argued in ch. 2 §4 that certain rules (those which permit binding agreements) allow collusion and the absence of these rules prevent it. As to the second question, the present chapter, as well as chs. 8 and 9, are directed toward providing an affirmative answer.

As a preliminary to defining the approximate best reply equilibrium, consider a vector of reaction functions, $\boldsymbol{\psi}$. Assume there is exactly one price vector, $\boldsymbol{p}^*$, which, if once chosen, is continually chosen; and, which is the eventual price vector chosen if $\boldsymbol{\psi}$ is used by the firms. Thus $\boldsymbol{p}^* = \boldsymbol{\psi}(\boldsymbol{p}^*)$ and, for any arbitrary price vector, $\boldsymbol{p}^1$, if a sequence of price vectors is defined by $\boldsymbol{p}^t = \boldsymbol{\psi}(\boldsymbol{p}^{t-1})$ (for $t = 2, 3, \ldots$), then $\boldsymbol{p}^* = \lim_{t\to\infty} \boldsymbol{p}^t$. $\boldsymbol{p}^*$ is a steady state equilibrium price vector in relation to the reaction functions $\boldsymbol{\psi}$. As an example, if $\boldsymbol{\psi} = \boldsymbol{w}$, the Cournot reaction functions, then $\boldsymbol{p}^*$ would be the Cournot price vector, $\boldsymbol{p}^c$.

Assume that $\boldsymbol{\phi}$ is the best reply against $\boldsymbol{\psi}$, in the sense that ϕ_i is the best reply against $\bar{\boldsymbol{\psi}}_i$ for $i = 1, \ldots, n$. $\boldsymbol{\phi}$ is an approximate best reply equilibrium if $\boldsymbol{p}^* = \boldsymbol{\psi}(\boldsymbol{p}^*) = \boldsymbol{\phi}(\boldsymbol{p}^*)$ and $\psi_i^j(\boldsymbol{p}^*) = \phi_i^j(\boldsymbol{p}^*)$ for $i, j = 1, \ldots, n$. In other words, $\boldsymbol{\phi}$ is an approximate best reply equilibrium if there is a vector of reaction functions $\boldsymbol{\psi}$ such that $\boldsymbol{\phi}$ and $\boldsymbol{\psi}$ have the same steady state prices, $\boldsymbol{p}^*$, and, at $\boldsymbol{p}^*$, the two vectors of reaction functions are tangent to one another. This is illustrated in fig. 5.1. It should be kept in

[3]i.e., a noncooperative equilibrium in the sense of Nash (1951).

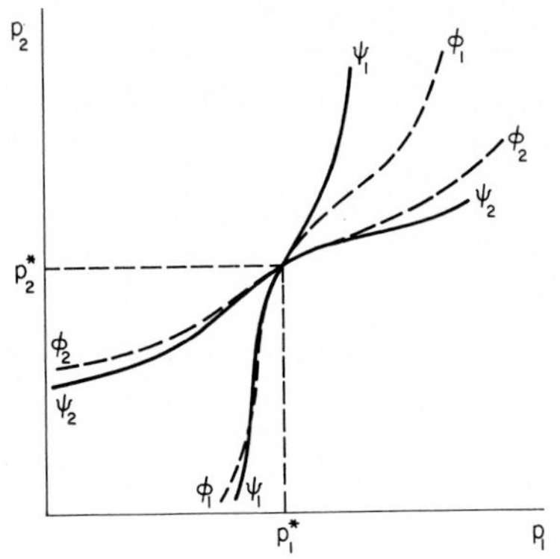

Fig. 5.1

mind that for $n = 2$, the reaction function of a firm, say ψ_1, is a function from $\mathscr{R}^2$ into $\mathscr{R}$. Thus, the graph of the function is in $\mathscr{R}^3$. Rather than try to draw three dimensions in fig. 5.1, the example assumes that ψ_1 depends only on p_2 and that ψ_2 depends only on p_1.[4]

2.1. *Optimal policy for a single firm with a finite time horizon*

In stating assumptions and theorems throughout §§2.1–2.3, most are written for firm 1; however it is understood that all such statements apply in the obvious way to all firms. Prior to proving results on optimal policy, additional assumptions are required. The first is that, no matter what the prices of other firms, firm 1 finds it profitable to choose a price at which its production and sales are greater than zero. Let p_1^0 be defined by the condition that $(p_1^0, \mathbf{0}) \in \alpha_1$. Thus $F_1(p_1^0, \bar{\mathbf{p}}_1) > 0$ for any $\bar{\mathbf{p}}_1 > 0$ and $F_1(p_1, \bar{\mathbf{p}}_1) > 0$ for any p_1 such that $0 \leq p_1 < p_1^0$ and $\bar{\mathbf{p}}_1 \geq \mathbf{0}$; hence, the firm's demand is necessarily positive if its price is below p_1^0 or if its price is at p_1^0 and at least one other firm has a positive price.

A10. $p_1^0 > C_1'(F_1(p_1^0, \mathbf{0})) = C_1'(0)$.

A10 assures that at an output level of zero, marginal revenue exceeds

[4]The results in this section (§2) come from Friedman (1971b), (1973b), (1976a) and, in an incidental way, from (1972).

marginal cost for the firm, no matter what prices are being charged by the other firms. An obvious implication of A10 is that $F_1(\boldsymbol{p}^c) > 0$. At the Cournot equilibrium, all firms have positive output levels – all are in the market.

The next assumption places restrictions on the class of reaction functions which are considered.

A11. *For the interior of the domain on which* $\boldsymbol{\psi}$ *is defined,* $\boldsymbol{\psi}$ *has continuous second partial derivatives. Furthermore,* $\psi_i^k(\boldsymbol{p}) > 0$ *and* $\sum_{k=1}^n \psi_i^k(\boldsymbol{p}) \leq \lambda < 1$. *Finally, for any* $\boldsymbol{p}$ *in the domain of* $\boldsymbol{\psi}$, $\boldsymbol{\psi}(\boldsymbol{p})$ *is in the domain.*

Unless otherwise stated, the domain of $\boldsymbol{\psi}$ is the set of price vectors, $\mathscr{D}$, no smaller than zero and no larger than $\boldsymbol{p}^+$, the maximal price vector in $\mathscr{A}$,

$$\mathscr{D} = \{\boldsymbol{p} | \boldsymbol{0} \leq \boldsymbol{p} \leq \boldsymbol{p}^+\}. \tag{5.3}$$

A11 requires that the price of a firm be an increasing function of the previous period's price of any firm, and that $\boldsymbol{\psi}$ be a *contraction* (i.e., that a pair of price vectors, $\boldsymbol{p}$ and $\boldsymbol{p}'$, are no "closer" to each other than their images, $\boldsymbol{\psi}(\boldsymbol{p})$ and $\boldsymbol{\psi}(\boldsymbol{p}')$).[5] Or, to put it another way, if all firms increase prices by Δp from period $t-1$ to period t, then, each firm will increase its price from period t to $t+1$ by no more than $\lambda \Delta p$. The contraction requirement implies that the reaction functions are stable. A sequence of prices generated by using the reaction functions over and over: $\boldsymbol{p}^t = \boldsymbol{\psi}(\boldsymbol{p}^{t-1})$, $t = 1, 2, \ldots,$ starting from an arbitrary $\boldsymbol{p}^0$, converges to a unique steady state price vector.

Turning now to the maximization problem of firm 1, it seeks to maximize

$$\sum_{t=1}^{T} \alpha_1^{t-1} \pi_1(\boldsymbol{p}_t), \tag{5.4}$$

subject to

$$\bar{\boldsymbol{p}}_{1t} = \bar{\boldsymbol{\psi}}_1(\boldsymbol{p}_{t-1}), \tag{5.5}$$

for $t = 1, 2, \ldots,$ and an initial condition, $\boldsymbol{p}_0$. It is assumed that when it is time to make the decisions of a given period, say t, that every firm knows all prices chosen in all periods up to and including period $t-1$. But period t prices are chosen simultaneously by the firms. Thus, when it is time to choose p_{1T}, firm 1 knows $\boldsymbol{p}_0, \ldots, \boldsymbol{p}_{T-1}$, and its profits for all periods prior to period T will have been totally determined. Therefore, at this time, the

[5]That is, for any $\boldsymbol{p}, \boldsymbol{p}' \in \mathscr{D}$, $\max_i |\psi_i(\boldsymbol{p}) - \psi_i(\boldsymbol{p}')| < \max_i |p_i - p_i'|$.

firm's maximization problem is simplified to

$$\max_{p_{1T}} \pi_1(p_{1T}, \bar{\boldsymbol{\psi}}_1(\boldsymbol{p}_{T-1})). \tag{5.6}$$

The first-order condition for an interior maximum for this problem is

$$G_{11}(\boldsymbol{p}_t) = \pi_1^1(p_{1t}, \bar{\boldsymbol{\psi}}_1(\boldsymbol{p}_{t-1})) = 0. \tag{5.7}$$

The subscript t appears rather than T to emphasize that there is nothing special in period T other than its being the last period to be taken into account. Thus, for any t, $G_{11} = 0$ is the appropriate first-order condition for firm 1 when firm 1 has an horizon of 1 period. If, for any $\boldsymbol{p}_{t-1} \in \mathscr{D}$, there is a unique p_{1t} satisfying the condition, then we may say that the first-order condition may be solved for p_{1t} as a function of $\boldsymbol{p}_{t-1}$, and that function may be represented by

$$p_{1t} = \phi_{11}(\boldsymbol{p}_{t-1}). \tag{5.8}$$

Note that the first-order condition for period t (and a one-period horizon) is a function of p_{1t} (the firm's own current price) and $\boldsymbol{p}_{t-1}$ (all firms' lagged prices, including firm 1).

Now consider the two-period horizon problem, when the last period decision is to be given by ϕ_{11},

$$\max_{p_{1,T-1}} \pi_1(p_{1,T-1}, \bar{\boldsymbol{\psi}}_1(\boldsymbol{p}_{T-2})) + \alpha_1 \pi_1(\phi_{11}(p_{1,T-1}, \bar{\boldsymbol{\psi}}_1(\boldsymbol{p}_{T-2})), \bar{\boldsymbol{\psi}}_1(p_{1,T-1}, \bar{\boldsymbol{\psi}}_1(\boldsymbol{p}_{T-2}))). \tag{5.9}$$

The first-order condition is

$$G_{12}(\boldsymbol{p}_t) = \pi_1^1(\boldsymbol{p}_t) + \alpha_1 \sum_{j=2}^{n} \pi_1^j(\boldsymbol{p}_{t+1}) \psi_j^1(\boldsymbol{p}_t) = 0. \tag{5.10}$$

Assuming this first-order condition can be solved, it may be represented as

$$p_{1t} = \phi_{12}(\boldsymbol{p}_{t-1}). \tag{5.11}$$

If reaction functions for firm 1 can be found in a similar way for horizons up to s periods, the first-order condition for profit maximization with a horizon of $s+1$ periods, given that the reaction functions $\phi_{1s}, \ldots, \phi_{11}$ are used in subsequent periods, is

$$G_{1,s+1}(\boldsymbol{p}_t) = \pi_1^1(\boldsymbol{p}_t) + \alpha_1 \sum_{j=2}^{n} \pi_1^j(\boldsymbol{p}_{t+1}) \psi_j^1(\boldsymbol{p}_t) + \sum_{l=2}^{s} \alpha_1^l \sum_{j_1=2}^{n} \pi_1^{j_1}(\boldsymbol{p}_{t+l}) \sum_{j_2=2}^{n} \psi_{j_1}^{j_2}(\boldsymbol{p}_{t+l-1}) \cdots \sum_{j_{l-1}=2}^{n} \psi_{j_{l-2}}^{j_{l-1}}(\boldsymbol{p}_{t+2}) \sum_{j_l=2}^{n} \psi_{j_{l-1}}^{j_l}(\boldsymbol{p}_{t+1}) \psi_{j_l}^1(\boldsymbol{p}_t) = 0, \tag{5.12}$$

subject to

$$p_{j,t+l} = \psi_j(\mathbf{p}_{t+l-1}), \qquad j = 2, \ldots, n, \quad l = 0, \ldots, s, \tag{5.13}$$

$$p_{1,t+l} = \phi_{1,s-l+1}(\mathbf{p}_{t+l-1}), \qquad l = 1, \ldots, s. \tag{5.14}$$

In both eqs. (5.10) and (5.12) there are many summations over firms with the index running from 2 to n, rather than running from 1 to n. The omitted terms are equal to zero due to the previous first-order conditions which are satisfied. For example, in eq. (5.10) the omitted term is $\alpha_1\pi_1^1(\mathbf{p}_{t+1})\psi_1^1(\mathbf{p}_t)$, which is known to be zero because $\pi_1^1(\mathbf{p}_{t+1})$ is the first-order condition for a one-period horizon. For brevity eq. (5.12) may be written

$$G_{1,s+1}(\mathbf{p}_t) = \pi_1^1(\mathbf{p}_t) + \sum_{l=1}^{s} \alpha_1^l \sum_{j=2}^{n} \pi_1^j(\mathbf{p}_{t+l}) M_{1,j,l}(\mathbf{p}_t) = 0, \tag{5.15}$$

where

$$M_{1,j,l}(\mathbf{p}_t) = \sum_{j_2=2}^{n} \psi_j^{j_2}(\mathbf{p}_{t+l-1}) \cdots \sum_{j_l-1}^{n} \psi_{j_l-2}^{j_l-1}(\mathbf{p}_{t+2})$$

$$\sum_{j_l=2}^{n} \psi_{j_l-1}^{j_l}(\mathbf{p}_{t+1}) \psi_{j_l}^1(\mathbf{p}_t), \qquad l \geq 2. \tag{5.16}$$

Another required assumption is:

A12. *For any* $\mathbf{p}_t \in \mathring{\mathscr{A}}_1^*$, $\mathbf{p}_{t+1}, \ldots, \mathbf{p}_{t+s-1} \in \mathscr{D}$ *and* $s \geq 1$, $G_{1s}^1(\mathbf{p}_t) \leq -\epsilon$; *for* $\epsilon > 0$.

A12 is an assumption that the second-order conditions for profit maximization are always met. It is usually undesirable to make such an assumption directly rather than to derive it from reasonable underlying conditions on the basic model. In the present instance, that would mean deriving A12 from conditions on the demand and cost functions, and perhaps some restriction of the discount parameter. All this is, in fact, done. Doing so is a particularly tedious business from which no worthwhile economic insights emerge. For this reason, it is postponed until §2.5. It is now possible to prove the existence of the reaction functions, ϕ_{1s}.

Theorem 5.1. *If A2–A5 and A8–A12 hold, the optimal sequence of reaction functions,* ϕ_{1s}, $s = 1, 2, \ldots$, *exists. The domain of each function is* $\mathscr{D}$ *and, for any* $\mathbf{p} \in \mathscr{D}$, $\phi_{1s}(\mathbf{p}) > C_1'(F_1(\mathbf{p}))$.

□ The proof proceeds by first showing that the optimal price of a firm is never at or below marginal cost (i.e. that $\phi_{1s}(\boldsymbol{p}) > C_1'(F_1(\boldsymbol{p}))$). This is done by, first, showing that a price which falls below marginal cost in the last period of the horizon is less profitable in that last period than a higher price. In addition, when p_{1t} is changed, profits from earlier periods are unaffected. Therefore, if the final period price is below marginal cost, raising it increases profit in the last period and leaves the profits of all earlier periods unchanged. In an earlier period, increasing a price which is below marginal cost has the same beneficial effect on current period profits, no effect on profits of earlier periods and a beneficial effect on the profits of later periods. Let $\boldsymbol{p}_0^0$ be an arbitrary initial vector of prices and let $p_{1t}^0 = \phi_{1,T-t+1}(\boldsymbol{p}_{t-1}^0)$, $p_{jt}^0 = \psi_j(\boldsymbol{p}_{t-1}^0)$, $j = 2, \ldots, n$, $t = 1, \ldots, T$. Then given $\boldsymbol{p}_0^0$ and the behavior of the other firms $(\bar{\boldsymbol{\psi}}_1)$, the choices $p_{11}^0, \ldots, p_{1T}^0$ maximize $\Sigma_{t=1}^T \alpha_1^{t-1}\pi_1(\boldsymbol{p}_t)$ with respect to the price choices open to firm 1. Assume that for some values of t, say $t_1, \ldots, t_k$ (where $t_1 < t_2 < \cdots < t_k$), $p_{1,t_l} \leq C_1'(F_1(\boldsymbol{p}_t^0))$, $l = 1, \ldots, k$. Consider first t_k,

$$\pi_1^1(\boldsymbol{p}_{t_k}^0) = F_1 + (p_{1,t_k} - C_1')F_1^1 > 0 \tag{5.17}$$

hence $\pi_1(\boldsymbol{p}_{t_k})$ increases if p_{1,t_k} is raised. If $t_k = T$, $\pi_1(\boldsymbol{p}_T)$ can be increased with no change in $\pi_1(\boldsymbol{p}_t^0)$ (for $t < T$); hence, $p_{11}^0, \ldots, p_{1T}^0$ would not be optimal. On the other hand, if $t_k < T$, then increasing p_{1,t_k} above p_{1,t_k}^0, leaving $p_{1t}(t \neq t_k)$ unchanged, causes all $p_{jt}(j \neq 1,\ t > t_k)$ to rise. As

$$\pi_1^j(\boldsymbol{p}_t^0) = (p_{1t}^0 - C_1'(F_1(\boldsymbol{p}_t^0))F_1^j(\boldsymbol{p}_t^0) \geq 0, \tag{5.18}$$

for $t > t_k$, future profits do not fall; hence if $p_{11}^0, \ldots, p_{1t}^0$ is optimal, $p_{1t}^0 > C_1'(F_1(\boldsymbol{p}_t^0))$ for all $t = 1, \ldots, T$.

The part of the proof which is above shows that, if an optimal price exists for a given period, it must be in $\mathscr{A}_1^*$. By A12, the objective function of the firm is concave with respect to p_1 on $\mathscr{A}_1^*$. Finally, it is also proved that the optimal price is above marginal cost, so it is above the lower boundary of $\mathscr{A}_1^*$. Taken together, the optimal price exists because maximization is over a compact set, is unique due to the strict concavity of the objective function, and is given by eq. (5.15) except for the possibility that $G_{1,s}(\boldsymbol{p}_t) > 0$ for all values of p_{1t} such that $(p_{1t}, \bar{\boldsymbol{\psi}}_1(\boldsymbol{p}_{t-1})) \in \mathring{\mathscr{A}}_1^*$. In the latter case, the optimal price is given by the conditions that $(p_{1t}, \bar{\boldsymbol{\psi}}_1(\boldsymbol{p}_{t-1})) \in \mathscr{A}_1^*$ and $F_1(p_{1t}, \bar{\boldsymbol{\psi}}_1(\boldsymbol{p}_{t-1})) = 0$. A12, together with an appeal to the implicit function theorem, implies that the optimal price of the firm in period t, when the horizon is s periods and optimal behavior is to be followed in subsequent periods, may be written $p_{1t} = \phi_{1s}(\boldsymbol{p}_{t-1})$. □

Theorem 5.1 states that the optimization problem for a single firm, when its horizon is finite, is solvable, and that the firm always chooses a price which, given its prediction for rival prices, exceeds its marginal cost. Both results are quite unremarkable, although they are essential steps in answering more interesting questions. Problems of backward induction, or dynamic programming, are generally solvable on much weaker assumptions than those made here; although, if the assumptions are very weak, existence of a solution could be all that may be obtained. It may prove impossible to know anything about the nature of the optimum. The reader seeking a very well written introduction to dynamic programming is referred to Howard (1960).

2.2. *Convergence of the optimal first-period reaction function as the time horizon lengthens*

With existence of the optimal policy sequence, ϕ_{1s}, $s = 1, 2, \ldots$, established, the next question is whether the sequence of reaction functions converges as $s \to \infty$. Under suitable additional assumptions, it is seen that the ϕ_{1s} do converge to a limiting reaction function, ϕ_1. Remember that ϕ_{1s} is the reaction function which is used in the first period of the firm's s period horizon; therefore, if the limiting reaction function exists, it means that when the horizon of the firm is very long, it uses, over the early periods, a reaction function which is quite close to ϕ_1. Statements using "very long" and "quite close" may be made precise, as is seen subsequently.

Associated with a vector of reaction functions for firms other than firm 1 is a special set of price vectors,

$$\mathcal{B}_1 = \{\boldsymbol{p} \mid \boldsymbol{p} \in \mathcal{D}, \bar{\boldsymbol{\psi}}_1(\boldsymbol{p}) = \bar{\boldsymbol{p}}_1\}. \tag{5.19}$$

A price vector is in $\mathcal{B}_1$ if, given that price vector, the ith firm would want to retain the same price, for all firms $i = 2, \ldots, n$. Thus if $\boldsymbol{p}^0$ were in $\mathcal{B}_1$ and firm 1 were willing to choose p_1^0 in response to $\boldsymbol{p}^0$, then $\boldsymbol{p}^0$ would be a steady state price vector; therefore, given that $\bar{\boldsymbol{\psi}}_1$ specifies the behavior of all firms other than firm 1, $\mathcal{B}_1$ contains all the price vectors which could conceivably be steady state price vectors. Several important facts about $\mathcal{B}_1$ should be noted. First, for any fixed p_1^0, where $0 \le p_1^0 \le p_1^+$, there is exactly one vector in $\mathcal{B}_1$ which has a first component of p_1^0. Second, if $\boldsymbol{p}'$, $\boldsymbol{p}'' \in \mathcal{B}_1$ and $p_1' < p_1''$, then $\boldsymbol{p}' \ll \boldsymbol{p}''$. This may be seen immediately by noting that if $\psi_i(\boldsymbol{p}') = p_i'$, then $\psi_i(p_1'', \bar{\boldsymbol{p}}_1') > p_i'$. The characteristics of $\mathcal{B}_1$ follow from A11.

For an arbitrary pair of price vectors in $\mathcal{B}_1$, call them $\boldsymbol{p}_*$ and $\boldsymbol{p}^*$, with $\boldsymbol{p}_* \ll \boldsymbol{p}^*$, let $\mathcal{D}_1^* = \{\boldsymbol{p} | \boldsymbol{p}_* \leq \boldsymbol{p} \leq \boldsymbol{p}^*\}$. A hypothetical $\mathcal{D}_1^*$, along with $\mathcal{B}_1$, is illustrated in fig. 5.2, where the large rectangle is $\mathcal{D}$ and the shaded area is $\mathcal{D}_1^*$. $\mathcal{B}_1$ is the curve which goes from the left to the right boundary of $\mathcal{D}$.

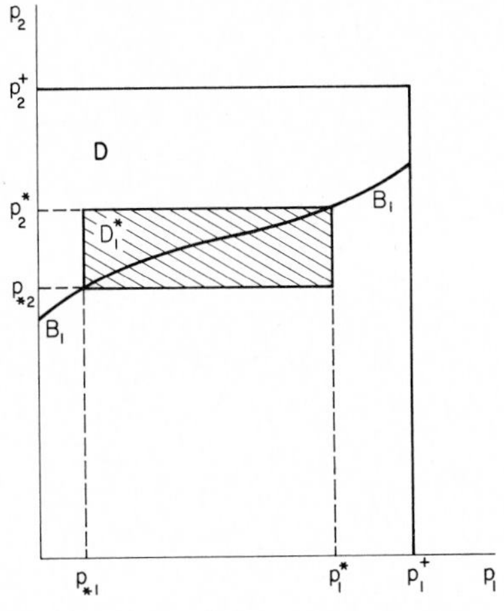

Fig. 5.2

An assumption is now introduced which specifies a set $\mathcal{D}_1^*$ within which the ϕ_{1s} have two special properties. For any ϕ_{1s}, the steady state price vector associated with $(\phi_{1s}, \bar{\boldsymbol{\psi}}_1)$ lies in $\mathcal{D}_1^*$ (in fact, it lies in $\mathcal{D}_1^* \cap \mathcal{B}_1$). In addition, for any $\boldsymbol{p} \in \mathcal{D}_1^*$, $\phi_{1s}(\boldsymbol{p}) < \phi_{1,s+1}(\boldsymbol{p})$.

A13. *There is a set* $\mathcal{D}_1^* = \{\boldsymbol{p} | \boldsymbol{p}_* \leq \boldsymbol{p} \leq \boldsymbol{p}^*\}$ *where* $\boldsymbol{p}_* \ll \boldsymbol{p}^*$ *and* $\boldsymbol{p}_*, \boldsymbol{p}^* \in \mathcal{B}_1$ *such that*

$$\pi_1^1(p_{*1}, \bar{\boldsymbol{\psi}}_1(\boldsymbol{p}_*)) = \pi_1^1(\boldsymbol{p}_*) \geq 0, \tag{5.20}$$

$$\pi_1^1(p_1^*, \bar{\boldsymbol{\psi}}_1(\boldsymbol{p}^*)) + \sum_{l=1}^{\infty} \alpha_1^l \sum_{j=2}^{n} \pi_1^j(p_1^*, \bar{\boldsymbol{\psi}}_1(\boldsymbol{p}^*)) M_{1,j,l}(\boldsymbol{p}^*)$$

$$= \pi_1^1(\boldsymbol{p}^*) + \sum_{l=1}^{\infty} \alpha_1^l \sum_{j=2}^{n} \pi_1^j(\boldsymbol{p}^*) M_{1,j,l}(\boldsymbol{p}^*) \leq 0, \tag{5.21}$$

and, for any $\boldsymbol{p}^1, \boldsymbol{p}^2 \in \mathcal{D}_1^*$ *with* $\boldsymbol{p}^1 \geq \boldsymbol{p}^2$,

$$\pi_1^j(\boldsymbol{p}^1) \geq \pi_1^j(\boldsymbol{p}^2), \qquad j = 2, \ldots, n, \tag{5.22}$$

$$\psi_j^k(\boldsymbol{p}^1) \geq \psi_j^k(\boldsymbol{p}^2), \qquad j = 2, \ldots, n, \quad k = 1, \ldots, n. \tag{5.23}$$

Eqs. (5.22) and (5.23) assure that ϕ_{1s} lies below $\phi_{1,s+1}$ in the sense previously indicated, while eq. (5.20) forces the steady state price vector associated with ϕ_{11} to be either $\boldsymbol{p}_*$ or above $\boldsymbol{p}_*$, and eq. (5.21) implies that $\boldsymbol{p}^*$ is an upper bound for the steady state price vectors associated with all of the ϕ_{1s}.[6] These results are proved below as part of theorem 5.2. A price vector which satisfies eq. (5.21) with equality is the steady state price vector associated with $(\phi_1, \bar{\psi}_1)$ (where ϕ_1 is the limit of the ϕ_{1s}). It is also proved in theorem 5.2 that the limiting function ϕ_1 obeys any Lipschitz condition which is satisfied by the ϕ_{1s}.

A Lipschitz condition is a condition which puts a bound on the rate of change of a function. For a differentiable function $g(\boldsymbol{x})$, with domain in $\mathcal{R}^m$ and in range $\mathcal{R}$, a *Lipschitz condition* could be stated: for any $\boldsymbol{x}$ in the domain of g, $\sum_{j=1}^m |g^j(\boldsymbol{x})| \leq k$ where k is a positive constant. If g is not necessarily differentiable, it is necessary to rephrase the condition to be in terms of a comparison between any two points. Thus let $\boldsymbol{x}$ and $\boldsymbol{y}$ be two points in the domain of g, and define the *distance from $\boldsymbol{x}$ to $\boldsymbol{y}$* as $d(\boldsymbol{x}, \boldsymbol{y}) = \max_j |x_j - y_j|$. Then, for a positive constant k, the function g is said to obey a Lipschitz condition with ratio k if $|g(\boldsymbol{x}) - g(\boldsymbol{y})| \leq k \cdot d(\boldsymbol{x}, \boldsymbol{y})$.

Theorem 5.2. *If A2–A5 and A8–A13 hold, then for $\boldsymbol{p} \in \mathcal{D}_1^*$, $\phi_{1s}(\boldsymbol{p}) > \phi_{1,s-1}(\boldsymbol{p})$, for $s = 2, 3, \ldots$, and $\lim_{s\to\infty} \phi_{1s}(\boldsymbol{p}) = \phi_1(\boldsymbol{p})$ exists. If the ϕ_{1s} obey a common Lipschitz condition, then ϕ_1 obeys the same Lipschitz condition.*

□ To see first that $\phi_{1s}(\boldsymbol{p}) > \phi_{1,s-1}(\boldsymbol{p})$ for any $\boldsymbol{p} \in \mathcal{D}_1^*$, consider again the first-order conditions,

$$G_{11}(\boldsymbol{p}_t^1) = \pi_1^1(\boldsymbol{p}_t^1) = 0, \tag{5.24}$$

$$G_{1s}(\boldsymbol{p}_t^s) = \pi_1^1(\boldsymbol{p}_t^s) + \sum_{l=1}^{s-1} \alpha_1^l \sum_{j=2}^{n} \pi_1^j(\boldsymbol{p}_{t+l}^s) M_{1,j,l}(\boldsymbol{p}_t^s) = 0, \qquad s = 2, 3, \ldots, \tag{5.25}$$

where $\boldsymbol{p}_t^s, \boldsymbol{p}_{t+1}^s, \ldots, \boldsymbol{p}_{t+s-1}^s$ is a sequence of price vectors which satisfy the first-order conditions for an s-period horizon, with the same initial condition, $\boldsymbol{p}_{t-1}^0$ for each. That is,

$$p_{jt}^s = \psi_j(\boldsymbol{p}_{t-1}^0), \qquad j = 2, \ldots, n, \quad s = 1, 2, \ldots, \tag{5.26}$$

$$p_{1t}^s = \phi_{1s}(\boldsymbol{p}_{t-1}^0), \qquad s = 1, 2, \ldots, \tag{5.27}$$

$$p_{j,t+l}^s = \psi_j(\boldsymbol{p}_{t+l-1}^s), \qquad j = 2, \ldots, n, \quad s = 2, 3, \ldots, \quad l = 1, \ldots, s-1, \tag{5.28}$$

[6]In short, eqs. (5.20) and (5.21) are seen below to insure that the steady state price vectors associated with all the ϕ_{1s} lie in $\mathcal{D}_1^*$.

$$p^{s}_{1,t+l} = \phi_{1,s-l}(\boldsymbol{p}^{s}_{t+l-1}), \qquad s = 2, 3, \ldots, \quad l = 1, \ldots, s-1. \tag{5.29}$$

It is immediate that $p^{1}_{1t} < p^{2}_{1t}$ because the terms in the expression for G_{1s} which involve π^{j}_{1} $(j \neq 1)$ are necessarily positive. The rest follows by induction. Assume for $s-1$ that the conclusion holds. Then $p^{s-l}_{jt} = p^{s-1}_{jt} = \psi_j(\boldsymbol{p}^{0}_{t-1})$ $(j \neq 1, l = 2, \ldots, s-1)$, $p^{s-k}_{j,t+l} > p^{s-k-1}_{j,t+l}$ $(j = 1, \ldots, n, l = 1, \ldots, s-2,$ $k = 1, \ldots, s-2)$, and $p^{s-k}_{1,t} > p^{s-k-1}_{1,t}$ $(k = 1, \ldots, s-2)$. Now compare $G_{1s}(\boldsymbol{p}^{s}_{t})$ with $G_{1,s-1}(\boldsymbol{p}^{s-1}_{t})$ on a term by term basis, and on the tentative assumption that $p^{s}_{1t} = p^{s-1}_{1t}$. Then $\boldsymbol{p}^{s}_{t} = \boldsymbol{p}^{s-1}_{t}$ and $\boldsymbol{p}^{s}_{t+l} \geqslant \boldsymbol{p}^{s-1}_{t+l}$ (for $l > 0$). Therefore, by A13, $\pi^{j}_{1}(\boldsymbol{p}^{s}_{t+l}) > \pi^{j}_{1}(\boldsymbol{p}^{s-1}_{t+l})$ and $M_{1,j,l}(\boldsymbol{p}^{s}_{t}) > M_{1,j,l}(\boldsymbol{p}^{s-1}_{t})$ $(l = 1, \ldots, s-2,\ j = 2, \ldots, n)$. Thus $p^{s}_{1t} = p^{s-1}_{1t}$ and $G_{1,s-1}(\boldsymbol{p}^{s-1}_{t}) = 0$ would imply $G_{1s}(\boldsymbol{p}^{s}_{t}) > 0$. Recalling A12, clearly $p^{s}_{1t} > p^{s-1}_{1t} \phi_{1s}(\boldsymbol{p}) > \phi_{1,s-1}(\boldsymbol{p})$, for $s = 2, 3, \ldots$, and $\boldsymbol{p} \in \mathscr{D}^{*}_{1}$.

The rest of the proof falls into place readily. From Choquet (1966, p. 136) it is seen that if a sequence of functions ϕ_{1s} is of Lipschitz class with ratio k and the upper envelope of the sequence is finite at one point, then the upper envelope is finite everywhere and is also of Lipschitz class with ratio k. Because all the ϕ_{1s} have the same domain, $\mathscr{D}^{*}_{1}$ which is compact and all have their images contained in a common compact set ($[p_{*1}, p^{*}_{1}]$), their upper envelope is clearly finite at all points. Because the functions are monotone (in the sense of $\phi_{1s}(\boldsymbol{p}) > \phi_{1,s-1}(\boldsymbol{p})$, i.e. each reaction function is *above* its predecessor), their upper envelope is $\lim_{s\to\infty} \phi_{1s} = \phi_1$. □

Figure 5.3 provides a schematic illustration of reaction functions which

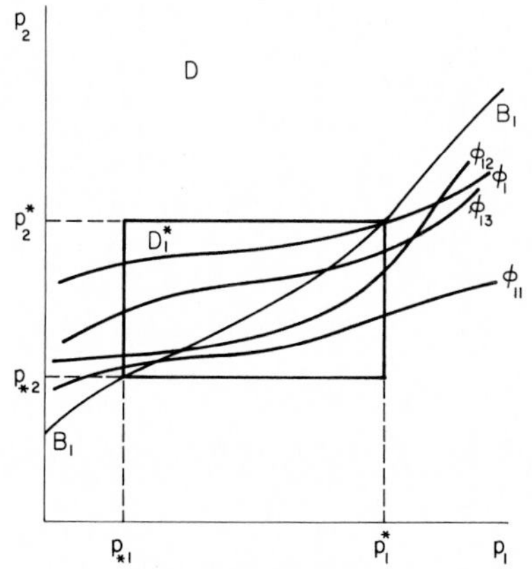

Fig. 5.3

are arrayed one above the other within the set $\mathscr{D}_1^*$, though they need not be so arrayed outside of $\mathscr{D}_1^*$. They are drawn to be monotone increasing, though as yet conditions have not been shown under which they are. That is easily done, as wherever ϕ_{1s} is defined by eq. (5.15), derivatives of ϕ_{1s} exist if the G_i^j are bounded for all $j > 1$. The derivatives are

$$\phi_{1s}^k(\mathbf{p}_{t-1}) = -\sum_{j=2}^{n} G_{1s}^j(\mathbf{p}_t)\psi_j^k(\mathbf{p}_{t-1})/G_{1s}^1(\mathbf{p}_t), \qquad k = 1, \ldots, n, \quad s = 1, 2, \ldots. \tag{5.30}$$

If all the G_{1s}^j $(j \neq 1)$ are nonnegative, then the ϕ_{1s}^k are all nonnegative for $k = 1, \ldots, n$. With respect to the existence of these derivatives, the formal assumption is:

A14. *For all* $\mathbf{p} \in \mathscr{D}$, $\Sigma_{j=2}^n |G_{1s}^j(\mathbf{p})|/G_{1s}^1(\mathbf{p})$ *is bounded in absolute value independently of s.*

It is important that the bound specified in A14 be independent of s, for, if it were not, it would be possible that $|\phi_{1s}^j| \to \infty$ as $s \to \infty$. Then each of the ϕ_{1s} would satisfy a Lipschitz condition, but not a common one, and the limit function ϕ_1 need not satisfy a Lipschitz condition. These remarks almost prove the following theorem:

Theorem 5.3. *If A2–A5 and A8–A14 hold, then for* $\mathbf{p} \in \mathring{\mathscr{D}}$ *(the interior of* $\mathscr{D}$*), first partial derivatives of the* ϕ_{1s} *exist almost everywhere and the* ϕ_{1s} *obey a common Lipschitz condition everywhere.*[7]

□ The theorem holds where the ϕ_{1s} are given by eq. (5.15). Where they are not given by eq. (5.15), $G_{1s} > 0$ and the optimal price is p_1^+. That is, for such a $\mathbf{p}$, $\phi_{1s}(\mathbf{p}) = p_1^+$. Inside a region where the optimal price is p_1^+ clearly the reaction function is constant; hence, derivatives, all of which are zero, exist. On the boundary of the two regions, directional derivatives exist and behave like one of the two cases, depending on direction. □

Somewhat stronger results than theorem 5.3 may be stated when, in addition to the other assumptions, the G_{1s} and ψ_j have derivatives of high order which are bounded.

[7]That the partial derivatives exist *almost everywhere* means they may fail to exist at only a countable set of points.

Theorem 5.4. *Let A2–A5 and A8–A14 hold. If the G_{1s} have continuous partial derivatives of kth order, bounded independently of s, and the ψ_j ($j \neq 1$) have bounded continuous partial derivatives on $\mathring{\mathcal{D}}$ of kth order; then the ϕ_{1s} have continuous partial derivatives of kth order on $\mathring{\mathcal{D}}$, and ϕ_1 has continuous partial derivatives of order $k-1$ on $\mathring{\mathcal{D}}$. The derivatives are the limits of the corresponding sequences of derivatives of the ϕ_{1s}.*

□ This theorem is an immediate consequence of two facts. First, because G_1^1 is bounded strictly away from zero and all other relevant derivatives are bounded, the derivatives of the ϕ_{1s} exist and are bounded. Boundedness of second-order derivatives of the ϕ_{1s} implies uniform convergence of the ϕ_{1s}^j. Thus the first partial derivatives are continuous and $\phi_1^j(p) = \lim_{s\to\infty} \phi_{1s}^j(p)$, by theorem (8.6.3) of Dieudonné (1960). The assertion for derivatives up to k-1st order follows by induction. □

Theorem 5.4 has a corollary when the payoff function, π_1, and the reaction functions, ψ_j ($j \neq 1$), are analytic. Then the ϕ_{1s} and ϕ_1 are also analytic.

Corollary 5.5. *If the ϕ_{1s} are analytic and their derivatives are bounded on $\mathring{\mathcal{D}}$ independently of s, then the ϕ_{1s} converge to ϕ_1 on $\mathcal{D}$ and ϕ_1 is analytic.*

The sense in which ϕ_1 approximates the optimal reaction functions for the early periods of a very long horizon is that for an arbitrary $\delta > 0$ and length of time T^*, there is a finite horizon, T, such that if the horizon of the firm is $T' \geq T$, then for any p, $|\phi_1(p) - \phi_{1s}(p)| < \delta$ for $s \geq T' - T^*$. In other words, for the first T^* periods of the long T' period horizon, the limiting reaction function will choose a price within δ of the price which is actually optimal.

2.3. Optimal behavior for the firm when it has an infinite horizon

The two preceding subsections contain conditions under which a firm in an oligopoly, having a finite horizon of T periods over which it wishes to maximize discounted profits, can solve its maximization problem and find that its optimal policy for the first period of the horizon converges as its horizon lengthens indefinitely. It is natural to ask whether the limiting reaction function, ϕ_1, is actually optimal for an infinite horizon. It would seem intuitively surprising if it were not, and, indeed, it is shown in this subsection that ϕ_1 is optimal for the firm when its horizon is infinite under relatively mild additional assumptions. This is done in two steps. The first is to prove that

$$\lim_{T\to\infty}\sum_{t=1}^{T}\alpha_1^{t-1}\pi_1(\phi_{1,T-t+1}(\mathbf{p}_{t-1}),\ \bar{\boldsymbol{\psi}}_1(\mathbf{p}_{t-1}))$$
$$=\sum_{t=1}^{\infty}\alpha_1^{t-1}\pi_1(\phi_1(\mathbf{p}_{t-1}),\ \bar{\boldsymbol{\psi}}_1(\mathbf{p}_{t-1})); \tag{5.31}$$

that is, as the horizon goes to infinity, the discounted profit of the firm, using the optimal sequence of reaction functions $\phi_{1T}, \ldots, \phi_{11}$, converges to the profit it would have if it were to only use the limiting reaction function, ϕ_1. The second step is to show that no reaction function or sequences of reaction functions can yield higher discounted profits over an infinite horizon than ϕ_1.

The additional assumption is:

A15. *The bounds of A14 are sufficient to insure that the Lipschitz condition obeyed by the ϕ_{1s} is the same as is obeyed by $\bar{\boldsymbol{\psi}}_1$, namely, $\lambda < 1$. The first partial derivatives of π_1 are bounded.*

The first part of A15 merely strengthens A14 in a self-evident fashion. The condition on π_1, which implies a Lipschitz condition on the profit function, still leaves a very large class of models which satisfy the list of assumptions. Denote the Lipschitz condition on π_1 by μ. Note that μ is not restricted in size, beyond having to be positive and finite.

Theorem 5.6. *If A2–A5 and A8–A15 hold and $\mathbf{p}_0 \in \mathcal{D}_1^*$, then eq. (5.31) holds.*

□ The proof makes important use of the boundedness of single-period profits and of uniform convergence of the reaction functions, ϕ_{1s}. To show that the discounted profits of the firm, using the sequence $\phi_{1T}, \ldots, \phi_{11}$, are for horizons of some given length and longer, within an arbitrary ϵ of what they would be using only ϕ_1, the proof is divided into two parts. First, it is shown that there is a finite time T_1 such that profits from that time onward, when discounted back to the present, are so small that the difference between any two conceivable policies is less than $\frac{1}{2}\epsilon$. Then it is shown that over the time from the present to T_1, the lengthening of the horizon, T^*, implies that the reaction functions used (i.e. $\phi_{1T^*}, \ldots, \phi_{1,T^*-T_1+1}$) converge uniformly to ϕ_1, which insures that the difference between the two policies in terms of discounted profits over the T_1 periods may be made arbitrarily small.

For a given value, T^*, define p_t^* as $p_{1t}^* = \phi_{1,T^*-t+1}(p_{t-1}^*)$, $\bar{p}_{1t}^* = \bar{\psi}_1(p_{t-1}^*)$, $t = 1, \ldots, T^*$ and define p_t^0: $p_{1t}^0 = \phi_1(p_{t-1}^0)$, $\bar{p}_{1t}^0 = \bar{\psi}_1(p_{t-1}^0)$, $t = 1, \ldots, T^*$. Furthermore let $p_0^* = p_0^0 \in \mathscr{D}_1^*$. The theorem is proved if it is shown that

$$\lim_{T^* \to \infty} \sum_{t=1}^{T^*} \alpha_1^{t-1} |\pi_1(p_t^0) - \pi_1(p_t^*)| = 0. \tag{5.32}$$

Choose an arbitrary $\epsilon > 0$. The minimum value of $\pi_1(p)$ is $-C_1(0)$ and the maximum value is less than $p_1^+ F_1(0, \bar{p}_1^+)$, which is finite. Let K denote the difference between the maximum and minimum possible values of profit per period. There is a finite T_1 such that

$$\sum_{t=T_1+1}^{\infty} \alpha_1^{t-1} K = \alpha_1^{T_1} K/(1-\alpha_1) < \tfrac{1}{2}\epsilon; \tag{5.33}$$

that is, the difference in discounted profits, discounted back to $t = 1$, between using the p^0 prices and the p^* prices, from T_1 onward is less than $\frac{1}{2}\epsilon$ for an arbitrary ϵ and any two price series.

On the other hand, uniform convergence of the ϕ_{1s} means that for any $\delta > 0$ there is T_2 such that for $s > T_2$, $|\phi_1(p) - \phi_{1s}(p)| < \delta$ for any $p \in \mathscr{D}_1^*$. Choose $T^* > T_1 + T_2$ and note that, using the Lipschitz condition on ϕ_1, the ϕ_{1s} and $\bar{\psi}_1$, then

$$|p_{jt}^* - p_{jt}^0| < \delta(1 + \lambda + \cdots + \lambda^{t-1}) = \delta(1-\lambda^t)/(1-\lambda), \qquad j = 1, \ldots, n, \quad t < T_1 \tag{5.34}$$

The Lipschitz condition on π_1 implies that $|\pi_1(p_t^*) - \pi_1(p_t^0)| < \mu\delta(1-\lambda^t)/(1-\lambda)$; therefore, for $T^* > T_1 + T_2$,

$$\begin{aligned} &\sum_{t=1}^{T_1} \alpha_1^{t-1} |\pi_1(p_t^0) - \pi_1(p_t^*)| \\ &< \mu\delta[(1-\alpha^{T_1}\lambda^{T_1})/(1-\alpha\lambda) - \alpha^{T_1}(1-\lambda^{T_1})/(1-\lambda)](1-\alpha) \\ &< \mu\delta(1-\alpha^{T_1}\lambda^{T_1})/[(1-\alpha\lambda)(1-\alpha)]. \end{aligned} \tag{5.35}$$

T_2 and T^* may be chosen sufficiently large that the final expression in eq. (5.35) is less than $\frac{1}{2}\epsilon$, assuring that

$$\sum_{t=1}^{T^*} \alpha_1^{t-1} |\pi_1(p_t^0) - \pi_1(p^*)| < \epsilon, \tag{5.36}$$

for arbitrary ϵ, which establishes the desired convergence. □

From theorem 5.6 it is known that in the limit using ϕ_1 is as profitable as using the ϕ_{1s}; however, the possibility remains that either of those courses of action are inferior for an infinite horizon to some other policy. That possibility is eliminated by the following theorem.

Theorem 5.7. *If A2–A5 and A8–A15 hold and $\mathbf{p}_0 \in \mathcal{D}_1^*$, then choosing $p_{1t} = \phi_1(\mathbf{p}_{t-1})$, for $t = 1, 2, \ldots$, maximizes $\Sigma_{t=1}^{\infty} \alpha_1^{t-1} \pi_1(\mathbf{p}_t)$.*

□ The proof proceeds by first noting that if ϕ_1 is not optimal, then some other policy yields discounted profits which are greater by some fixed, positive amount ϵ. Because π_1 is bounded, most of the extra profit must be earned within a finite time; however, for that to be possible contradicts the known optimality of the ϕ_{1s}.

Surely if ϕ_1 is not an optimal policy for the infinite horizon, there is some sequence p_{1t}^1 $(t = 1, 2, \ldots)$ which is better. Let $\mathbf{p}_t^0$ and $\mathbf{p}_t^*$ $(t = 1, 2, \ldots)$ be defined as in the proof of theorem 5.6, and define $\mathbf{p}_t^1$ as $p_{jt}^1 = \psi_j(\mathbf{p}_{t-1}^1)$, $j \neq 1$, $t \geq 1$, and p_{1t}^1 is the optimal sequence mentioned above, with $\mathbf{p}_0^1 = \mathbf{p}_0^0 = \mathbf{p}_0^*$. If the $\mathbf{p}^1$ sequence is superior to the $\mathbf{p}^0$ sequence, then

$$\sum_{t=1}^{\infty} \alpha_1^{t-1} \pi_1(\mathbf{p}_t^1) - \sum_{t=1}^{\infty} \alpha_1^{t-1} \pi_1(\mathbf{p}_t^0) = \epsilon > 0. \tag{5.37}$$

Now choose T^* so that

$$\sum_{t=1}^{T^*} \alpha_1^{t-1} \pi_1(\mathbf{p}_t^*) - \sum_{t=1}^{T^*} \alpha_1^{t-1} \pi_1(\mathbf{p}_t^0) < \epsilon/4$$

and

$$\alpha_1^{T^*} K/(1 - \alpha_1) < \epsilon/4. \tag{5.38}$$

Then $\Sigma_{t=1}^{T^*} \alpha_1^{t-1} \pi_1(\mathbf{p}_t^1) - \Sigma_{t=1}^{T^*} \alpha_1^{t-1} \pi_1(\mathbf{p}_t^*) > \epsilon/2$, which contradicts the known optimality $\Sigma_{t=1}^{T^*} \alpha_1^{t-1} \pi_1(\mathbf{p}_t^*)$ for an horizon of T^* periods; hence, p_{1t}^1 $(t = 1, \ldots)$ cannot be superior to ϕ_1 for the infinite horizon. □

To summarize the results of §2.1–2.3, conditions are given under which the optimal sequence of reaction functions for firm 1, when its horizon is T periods, exists, converges to ϕ_1 and satisfies a Lipschitz condition like the condition on $\bar{\psi}_1$. The ϕ_{1s} are also differentiable, as is ϕ_1, as well as ϕ_1 being the optimal behavior rule for the firm when its horizon is infinite. The convergence and optimality results for ϕ_1 are confined to a subset, $\mathcal{D}_1^*$ of $\mathcal{D}$, the price domain. It is possible that for many models of interest, the conditions which are assumed on $\mathcal{D}_1^*$ hold everywhere on $\mathcal{A}_1^*$, in which case, the results are of a global character. One other interesting point about the ϕ_{1s} is that they need not have positive first partial derivatives. It could be assumed that the G_1^j $(j \neq 1)$ are all positive, which would imply the ϕ_{1s}^k $(k = 1, \ldots, n)$ are also positive. To insure a market equilibrium in §2.4, this is, in fact assumed.

2.4. *Equilibrium in the market*

The results for a single firm which are developed to this point yield, when applied to all n firms in the market, a means of calculating a vector of reaction functions, $\boldsymbol{\phi} = (\phi_1, \ldots, \phi_n)$, which are related to another vector of reaction functions, $\boldsymbol{\psi} = (\psi_1, \ldots, \psi_n)$. For each firm, i, ϕ_i is the best reply to $\bar{\boldsymbol{\psi}}_i$, given that the ith firm has as its objective the maximum of $\sum_{t=1}^{\infty} \alpha_i^{t-1} \pi_i(\boldsymbol{p}_t)$. The maximization process described in §2.2 may be thought of as a mapping or function which maps $\boldsymbol{\psi}$ into $\boldsymbol{\phi}$. An individual component of the mapping takes $\bar{\boldsymbol{\psi}}_i$ into ϕ_i. These may be written in the form $\boldsymbol{\phi} = \boldsymbol{W}(\boldsymbol{\psi})$ and $\phi_i = W_i(\bar{\boldsymbol{\psi}}_i)$. Returning to the single-period Cournot quantity model, these best reply functions are analogous to the best reply functions $\boldsymbol{q}' = \boldsymbol{w}(\boldsymbol{q})$ and $q_i' = w_i(\bar{\boldsymbol{q}}_i)$. The difference is that in the present model, the firm chooses a function, whereas in the single-period model the firm chooses a number. Best reply equilibria are characterized by $\boldsymbol{q} = \boldsymbol{w}(\boldsymbol{q})$ and $\boldsymbol{\psi} = \boldsymbol{W}(\boldsymbol{\psi})$, respectively. In the former instance, all there is to know is whether $\boldsymbol{q}$ and $\boldsymbol{w}(\boldsymbol{q})$ are the same; but in the reaction function model, the situation is more complex. There are intermediate possibilities between $\boldsymbol{\psi} = \boldsymbol{W}(\boldsymbol{\psi})$ and $\boldsymbol{\psi} \neq \boldsymbol{W}(\boldsymbol{\psi})$. These have to do with the extent to which $\boldsymbol{\psi}$ and $\boldsymbol{W}(\boldsymbol{\psi})$ approximate each other. The least which could reasonably be required of an "equilibrium" $\boldsymbol{\psi}$ is that both $\boldsymbol{\psi}$ and $\boldsymbol{W}(\boldsymbol{\psi})$ have the same steady state price vector:

$$\boldsymbol{p}^r = \boldsymbol{\psi}(\boldsymbol{p}^r) = \boldsymbol{\phi}(\boldsymbol{p}^r) = \boldsymbol{W}(\boldsymbol{\psi}(\boldsymbol{p}^r)). \tag{5.39}$$

If eq. (5.39) holds, then, at least if the current price vector is $\boldsymbol{p}^r$ and the firms are behaving according to the ψ_i, then no firm can, by itself, change its behavior rule (i.e. its reaction function) and increase its profit. This could be stated two ways. If the firms have the perfect initial price vector, $\boldsymbol{p}_0 = \boldsymbol{p}^r$, then $\boldsymbol{\phi}$ is a best reply against $\boldsymbol{\psi}$, though it is not a best reply for an arbitrary (indeed for any other) price vector. Or, if one can countenance the firms each using ϕ_i while believing their rivals are using $\bar{\boldsymbol{\psi}}_i$, in the limit, as $t \to \infty$, $\boldsymbol{p}_t \to \boldsymbol{p}^r$, and, in the limit, the firms see what they expect to see.[8]

A condition which could be required in addition to eq. (5.39) is

$$\psi_i^k(\boldsymbol{p}^r) = \phi_i^k(\boldsymbol{p}^r), \qquad i, k = 1, \ldots, n. \tag{5.40}$$

Equation (5.40) states that $\boldsymbol{\psi}$ and $\boldsymbol{\phi}$ are tangent to one another at the steady state price vector, making price changes in one be very close to price changes in the other in the neighborhood of $\boldsymbol{p}^r$. If eq. (5.40) holds with the firms using the ϕ_i and assuming others use the $\psi_j (j \neq i)$, the "right for the wrong reasons" problem is still present, but in a less severe form than if only eq. (5.39) held. At least in a neighborhood of $\boldsymbol{p}^r$ the firms' expectations about both prices and price changes are approximately correct. Though an equilibrium based upon eqs. (5.39) and (5.40) is not entirely satisfactory because it is not a best reply equilibrium, such an equilibrium is the furthest formal development of the simultaneous decision reaction function models of which I am aware. Proof of the main result of this section requires two additional assumptions.

A16. *For all* $\boldsymbol{p} \in \mathscr{D}$ *and* $s \geq 1$, $G_{is}^j(\boldsymbol{p}) \geq 0$, $j \neq i$, $i = 1, \ldots, n$.

A17. $\mathscr{D}^* = \cap_{i=1}^n \mathscr{D}_i^*$ *has a nonempty interior which contains* $\boldsymbol{p}^c$.

A glance at eq. (5.30) shows that A16 forces the ϕ_i^k to be nonnegative. If A16 required the G_{is}^j to be bounded strictly away from zero and G_{1s}^1 to be bounded in absolute value, then a positive lower bound is implied for the first derivatives of the ϕ_i. There is a problem with A17 which should be pointed out. $\mathscr{D}_i^*$ depends upon characteristics of $\bar{\boldsymbol{\psi}}_i$; however, we want to be in a position to define a set $\mathscr{D}^*$ on which A13 holds (i.e. on which eqs. (5.20)–(5.23) hold) whose size and location are not determined by characteristics of some particular $\boldsymbol{\psi}$. Therefore $\mathscr{D}^*$ is required to be a set with $\boldsymbol{p}^c$ on its main diagonal: $\mathscr{D}^* = \{\boldsymbol{p} | p_i^c - a \leq p_i \leq p_i^c + b,\ i = 1, \ldots, n\}$, where $a > 0$, $p_i^c - a \geq 0$, $b > 0$, $p_i^c + b \leq p_i^+$, and $(p_1^c - a, \ldots, p_n^c - a) \in \mathscr{A}^*$.

Theorem 5.8. *Let A1–A5 and A8–A17 hold. Then discount parameters* α_i *and reaction functions* ψ_i *can be chosen so that: (a)* $\boldsymbol{p}^r \in \mathring{\mathscr{D}}^*$, *with* $\boldsymbol{\psi}(\boldsymbol{p}^r) = \boldsymbol{p}^r$; *and (b) the best response reaction functions,* ϕ_i, *have the same*

[8] This is a more sophisticated instance of what Fellner calls being "right for the wrong reasons" (see ch. 4 §6).

price equilibrium and first partial derivatives at the steady state prices – that is, $\boldsymbol{\phi}(\boldsymbol{p}^r) = \boldsymbol{\psi}(\boldsymbol{p}^r)$ and $\phi_i^j(\boldsymbol{p}^r) = \psi_i^j(\boldsymbol{p}^r)$, $i, j = 1, \ldots, n$.

□ The proof is by construction. Before choosing α_i and constructing the reaction functions, $\boldsymbol{\psi}$, which satisfy the conditions of the theorem, the conditions which the α_i and $\boldsymbol{\psi}$ must obey are reviewed. Let $\boldsymbol{p}^r$ denote the steady state price vector associated with $\boldsymbol{\psi}$. The conditions insuring that $\boldsymbol{\phi}$, the best response to $\boldsymbol{\psi}$, have the same steady state price vector are

$$G_i(\boldsymbol{p}^r) = \pi_i^i(\boldsymbol{p}^r) + \sum_{l=1}^{\infty} \alpha_i^l \sum_{j \neq i} \pi_i^j(\boldsymbol{p}^r) M_{i,j,l}(\boldsymbol{p}^r) = 0, \qquad i = 1, \ldots, n. \tag{5.41}$$

As the proof is by construction, there is considerable freedom in the choice of the values of $\psi_i^j(\boldsymbol{p}^r)$. In particular, they may be all chosen equal; so if $\psi_i^j(\boldsymbol{p}^r) = b^*$ for $i, j = 1, \ldots, n$, with $b^* < 1/n$, then $M_{i,j,l}(\boldsymbol{p}^r) = b^{*l}(n-1)^{l-1}$ and eq. (5.41) reduces to

$$G_i(\boldsymbol{p}^r) = \pi_i^i(\boldsymbol{p}^r) + \alpha_i b^*/[1-(n-1)b^*\alpha_i] \sum_{j \neq i} \pi_i^j(\boldsymbol{p}^r) = 0. \tag{5.42}$$

Similarly, the conditions insuring that $\phi_i^j(\boldsymbol{p}^r) = \psi_i^j(\boldsymbol{p}^r)$, $i, j = 1, \ldots, n$, are readily seen from eq. (5.30) to be

$$\begin{bmatrix} G_1^1(\boldsymbol{p}^r) \cdots G_1^n(\boldsymbol{p}^r) \\ \cdot \qquad\qquad \cdot \\ \cdot \qquad\qquad \cdot \\ \cdot \qquad\qquad \cdot \\ G_n^1(\boldsymbol{p}^r) \cdots G_n^n(\boldsymbol{p}^r) \end{bmatrix} \begin{bmatrix} \psi_1^1(\pi^r) \cdots \psi_1^n(\boldsymbol{p}^r) \\ \cdot \qquad\qquad \cdot \\ \cdot \qquad\qquad \cdot \\ \cdot \qquad\qquad \cdot \\ \psi_n^1(\boldsymbol{p}^r) \cdots \psi_n^n(\boldsymbol{p}^r) \end{bmatrix} = \begin{bmatrix} 0 \cdots 0 \\ \cdot \quad \cdot \\ \cdot \quad \cdot \\ \cdot \quad \cdot \\ 0 \cdots 0 \end{bmatrix} \tag{5.43}$$

If the first partials at $\boldsymbol{p}^r$ are all equal to b^*, then eq. (5.43) reduces to

$$\sum_{j=1}^{n} G_i^j(\boldsymbol{p}^r) = 0, \qquad i = 1, \ldots, n. \tag{5.44}$$

The first step in the construction is to choose a suitable $\boldsymbol{p}^r$. This is done in conjunction with finding allowable intervals within which the α_i may be chosen, and under the assumption that $\psi_i^j(\boldsymbol{p}^r) = b^*$, for all i, j. By A8, $\pi_i^{ii}(\boldsymbol{p}) < 0$ for all $\boldsymbol{p} \in \mathscr{D}^*$; hence either A12 is satisfied for any $\alpha_i \in (0, 1)$ or there is a least upper bound between zero and one below which A12 is satisfied. Let α_i^* be this lower bound, if it applies or one if it does not. Now consider points in $\mathscr{D}^*$ which satisfy $\boldsymbol{p} = (p_1^c + v, \ldots, p_n^c + v)$ for $v \geq 0$. Call this set $\mathscr{B}^*$ and choose $\boldsymbol{v}^i = (p_1^c + v_i, \ldots, p_n^c + v_i)$ to be the largest element of $\mathscr{B}^*$ for which

$$G_i(\boldsymbol{v}^i) = \pi_i^i(\boldsymbol{v}^i) + \sum_{i=1}^{\infty} \alpha_i^{*l} \sum_{j \neq i} \pi_i^j(\boldsymbol{v}^i) M_{i,j,l}(\boldsymbol{v}^i) \geq 0. \tag{5.45}$$

Obviously $G_i(\boldsymbol{p}^c) > 0$, so the required $\boldsymbol{v}^i$ must exist and be larger than $\boldsymbol{p}^c$. Do the preceding exercise for all firms. Now choose a point, $\boldsymbol{p}^r \in \mathscr{B}^*$ which satisfies $\boldsymbol{p}^r \gg \boldsymbol{p}^c$ and $\boldsymbol{p}^r \ll \boldsymbol{v}^i$ for $i = 1, \ldots, n$. $\boldsymbol{p}^r$ is the steady state price vector. The α_i may now be selected so that eq. (5.42) is satisfied, and values of $\psi_i^{jk}(\boldsymbol{p}^r)$ may be assigned which insure that eq. (5.44) is satisfied. It is trivial to finish the specification of $\boldsymbol{\psi}$ so that A11 is satisfied everywhere and $\boldsymbol{\psi}(\boldsymbol{p}^r) = \boldsymbol{p}^r$. □

The possibilities for cooperation and the meaning of *tacit cooperation* may arise in connection with the reaction function models. The method of construction used in the proof of theorem 5.8 caused the steady state price vector $\boldsymbol{p}^r$ to be larger than $\boldsymbol{p}^c$, the Cournot equilibrium price vector, by an equal amount in each component. It is seen in the proof of theorem 3.2 that if all firms increase their prices at an equal rate, starting from $\boldsymbol{p}^c$, all have their profits increased. Therefore, in moving on a straight line from $\boldsymbol{p}^c$ to $\boldsymbol{p}^r$, all firms experience an increase in profits for at least part of the way. This shows, in turn, that the steady state prices, $\boldsymbol{p}^r$, associated with the reaction function equilibrium whose existence is proved in theorem 5.8 can be prices at which $\boldsymbol{\pi}(\boldsymbol{p}^r) \gg \boldsymbol{\pi}(\boldsymbol{p}^c)$.

Another question of interest is whether the steady state profits, $\boldsymbol{\pi}(\boldsymbol{p}^r)$, can be on the profit frontier; however, to the best of my knowledge that remains an open question.

With respect to cooperation, the reader might give some thought to whether earning profits in excess of $\boldsymbol{\pi}(\boldsymbol{p}^c)$ either implies cooperation or ought to serve as a definition of cooperation. This question recurs in ch. 8 where a model is presented in which there is a best reply equilibrium giving profits which are Pareto optimal.[9]

2.5. *On replacing some assumptions with more fundamental conditions*

The assumptions A12, A14–A16 place various bounds on the derivatives G_{1s}^j. All the bounds are independent of s, the horizon, and may be summarized as (a) $G_{1s}^j \geq 0$ for $j \neq 1$ and (b) $\Sigma_{j=1}^n G_{1s}^j \leq -\epsilon$. These condi-

[9]See particularly §4.

tions insure that $G^1_{1s} \le -\epsilon$, $\phi^j_{1s} \ge 0$ and $\Sigma^n_{j=1}\phi^j_{1s} \le \lambda$, which are precisely the conditions they were introduced to insure.

Consider G^j_{1s}:

$$G^j_{1s}(\boldsymbol{p}_t) = \pi^{1j}_{1t} + \sum_{l=1}^{s-1} \alpha^l_1 \sum_{k=2}^{n} [\pi^k_{1,t+l}\partial M_{1,k,l}/\partial p_{jt} + M_{1,k,l}\partial\pi^k_{1,t+l}/\partial p_{jt}]. \tag{5.46}$$

Without getting buried in details, it is easy to see that the term multiplied by α^l_1 involves first derivatives of ϕ_{1s} (which occur in derivatives of $M_{1,k,l}$), as well as first and second derivatives of $\bar{\boldsymbol{\psi}}_1$ and π_1. These may all be bounded to provide a bound for the term in eq. (5.46) appearing within the square brackets. If that bound is called V^*, then

$$G^j_{1s} \ge \pi^{1j}_{1t} - \sum_{l=1}^{s-1} \alpha^l_1(n-1)V^* > \pi^{1j}_{1t} - \alpha_1(n-1)V^*/(1-\alpha_1). \tag{5.47}$$

Thus if π^{1j}_{1t} has a strictly positive lower bound, there is a range of values of α_1, extending from zero to some strictly positive number, for which $G^j_{1s} \ge 0$. Similarly, summing the G^j_{1s}, noting that the terms multiplied by α^l_1 may be bounded in absolute value, and calling the bound V^{**} yields

$$\sum_{j=1}^{n} G^j_{1s} \le \sum_{j=1}^{n} \pi^{1j}_1 + \alpha_1 n(n-1)V^{**}/(1-\alpha_1). \tag{5.48}$$

Satisfaction of $\Sigma^n_{j=1} G^{1j}_{1s} \le -\epsilon$ is guaranteed if $\Sigma^n_{j=1} \pi^{1j}_1$ has a strictly negative upper bound smaller than $-\epsilon$. Then a range can be found for α_1 within which the required condition on $\Sigma^n_{j=1} G^j_{1s}$ is met. From

$$\pi^{1j}_1 = (1 - C''_1F^1_1)F^j_1 + (p_1 - C'_1)F^{1j}_1, \tag{5.49}$$

$$\sum_{j=1}^{n} \pi^{1j}_1 = F^1_1 + (1 - C''_1F^1_1)\sum_{j=1}^{n} F^j_1 + (p_1 - C'_1)\sum_{j=1}^{n} F^{1j}_1, \tag{5.50}$$

it is easy to see that the required conditions are met if $C''_1 \ge 0$, $F^{1j}_1 \ge 0$ $(j \ne 1)$, $\Sigma^n_{j=1} F^{1j}_1 \le 0$ and F^j_1 has a strictly positive lower bound, h. Then $\pi^{1j}_1 \ge h$, and because $\Sigma^n_{j=1} F^j_1 \le 0$, $\Sigma^n_{j=1} \pi^{1j}_1 \le -(n-1)h$.

3. Duopoly models with homogeneous goods and quantity as the decision variable of the firm

3.1. Simultaneous decision

In this section a model is briefly sketched which is worked out in detail in Friedman (1968). Imagine two firms in a market satisfying the assumptions for quantity models, A1, A2, A6 and A7. Also suppose f'' and C_1'' to be restricted so that $\pi_1^{12} = f' + q_1 f'' - C_1'' \leq 0$. If the first firm seeks to maximize $\Sigma_{t=1}^{T} \alpha_1^{t-1} \pi_1(q_t)$ and the second firm is supposed to behave according to $q_{2t} = \psi_2(q_{1,t-1})$, then the problem is similar to that of the dynamic Stackelberg leader–follower model examined in ch. 4 §5 and to that of the firm in §2.1 of this chapter. The only restrictions put onto ψ_2 are that it is defined for all q_1 between zero and $\bar{Q}$, and $-1 < \psi_2' < 0$. Note that, like the Stackelberg model and unlike the price model, the reaction function of firm 2 depends only on the previous output of firm 1, rather than on both output levels.

Solution of the firm's maximization problem leads, as in the Stackelberg model, to a sequence of behavior rules of the form $q_{1t} = v_{1s}(q_{1,t-1})$, $s = 1, \ldots, T$. v_{1s} is the behavior rule when the remaining number of periods with which the firm is concerned is s. These are related to one another as shown in fig. 5.4, with $v_{1s}' > 0$ and $v_{1s}(q_1) > v_{1,s-1}(q_1)$. That is, the longer the firm's horizon, the larger its optimal level of output, and the larger its previous period output level (the smaller its rival's current output so), the larger its current output will be. The v_{1s} are shown to converge, as $s \to \infty$, to v_1. The output level at which a function v_{1s} crosses the 45° line in fig. 5.4 gives the steady state output level associated with v_{1s} and ψ_2. For example, the steady state associated with v_1, the limiting behavior rule, and ψ_2 is q_1^* for firm 1 and $\psi_2(q_1^*)$ for firm 2.

As shown in ch. 4 §5, v_1 may be converted to a reaction function by setting $\phi_1(q_{2,t-1}) = v_1(v_1(\psi_2^{-1}(q_{2,t-1})))$. The process sketched for getting ϕ_1 from ψ_2 may also be used to get ϕ_2 from ψ_1. Thus it is possible to begin with a pair of arbitrary reaction functions for the firms and find best responses to them, just as was done in §2. The equilibrium developed is one for which ϕ_1 and ϕ_2 have the same steady state output levels (q_1^*, q_2^*) as ψ_1 and ψ_2, as well as the same first derivatives at those output levels. These results are parallel to the results of §2; however, it has not proved possible to generalize the model to more than two firms. The model, apart from the difficulty of generalization, is less satisfactory than the price model on two grounds. The price model with differentiated products is

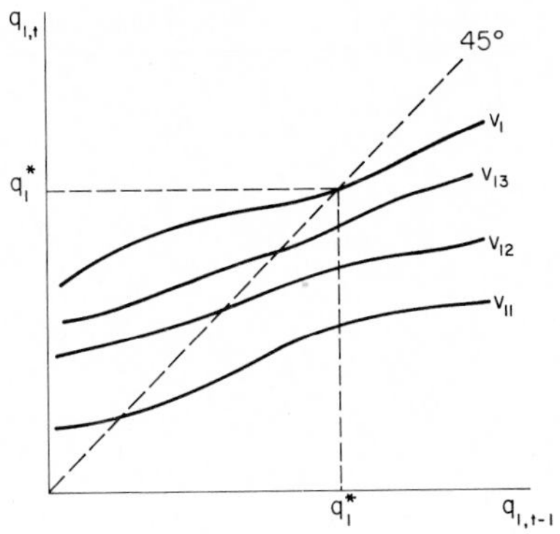

Fig. 5.4

nearer reality, and the optimal behavior of a firm responding to reaction function behavior of others is immediately a reaction function rather than another behavior rule which is not in quite the same form as the rules governing the behavior of the others.

3.2. Alternate decision

Cyert and deGroot (1970) use a homogeneous products quantity model of duopoly in which they assume a linear market demand function and zero costs. Time is discrete with one firm choosing a new output level in each of the even numbered periods and the other firm doing so in each of the odd-numbered periods. By this trick, a firm always knows that when it chooses a new output level, it can invariably count on its rival to keep output unchanged (because the rival is disallowed by the rules from changing its output). Analytically this is a great boon; for it completely removes the simultaneity from the problem of finding a mutually optimal pair of reaction functions. In effect, firm 1 finds the reaction function which is optimal against $q_{2t} = q_{2,t-1}$ and firm 2 finds the function which is best against $q_{1t} = q_{1,t-1}$. Half the time, one firm is forced to a particular action (that of repeating what was previously chosen), while the other reacts to that forced action in the best way.

Rather than prove results for a class of models, Cyert and deGroot calculate a particular example for which limiting reaction functions appear to exist. The steady state output levels associated with the limiting behavior call for lower output and yield higher profits than would be obtained at the Cournot equilibrium. Because the model is symmetric, it is not at all clear whether both firms would have higher profits if they had different cost functions.

4. Final observations on the reaction function models

Almost the whole of the present chapter is devoted to developing reaction function equilibria for an explicitly dynamic differentiated products price model in which the firms are maximizers of a discounted stream of profits. The model and everything done with it is a natural outgrowth of the work Cournot (1960), Bowley (1924) and Stackelberg (1934) and of the critical suggestions and perceptions of Fellner (1949). Though these writers did not make explicit the time relationships in their models, nor explicitly assume firms to have maximization of discounted profits as goals, their work pointed strongly in these directions.

Indeed, there is a second way of viewing the implicit time structure they had in mind. It involves assuming that firms are able to respond instantaneously to changes in market conditions. Then each firm, before making any changes in its price, contemplates what is the likely outcome of its proposed change after rivals have made appropriate adjustments. This line is followed and developed further by Marschak and Selten (1974) who write (p. 5):

> Very broadly speaking, our approach follows the "conjectural" tradition of classical oligopoly theory. In the equilibria we study each firm is content with its actions in the sense that it prefers them to what it believes would be the *end result* of deviating from them. It conjectures what others' responses to its deviations would be and effectively assumes the others' responses to each deviation to be *instantaneous*, so that it cannot derive any transitional benefit from the deviation before the others respond.

It is noted above in ch. 4 §5 that as the discount parameter of the firm gets closer to 1, the leader–follower equilibrium developed there approaches the leader–follower equilibrium given by Stackelberg. Letting the discount parameter go to 1 is analogous to letting the length of the time period go to zero – which means, in the limit, response is instantaneous. This raises the question whether the Marschak and Selten model and results are a limiting case of an appropriately specified model having the sort of response lags which characterize the models in this chapter.

The principal model studied in this chapter is, then, one in which the aim is to develop Cournot type best reply equilibria when the firms choose reaction functions, which are rules by which prices are chosen, rather than by choosing prices.

Success is partial, in the sense that best reply reaction function equilibria are not found to exist (or proved to not exist); but, instead, existence of an approximation to a best reply equilibrium is proved. This leaves open the question whether such equilibria, other than the degenerate $\psi(p_t) = p^c$ can exist. It is true that only the degenerate best reply equilibrium is possible in the model of §2 when the firms have finite horizons. This may be easily seen from a very familiar backward induction argument.[10] In the last time period, it is clear that the only equilibrium behavior which is best reply simultaneously for all firms is $p_T = p^c$. In the second to last period, the firms' choices do not influence behavior in the last period because last-period prices are p^c in any case; therefore, it is as if period $T-1$ is the last period, and the only best reply actions which the firms can take is $p_{T-1} = p^c$. Carrying on in this way, assume it is period $T-k$, and for all the periods $T-k+1, \ldots, T$ the firms choose p^c in each period. Then the choices made in period $T-k$ can have no future repercussions; hence, the optimal choice all around is $p_{T-k} = p^c$. Among other places, Cyert and deGroot (1970) cite this argument. They use it to provide evidence that models of simultaneous decision can provide no other noncooperative equilibria of interest; however, if the horizon is infinite, the situation changes dramatically. In ch. 8 a best reply equilibrium is presented which exists under quite general conditions and is not one which has the firms choosing analogously to p^c in each period. Its existence depends crucially on the infinite horizon. Were the same model formulated with a finite horizon, the backward induction argument would apply.

[10]See ch. 2 §5 for other comments and references concerning the backward induction argument.

CHAPTER 6

A MODEL OF THE FIRM WITH FIXED CAPITAL

1. Introduction

Most of this chapter is taken up with a model of a monopolistic firm in a world of certainty. The decision problem for the firm is to choose a price and investment policy. It is usual in much of the literature of economics to suppose that the capital stock of a firm is a homogeneous mass which decays exponentially and which may be increased in a simple straightforward way at any time. It is as if capital is a block of ice which, left to itself, melts at a given rate. Investment consists of retarding the rate of melting or causing the block to grow rather than diminish.

This view, which is recognized by any sensible person as being unreasonable (including by those who adopt this view for purposes of theory construction), oversimplifies the basic nature of capital. The *fixedness* of capital impinges on the firm in two ways. The first is that the illiquidity of capital, together with its productivity, causes downward adjustment in the size of the capital stock to be slow. Of itself the ice melts at a fixed rate (the rate of depreciation). The alternative to depreciating the capital stock is to sell off some of it; however, it generally cannot be sold for as much as its replacement cost (net of depreciation). There are several possible contributing causes to the poor used market for capital. The cost of removal and reinstallation may be high. Some capital may have been specially designed for the original user, and modification to suit another may be expensive. If the market is thin, acquiring information on what is available used may be difficult, which discourages prospective buyers from that market due to high search costs. For simplicity, assume the firm cannot sell any capital at a positive price; and, because its marginal product is positive, it does not give capital away to get quickly to the optimal level. Although the level of the capital stock cannot profitably be adjusted downward at any desired speed, capital does have perfect malleability and flexibility in another respect. Capital consists of a single, homogeneous commodity which can be made to expand or shrink as needed.

The second way that the fixed nature of capital enters points up the fault of the ice analogy used above. When a firm obtains a particular capital stock, it has limited the set of technologies which it can actually use. Items of physical capital tend to be specific in the uses to which they might be put. Imagine a firm which has no physical capital and is deciding what to buy. Assume it spends a sum of A on particular plant and equipment. Once the expenditure is made and the capital actually acquired, assume the firm regrets its choice. It wishes it had put $1.5A$ into capital. It has now foregone the chance to have the optimal plant–equipment configuration which can be bought for $1.5A$. There are several possibilities. One is that for a price of something greater than $0.5A$ it can convert from what it purchased to what it wishes it had purchased. Another is that the optimal adjustment is one which leaves the firm with something different than it wishes it originally bought. This adjustment could cost more or less than $0.5A$.

The original plant may turn out to be inappropriate for any one (or combination) of several reasons. The demand conditions facing the firm may change. For example if the demand function facing the firm should "shift outward," so that the amount demanded of the firm, at any given price, were to rise, it would probably want a larger plant than what would previously have been optimal. Second, input prices may change, in which case, the input mix which would be optimal, given free adjustment of all inputs, would also change. Finally, with the passage of time the plant can deteriorate and also new techniques can come into existence.

Consider a firm which equips itself, anticipating particular demand conditions over some extended horizon of time. Say that shortly after the firm's capital is in operation, it is predicting much different levels of demand or of (variable) input costs. It would, if choosing plant and equipment at the later time with no fixed commitment previously made, select a different configuration than it presently owns. Meanwhile, it is highly unlikely that the optimal plant–equipment configuration, given its sunk costs, is the same as it would have been if no capital had previously been purchased. Clearly, the way in which the firm adjusts is sensitive to its pre-existing fixed commitments at any point in time.

Indeed, these same considerations apply when invention takes place over time. Compare a firm which has been in business for a long time with one which is about to begin operations. The older firm may not find it in its best interests to adopt the most advanced technology because of the capital it owns and cannot sell at a "reasonable" price. The new firm, having no capital of zero market value, but positive marginal productivity, buys the plant–equipment configuration which produces at least cost given the present purchase price of all inputs (capital included).

The plant–equipment combination that a firm would purchase is, in general, a joint decision. That is, the particular types and configuration of equipment, as well as the precise size and nature of the plant, are jointly determined. If, after the passage of time, the firm should decide that it has less than the optimal amount of capital, again, it is not likely to expand to what it would currently buy if it were starting from scratch. The latter may involve quite a different selection of equipment, as well as a different size and layout of plant. Given the sunk costs, it is likely to adjust very differently than it would if it started from having nothing. It is possible that, at the higher level of capital, the firm would want to use quite a different technology than the one it actually adopted. Meanwhile, given the earlier commitment, it is wiser (cheaper) for it to add more capital in a way which expands its capacity with the old technology rather than converting to the technology which is more efficient at the scale of output which it anticipates. Or, alternatively, at the later time a new technology may become available which it would adopt if it were a new firm just setting up; but which it does not pay to adopt given the plant and equipment which it presently owns. In general, a firm probably never possesses the (free adjustment) optimal capital stock at any time of its existence after the point in time when its initial investment is made.

The considerations and examples reviewed above are no doubt neither novel nor surprising to most readers, though, so far as I am aware, they have not been made the subject of systematic, formal analysis. A start is made on this task in this chapter.

The simplest way to handle the nonmalleable nature of capital is to assume that if a firm wishes to change its amount of capital, it must discard all of the old and buy entirely new. A model of this sort is outlined in §2. In §3 optimal behavior for a monopolist is investigated; and in §4 some generalizations to the model developed in §2 and §3 are considered. §5 contains a discussion of the relative merits of continuous and discrete time models. §6 generalizes the monopolistic market of §2 and §3 to an oligopoly of n firms; and §7 contains concluding comments.

2. The model of the monopolistic firm

In the model developed in this and the following section time is a continuous variable, although time does, in the end, behave more like it were discrete. Letting p_t denote the firm's price at time t, the firm is free to choose any price policy whatsoever over the time interval $[0, \infty)$, subject only to the restriction that price be nonnegative. Thus, abrupt changes in price can occur at any time. It turns out, for obvious reasons, that the firm does not want to continuously alter its price and that, between price changes, there is a certain minimum time interval. The reasons are easily stated: the zero scrap value for old plant assures that there is a minimum and strictly positive time interval during which a given plant is used. In addition, the optimal price for a firm at any moment depends only on the plant in use; hence, price is only changed when a new plant is acquired.

In §2.1 the assumptions to be used are presented, and in §2.2 a few minor results are derived. Notation is similar to that of chs. 2–5.

2.1. The underlying assumptions for the firm and market

The assumptions made in this section parallel assumptions made in ch. 3, with obvious modifications. First, the market has only one seller ($n = 1$), and the costs of production of the firm depend on both the amount produced and the capital stock which the firm possesses. In addition to costs of production there are capital costs. As in earlier chapters, it is assumed that the rate of sales and production are equal, and that demand at the ruling price equals actual sales. The basic conditions on demand are:

E1. *The demand function, $q_t = F(p_t)$, is defined, nonnegative and continuous on the interval $[0, \infty)$. For a positive, finite p^+, $F(p) = 0$ when $p \geq p^+$. For $p \in (0, p^+)$, $F(p)$ is bounded and twice differentiable with $F' < 0$.*

E1 requires that the demand function is differentiable, downward sloping and cuts both axes.

With a stationary demand function, as E1 specifies, if costs as a function of output and the stock of capital are also stationary, investment policy comes trivially to choosing the optimal capital stock at the outset and nothing more. Two obvious routes are open to make investment policy slightly more realistic and interesting. One is to assume that capital deteriorates and the other is to assume that newer capital is more productive than old. The latter course is followed.

Most of the conceptual problems related to capital are sidestepped here.[1] It is assumed, implicitly, that the firm has a known production function, that all prices of its inputs into the indefinite future are known and that, as a result, the firm has a total cost function which is specified in E2. With newer capital assumed to be more productive than old, it is necessary to always keep two dates in mind – the present date, t, and the time when the present capital stock was put into service, τ. Of course, $\tau \leq t$. Without loss of generality, the price of capital may be taken to be unity. A unit of capital represents a number of *efficiency units* which depends on the vintage of the capital. One unit of capital from time τ represents $e^{\lambda\tau}$ efficiency units, where λ is a fixed positive constant. Thus if K_t^* is the number of efficiency units represented by the capital stock in service at time t, and the capital stock dates from τ, then $K_t^* = e^{\lambda\tau}K_\tau$. The total cost function is $C(q_t, K_t^*)$. In this case, *total cost* means total cost of production and does not include outlays to purchase capital.

E2. *$C(q_t, K_t^*)$ is defined and continuous for $q_t \in [0, F(0)]$ and $K_t^* \in [0, \infty)$, and has continuous second partial derivatives on the interior of its domain. $C(0, 0) = 0$, $C^1(q, K^*) > 0$, $C^2(q, K^*) < 0$, $C^{11}(q, K^*) \geq 0$ and $C^{12}(q, K^*) < 0$. There is a minimal $\underline{K}^* < \infty$ such that $C^1(0, \underline{K}^*) \leq p^+$.*

These conditions assert that the marginal cost of production, C^1, is positive, more (efficiency units of) capital lower both marginal and total variable cost at any given level of output and there is a finite capital stock such that, for larger capital stocks, strictly positive rates of production can be found which are more profitable than zero production. Indeed, unless marginal cost is constant (i.e. $C^{11} = 0$), revenue exceeds variable cost whenever price equals marginal cost and production is positive. Let

$$\mathcal{D} = \{(p, K^*) | (p, K^*) \geq 0 \quad \text{and} \quad p \geq C^1(F(p), K^*)\}. \tag{6.1}$$

E1 and E2 imply that $\mathcal{D}$ is nonempty and, in fact, has a nonempty interior.

[1]The interested reader is directed to Bliss (1975) for an interesting treatment of questions related to the incorporation of capital into economic models. Though Bliss' discussion is directed toward models of the economy, much, if not all, of what he has to say pertains to modelling the firm.

It is assumed that the profit function,

$$\pi(p, K^*) = pF(p) - C(F(p), K^*), \tag{6.2}$$

is concave on $\mathscr{D}$:

E3. *$\pi(p, K^*)$ is concave with respect to $(p, K^*) \in \mathscr{D}$. The profit function $\pi(p, K^*)$ is the rate of flow of profit per unit of time, not taking into account the cost of acquiring capital.*

Contrary to all the other chapters in this book, time is treated here as a continuous variable. Thus the firm is allowed to acquire new plant at any point in time, indeed to continuously change its plant if it wishes; and to continuously change its price. The price path of the firm is denoted $\{p(t)\}$, with the understanding that, for any $t \geq 0$, $p(t) \in [0, p^+]$. It is seen in the following sections that the firm does not wish to alter its capital stock continuously; because the adjustment cost generated by such action is unbounded. It is further seen that each value of the capital stock has an associated optimal price; therefore, price changes are only made when capital is changed. To form the discounted profit function of the firm, let the discount rate of the firm be ρ and assume that the times at which the capital stock is changed are $t_0, t_1, t_2, \ldots$, where it is understood that $t_0 = 0$ and $t_k < t_{k+1}$. The profit function in eq. (6.2) gives a rate of profit per unit time when capital is fixed. In addition to the discounted flow of revenue minus production costs from eq. (6.2), the firm pays the cost of each new plant it acquires. Recall, it is assumed that any adjustment to the capital stock implies the complete scrapping of the old plant, with zero scrap value, and the purchase of a new plant. Thus, the discounted profits of the firm are given by

$$G(p(t), K_{t_0}, K_{t_1}, \ldots)$$
$$= \sum_{k=0}^{\infty} \left[-e^{-\rho t_k} K_{t_k} + \int_{t_k}^{t_{k+1}} e^{-\rho\tau} \pi(p(\tau), e^{\lambda t_k} K_{t_k}) d\tau \right]. \tag{6.3}$$

2.2. *The firm changes its price only when it changes its capital stock*

There are two results which follow quite easily from the assumptions, and which simplify the analysis which comes later. These are that it is not in the best interest of the firm to refrain from ever acquiring a positive capital stock and entering the market; and, optimal price policy requires that the firm's price remain constant over any period during which the capital stock is unchanged. These results are easily explained. There are price–capital combinations for which revenue exceeds the cost of production. For any one such combination, if capital were sufficiently cheap, it would be profitable to purchase the given amount of capital and adopt the indicated price. Because the purchase price of capital falls continuously, going to zero in the limit, a time is eventually reached when capital has become sufficiently cheap. That the optimal price does not change as long as the capital stock is fixed results from the stationary demand function, together with the condition that production costs are not a function of time.

Let $\{p(t), K^*(t)\}$ denote a *plan* for the firm. The plan is a path for the values of price and plant. It is shown in lemma 6.1 that, from some finite time onward, $(p(t), K^*(t)) \in \mathcal{D}$ for a plan which is optimal. That is, eventually the capital stock is greater than zero and the firm *enters the market.*

Lemma 6.1. *Assume E1–E3 to be satisfied and let $\{p(t), K^*(t)\}$ be an optimal plan. Then there is a finite $\underline{t}$ such that for $t \geq \underline{t}$, $(p(t), K^*(t)) \in \mathcal{D}$.*

□ The last part of E2 asserts that there is a threshold number of efficiency units of capital, $\underline{K}^*$, such that for $K^* \geq \underline{K}^*$, $p^+ \geq C^1(0, K^*)$. *If* $\underline{K}^* > 0$, then for $K^* < \underline{K}^*$, $p^+ < C^1(0, K^*)$. Given that the marginal cost of production (C^1) rises with the rate of output and falls as the size of the plant increases, it is clear that plant sizes larger than zero and smaller than $\underline{K}^*$ are devoid of interest; because the rate of flow of profit associated with them is negative. With plants of such size, the optimal price is p^+ and the optimal output level is zero. Clearly, if optimality entails zero production, this is most profitably accomplished with a plant of zero size. Thus if $\{p(t), K^*(t)\}$ is an optimal plan, then, $K^*(t) \notin (0, \underline{K}^*)$ for any t – that is, $K^*(t)$ may be zero or at least as large as $\underline{K}^*$, but it cannot be between. It is immediate that if $K^*(t) = 0$, then $p(t) = p^+$ and if $K^*(t) \geq \underline{K}^*$ when $p(t) \geq C^1(0, K^*(t))$.

Thus the lemma is proved if it can be shown that (a) once $K^*(t)$ rises to at least $\underline{K}^*$, it never falls below; and (b) by some finite time $\underline{t}$, $K^*(\underline{t})$ must be at least $\underline{K}^*$. That (a) is true for an optimal plan follows from two facts: lowering $K^*(t)$ causes an increase in the marginal and total production costs associated with any given production level, and requires expenditure on a new plant. Thus $K^*(t)$ is a monotone nondecreasing function. Showing (b) now means showing that $K^*(t) = 0$ for all t cannot be optimal, which, in turn, requires that there be some plan for which the discounted profit is strictly positive. Choose an arbitrary $K_0^* > \underline{K}^*$. Let $K^*(t) = 0$ for $t < t^*$ and $K^*(t) = K_0^*$ for $t \geq t^*$. Then let $p(t) = p^+$ for $t < t^*$ and $p(t) = p_0$ for $t \geq t^*$, where p_0 is the price which maximizes the rate of profit, given K_0^*. Under this plan, discounted profits are

$$e^{-\rho t^*}[-e^{-\lambda t^*}K_0^* + \pi(p_0, K_0^*)/\rho]. \tag{6.4}$$

With K_0^* and p_0 given, eq. (6.4) is maximized for finite

$$t^* = \lambda^{-1} \ln [(\rho + \lambda)K_0^*/\pi(p_0, K_0^*)]. \tag{6.5}$$

At $\underline{t}^*$ discounted profits are $\lambda e^{-\rho \underline{t}^*}\pi(p_0, K_0^*)/(\rho + \lambda) = \pi_0 > 0$.

Thus any optimal plan must yield a discounted profit of at least π_0. Now consider that no matter what the size of the capital stock, the rate of profit cannot exceed $p^+F(0)$; therefore if $K^*(t) = 0$ until $\underline{t}$, discounted profits cannot exceed $e^{-\rho \underline{t}}p^+F(0)/\rho$. Choose $\underline{t}$ so that $e^{-\rho \underline{t}}p^+F(0)/\rho = \pi_0$. Clearly $\underline{t}$ is finite and it is impossible for any plan to give discounted profits of π_0 or greater if capital is zero over the finite interval $[0, \underline{t}]$. Therefore under an optimal plan, $K^*(t)$ becomes greater than $\underline{K}^*$ by no later than $\underline{t}$. □

Another way to state the conclusion of lemma 6.1 is that the monopolist cannot maximize discounted profits if it fails to *get into the market* later than time $\underline{t}$. Without loss of generality, it may be assumed that the initial price of capital is low enough that the monopolist chooses to acquire a positive amount of capital at $t = 0$.

Over an interval of time during which the capital stock is unchanged, the profit function is stationary. This, coupled with concavity of the payoff function on $\mathcal{D}$ and the knowledge that K^* is always chosen so that $\{p \mid (p, K^*) \in \mathcal{D}\}$ is not empty implies the following lemma.

Lemma 6.2. *Let E1–E3 hold and assume that K_t^* is constant and greater than $\underline{K}^*$ over the time interval (t_k, t_{k+1}). Then the optimal price for the firm over this time interval is given by the (unique) solution to*

$$\pi_t^1 = F(p_t) + (p_t - C^1(F(p_t), K_t^*))F'(p_t) = 0. \tag{6.6}$$

The final result for this section is that the optimal price and the size of the capital stock, in efficiency units, are inversely related.

Lemma 6.3. *Under E1–E3, for $K_t^* \geq \underline{K}^*$, the solution to eq. (6.6) may be* *written*

$$p_t = \phi(K_t^*); \tag{6.7}$$

ϕ is differentiable with $\phi' < 0$ for $K_t^* > \underline{K}^*$.

□ For $K_t^* > K^*$, concavity of π_t allows that the implicit function theorem may be applied to assure that ϕ exists. Furthermore, $\phi(K_t^*)$ must be strictly greater than zero and less than p^+. The former must hold because profits would be negative otherwise. The latter implies zero sales with marginal revenue (equal to price) greater than marginal cost. Thus, profit maximization occurs at a point interior to $\mathcal{D}$, implying that eq. (6.6) must be satisfied. Differentiating eq. (6.6) to obtain an expression for ϕ' yields

$$\phi'(K_t^*) = dp_t/dK_t^* = C^{12}(F(p_t), K_t^*)F'(p_t)/\pi^{11}(p_t, K_t^*) < 0. \tag{6.8}$$

The sign of ϕ' is necessarily negative because $\pi^{11} < 0$ (by E3), $F' < 0$ (by E1) and $C^{12} < 0$ (by E2). Finally, $\phi(\underline{K}^*) = \lim_{K^* \to \underline{K}^*} \phi(K^*)$. □

3. *The monopolist's optimal policy*

The principal result of this section, theorem 6.8, is that there is an optimal policy for the monopolist. This is shown in §3.3; however, in the process of deriving the result, a number of characteristics of the optimal policy emerge in §3.1 and §3.2. In §3.1, optimal policy is examined for a firm which can change price continuously, but which can acquire new capital only at certain previously specified dates. It is seen that over any span of time during which the capital stock is unchanged, the firm keeps price unchanged also. Then, it is shown that the capital stock which is acquired at any date is larger in terms of efficiency units than the preceding capital stock, and that the optimal capital to acquire at a given time depends on the date of acquisition and the date at which capital is next to be acquired (i.e. on the date of acquisition and the length of time the capital is to be in service).

In §3.2 it is proved that there is a minimum length of time, strictly positive, during which the firm must use a given plant, if it is to have nonnegative profits. Thus, new plant, under an optimal policy, is acquired a countable number of times. Finally, it is shown in §3.2 that an optimal plan requires that a new plant be acquired a countably infinite number of times. The results of §3.1 and §3.2 summarize to: (a) optimal price is a function of the capital stock alone, (b) the capital stock is changed a countably infinite number of times, (c) it is increased each time it is changed, and (d) at each time the capital stock is changed, the size of the new capital stock depends only on the date of acquisition and the length of time the new capital is to be used. With these results in hand, it is an easy matter to prove that there is an optimal policy.

3.1. *Optimal behavior when new plant may only be acquired at predetermined points in time*

Imagine a (countable) list of times t_0, t_1, $t_2, \ldots$, with $t_k < t_{k+1}$ for $k = 1, 2, \ldots$. Say that only at these specific dates is it possible for the firm to purchase new plant. At any of these times, the firm may elect to leave its plant unchanged. In view of lemma 6.3, the model becomes a discrete time model with a sequence of time periods of length $t_1 - t_0$, $t_2 - t_1$, etc. These lengths may be nonuniform; however, that does not alter the discrete time character of the model.

Lemma 6.4. *Assume that E1–E3 hold and the times t_0, t_1, $t_2, \ldots$ are the only times at which new capital may be purchased. At any of these t_k the firm may decide not to purchase new capital. Then, given the constraint on the timing of new plant acquisition, if the firm follows an otherwise optimal plan, any time it purchases a new plant, it raises its holding of efficiency units of capital. That is, K^* rises each time a new plant is acquired. The optimal value for K_{t_k}, if a new plant is to be acquired at t_k and another at t_{k+1}, depends only on the dates t_k and t_{k+1}, given an optimal price policy.*

□ The reason why the monopolist only obtains a larger plant when a new plant is purchased is quite obvious. Reduction of plant size brings no possible advantage. Not only is marginal cost, for a given output level, a decreasing function of plant size, the plant causes no flow of fixed costs ($C(0, 0) = 0$). Were there a positive flow of fixed costs, increasing with plant size, then with shrinking demand over time, optimal policy might well involve acquiring new, smaller plants from time to time; however, neither of these conditions are part of the present model. Buying a new plant, smaller than the old, means paying a positive fee to increase the flow of production costs over the future. Thus all but the last sentence of the lemma has been shown, so, without loss of generality it may be supposed that $t_0 = 0$ and that capital is purchased at each of the t_k. Recalling that the firm's optimal price is given by $\phi(K^*)$, and letting K_k denote the number of units of capital purchased at time t_k, the number of efficiency units of capital put into service at t_k is

$$K_k^* = e^{\lambda t_k} K_k \tag{6.9}$$

and the discounted profit of the firm is

$$\sum_{k=0}^{m} e^{-\rho t_k}[-K_k + \pi(\phi(e^{\lambda t_k}K_k), e^{\lambda t_k}K_k)(1 - e^{-\rho(t_{k+1}-t_k)})/\rho], \tag{6.10}$$

where m may, but need not be, finite. Given the timing of plant changes, the condition which determines the optimal size of a new plant purchased at t_k is

$$\begin{aligned} -1 &+ e^{\lambda t_k}\phi'(e^{\lambda t_k}K_k)\pi^1(\phi(e^{\lambda t_k}K_k), e^{\lambda t_k}K_k) \\ &+ \pi^2(\phi(e^{\lambda t_k}K_k), e^{\lambda t_k}K_k)] [1 - e^{-\rho(t_{k+1}-t_k)}]/\rho = -1 \\ &+ e^{\lambda t_k}\phi'(K_k^*)\pi^1(\phi(K_k^*), K_k^*) + \pi^2(\phi(K_k^*), K_k^*)] \\ &\quad [1 - e^{-\rho(t_{k+1}-t_k)}]/\rho = 0, \qquad k = 0, 1, 2, \ldots, m. \end{aligned} \tag{6.11}$$

Eq. (6.11) is the partial derivative of eq. (6.10) with respect to K_k. Due to concavity of eq. (6.11), the implicit function theorem may be called upon. The optimal value for K_k may be written

$$K_k = \psi(t_k, t_{k+1}). \tag{6.12}$$ □

The results on optimal behavior to this point may be summarized:

Theorem 6.5. *Under E1–E3, if the firm is constrained to purchase new plant only at times $t_0, t_1, t_2, \ldots$, then optimal behavior is described by eqs. (6.7) and (6.12):*

$$p_t = \phi(e^{\lambda t_k}K_k), \tag{6.7}$$

$$K_k = \psi(t_k, t_{k+1}). \tag{6.12}$$

Furthermore, $\phi' < 0$, and $K_k < K_{k+1}$ for all k. Thus optimal behavior is characterized fully if the firm cannot choose the times at which it is to acquire new plant.

3.2. *Under an optimal plan, the firm invests a countably infinite number of times*

It is not surprising that any optimal policy requires that the firm acquire a new plant a countably infinite number of times. On the one hand, purchase of a new plant causes a lump-sum outlay. With an upper bound on the rate of profit, there is a minimum length of time which must pass before the firm could conceivably recoup enough additional profits to pay for the plant. Thus, the number of times when a new plant is purchased is at most countable. On the other hand, if the number is finite, there is a time after which no new plant is again purchased. The reason this turns out to be impossible under an optimal plan is that, no matter what the size of the plant, if it were larger, the rate of profit flow would be higher. Meanwhile, the cost of an efficiency unit of capital declines with time, going to zero in the limit. Therefore, a given size of plant eventually becomes arbitrarily cheap to buy; so for any existing plant size, a larger one of a given size, after enough time elapses, becomes profitable to acquire.

Lemma 6.6. *Under E1–E3, if a new plant is purchased at time t, there is a minimum length of time during which it must be kept in use if its acquisition is to increase discounted profits. A lower bound can be found for this minimum which is strictly positive and depends only on t and* $\underline{K}^*$.

□ Let K be the amount of capital to be purchased at time t, $e^{\lambda t}K$, the associated number of efficiency units, π the rate of profit with the capital in use just before t, and $\pi + \Delta\pi$ the rate of profit after the new capital is set in place. The addition to discounted profits resulting from the new plant, discounted to t, is

$$-K + \Delta\pi(1 - e^{-\rho T})/\rho, \tag{6.13}$$

on the assumption that the new plant is replaced at $t + T$. Equation (6.13) gives the addition to profit resulting from acquiring the new plant at t and replacing it at $t + T$ as compared with keeping the old plant until $t + T$. An admissible lower bound on eq. (6.13) is zero, if the new plant is to add to the firm's payoff. Nonnegativity of eq. (6.13) is equivalent to

$$T \geq \ln\,[\Delta\pi/(\Delta\pi - \rho K)]/\rho. \tag{6.14}$$

The expression to the right of the inequality in eq. (6.14) falls with larger $\Delta\pi$ and smaller K. $\Delta\pi$ is bounded above by $p^+F(0)$ and K is bounded below by $e^{-\lambda t}\underline{K}^*$. Thus

$$T \geq \ln\left[p^+F(0)/(p^+F(0) - \rho e^{-\lambda t}\underline{K}^*)\right]/\rho = B(t) > 0, \tag{6.15}$$

which satisfies the lemma. □

Lemma 6.7. *Under E1–E3, with an optimal plan, the number of times that a new plant is acquired is countably infinite.*

□ From lemma 6.6, the number of new plant acquisitions is known to be countable. It remains to show that the number cannot be finite. This is done by proving that for any size of the capital stock, it is eventually profitable to replace it. Let K_0^* be the present capital stock and let $K_1^* > K_0^*$. It is seen below that irrespective of the size of K_0^*, there is a time t^* such that, replacing the K_0^* plant with a K_1^* at any time beyond t^* results in increased discounted profit. After the switch is made, the rate of flow of profit increases by $\Delta\pi > 0$. If the change were made at time t, the change in discounted profits, discounted to t, would be

$$-e^{-\lambda t}K_1^* + \Delta\pi/\rho, \tag{6.16}$$

which exceeds zero if

$$\Delta\pi/\rho K_1^* > e^{-\lambda t}. \tag{6.17}$$

The expression on the left is a strictly positive constant, and that on the right is a decreasing function of t which goes to zero; therefore, there is a finite t^* such that for $t > t^*$, eq. (6.17) holds. There can, then, be no final plant purchase. □

3.3. Existence of an optimal policy

The role played in the preceding proof by the price of an efficiency unit of capital suggests that if that price had a positive lower bound, the optimal number of new plants would be finite. Taking together the various results obtained, it becomes useful to reformulate the discounted payoff function as a special sort of function with a discrete time structure. It is known that the optimal price does not change as long as the capital stock is unchanged, and that the capital stock is changed a countably infinite number of times. Therefore, if attention is restricted to policies with these characteristics, then the optimal policies, if any exist, are in this smaller set. The discounted payoff function may be written

$$G(p_0, K_0, p_1, K_1, \ldots)$$
$$= \sum_{k=0}^{\infty} e^{-\rho t_k}[-K_k + \pi(p_k, e^{\lambda t_k}K_k)(1 - e^{-\rho(t_{k+1}-t_k)})/\rho]. \tag{6.18}$$

Taking into account the knowledge that the optimal plant to choose at t_k is $\psi(t_k, t_{k+1})$ and the optimal price to use with that plant is $\phi(\psi(t_k, t_{k+1}))$, the discounted payoff function can be written as a function of the investment dates alone. The firm's problem reduces to choosing the optimal set of times to acquire new plant, and eq. (6.18) becomes, denoting $\psi(t_k, t_{k+1})$ by ψ_k and $\phi(\psi(t_k, t_{k+1}))$ by ϕ_k,

$$\sum_{k=0}^{\infty} e^{-\rho t_k}[-\psi_k + \pi(\phi_k, e^{\lambda t_k}\psi_k)(1 - e^{-\rho(t_{k+1}-t_k)})/\rho]. \tag{6.19}$$

The existence of an optimal policy reduces to the existence of a set of numbers, $t_0, t_1, t_2, \ldots$, which maximize eq. (6.19).

Theorem 6.8. *Under E1–E3 an optimal policy exists for the firm. Under an optimal policy, price and capital are chosen according to ϕ and ψ respectively, and the number of dates at which new plant is acquired is countably infinite.*

□ From the results which have been proved, the firm's payoff may be regarded as a function of the investment dates alone, as in eq. (6.19). Then if eq. (6.19) is a bounded and continuous function of the investment dates $(t_0, t_1, \ldots)$ and the vector of dates comes from a compact set, eq. (6.19) has a maximum. The vector of dates associated with the maximum, which need not be unique, is the optimal policy which we seek. The proof proceeds by defining a distance measure for the set of investment dates with respect to which the payoff function is continuous, and for which the set of investment dates is compact. Then an optimal policy must exist; because a continuous function defined on a compact set has a maximum on that set.[2]

To define a set of vectors of investment dates or *plans* let $\beta_0 = B(0)$ and $\beta_l = B(\Sigma_{k=0}^{l-1} \beta_k)$ for $l = 1, 2, \ldots$. From lemma 6.6, a plan t cannot be optimal if $t_l - t_{l-1} < \beta_l$; therefore attention may be restricted to the set of plans

$$\mathcal{T} = \{t | t_0 \geq 0, t_l - t_{l-1} \geq \beta_l, \quad l = 1, 2, \ldots\}. \tag{6.20}$$

The members of $\mathcal{T}$ each have a countable number of elements; but the

[2]I am grateful to Peter Rice for pointing out an error in this proof.

number may be finite. Now choose $t^0, t^1 \in \mathcal{T}$ and let $\mathcal{T}^i = \{t_0^i, t_1^i, t_2^i, \ldots\}$. $\mathcal{T}^i$ is the set of dates which appear in the investment date vector t^i. ($i = 0, 1$). Further, let $\mathcal{T}_0 = \mathcal{T}^0 \cup \mathcal{T}^1$, the distinct set of dates appearing in both sets; so if $t^0 = t^1$, then $\mathcal{T}_0 = \mathcal{T}^0 = \mathcal{T}^1$. Next, the members of $\mathcal{T}_0$ are arranged into order. They are labelled so that $\mathcal{T}_0 = \{\tau_0, \tau_1, \tau_2, \ldots\}$ with $\tau_{l+1} > \tau_l$, for $l = 1, 2, \ldots$. From the elements of $\mathcal{T}_0$, two vectors of dates are constructed corresponding to t^0 and t^1. $\tau^i = (\tau_0^i, \tau_1^i, \tau_2^i, \ldots)$ where

$$\tau_l^i = \max_{t \in T^i, t \leq \tau_l} t, i = 0, 1, \qquad l = 0, 1, 2, \ldots. \tag{6.21}$$

Finally, let

$$\delta_l = (e^{-\rho\tau_l} - e^{-\rho\tau_{l+1}})/\rho, \qquad l = 0, 1, 2, \ldots. \tag{6.22}$$

The construction completed above can be explained with the help of fig. 6.1. Each time interval $[\tau_l, \tau_{l+1})$ is an interval over which no new plant is purchased under either t^0 or t^1; however, at each of the times τ_l new plant is purchased under one or both of the plans. Associated with any particular point in time, t, is the *vintage* (purchase date) of the plant in use under a given plan. Thus if $t \in [\tau_l, \tau_{l+1})$ the plant in use under plan t^0 is τ_l^0 and under t^1 is τ_l^1. For example, in fig. 6.1 for $l = 3$, $\tau_3^0 = t_2^0$ and $\tau_3^1 = t_1^1$.

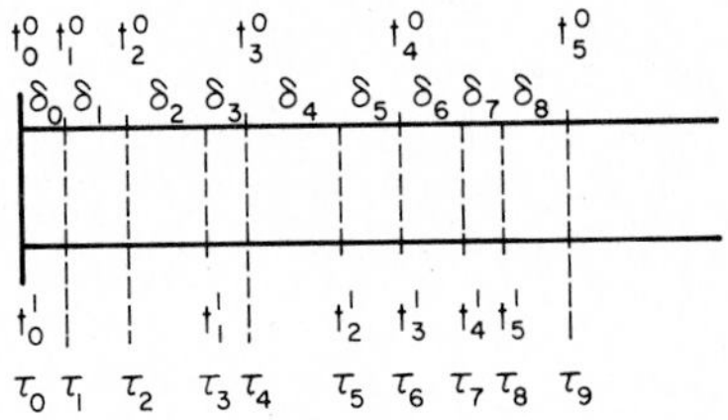

Fig. 6.1

Given the discount rate ρ, the importance, or weight, assigned to this time interval is

$$\int_{\tau_3}^{\tau_4} e^{-\rho t} dt = (e^{-\rho\tau_3} - e^{-\rho\tau_4})/\rho = \delta_3, \tag{6.23}$$

and the difference between the two plans at such a time, t, is the difference between the vintages, $|\tau_3^0 - \tau_3^1| = |t_2^0 - t_1^1|$. Consequently, the distance between the two plans, $d(t^0, t^1)$, is defined as

$$d(t^0, t^1) = \sum_{l=0}^{m} \delta_l |\tau_l^0 - \tau_l^1|, \tag{6.24}$$

where $m < \infty$ if the plans t^0 and t^1 both involve only a finite number of investment times. Using the distance defined in eq. (6.24), it is clear that the discounted payoff function is continuous; and that the set of plans, $\mathcal{T}$, is closed. It remains to verify that the set of plans, $\mathcal{T}$ is bounded. With t^0 and t^1 representing two arbitrary elements of $\mathcal{T}$ and $\tau_0 = 0$, this may be seen as follows:

$$\begin{aligned} d(t^0, t^1) &= \sum_{l=0}^{m} (e^{-\rho\tau_l} - e^{-\rho\tau_{l+1}})|\tau_l^0 - \tau_l^1|/\rho \\ &\leq \sum_{l=0}^{m} (e^{-\rho\tau_l} - e^{-\rho\tau_{l+1}})\tau_l/\rho \\ &= \sum_{l=0}^{m} (\tau_l - \tau_{l-1})e^{-\rho\tau_l}/\rho \\ &\leq \int_0^{\tau_m} (e^{-\rho\tau}/\rho)d\tau \leq \int_0^{\infty} (e^{-\rho\tau}/\rho)d\tau = 1/\rho^2. \end{aligned} \tag{6.25}$$

Compactness may be seen by choosing an arbitrary sequence of members t^k of $\mathcal{T}$. Assume there is no $t^0 \in \mathcal{T}$ such that any subsequence of $\{t^k\}$, converges pointwise to t^0. Then, regarding $\{t^k\}$ as a countable family of sequences, $\{t_l^k\}$, for $l = 0, \ldots, l^* - 1$, $\{t_l^k\}$ has convergent subsequences and for $l \geq l^*$, $\{t_l^k\}$ has no convergent subsequences. Hence t_l^k goes to ∞ as k goes to ∞ for $l \geq l^*$. Now let $t^0 \in \mathcal{T}$ and let $\{t^{k_r}\}$ denote a subsequence of $\{t^k\}$ for which $\{t_l^{k_r}\}$ converges to $t_l^0 (l < l^*)$. $l^* \geq 1$ due to the convention that $t_0 = 0$ for any member of $\mathcal{T}$. From eq. (6.25) it is clear that

$$d(t^{k_r}, t^0) \leq \sum_{l=0}^{l^*-1} |t_l^{k_r} - t_l^0| + e^{-\rho t_{l^*}^{k_r}}/\rho^2;$$

however, $|t_l^{k_r} - t_l^0|$ goes to zero as r goes to ∞ for $l = 0, \ldots, l^* - 1$, and $t_{l^*}^{k_r}$ goes to ∞. Thus $d(t^{k_r}, t^0)$ converges to zero, which implies that $\{t^k\}$ has a convergent subsequence. Hence $\mathcal{T}$ is compact and the theorem is proved. □

4. On generalizing the model

There are some relatively easy and obvious avenues of generalization for the model which has been presented. These are sketched rather than worked out in detail; for the interested reader can undoubtedly fill in the gaps quite easily himself.

4.1. The plant has scrap value or can be updated

It has been assumed that any change whatsoever in the size of plant requires abandonment of the old plant and purchase of a new one. A modified version of this condition is to assume that the old plant can be either modernized or sold for a positive price. These may be simply incorporated into the model in the following way. Say the firm is considering at time t_2 whether to alter its plant, which was acquired at time t_1. At time t_2 the price of an efficiency unit of capital is $e^{-\lambda t_2}$. If there were no costs of conversion from a plant of one size to a plant of another, the old plant would be worth

$$e^{-\lambda t_2}K_1^* = e^{\lambda(t_1-t_2)}K_1. \tag{6.26}$$

The simplest mode of imposing costs is to assume that the cost of conversion is proportional to the size of the old plant. Thus, for $\gamma \in [0, 1)$, the value of the old plant, given conversion, is

$$\gamma e^{\lambda(t_1-t_2)}K_1, \tag{6.27}$$

which makes the net cost of the new plant

$$K_2 - \gamma e^{\lambda(t_1-t_2)}K_1. \tag{6.28}$$

$\gamma = 0$ corresponds to the case analyzed in §3 and §4; while $\gamma = 1$ is a conventional model in which there are no costs of adjustment. An intermediate value for γ must yield results essentially like the results of §2 and §3. It is obvious that in the total absence of adjustment costs, the firm would have its capital stock changing continuously over time. As soon as any adjustment cost is introduced, however small, it is equally obvious that continuous adjustment brings unbounded costs; hence, qualitatively, the firm must behave as theorem 6.8 indicates. Investment takes place at discrete intervals, though the intervals may be unequal in length.

4.2. More than one kind of capital

For the moment, assume two kinds of capital, *plant* and *equipment*. Assume that equipment is continuously adjustable while plant behaves as in §2 and §3. Then the optimal level of equipment would be determined in an analogous fashion to the way that optimal price is arrived at. Indeed, they would be determined simultaneously, each as a function of the size of plant. If adjustment costs should attach to changes in equipment or the adjustment costs associated with changing plant depend on both plant and equipment stocks, the model remains, basically, like the one studied in detail. No changes take place in any type of capital or in price, except at a countable number of points in time.

For a final example, say that there are no adjustment costs associated with changes in equipment alone, but that equipment depreciates. Then, equipment would undoubtedly be purchased continuously at a rate which would change only at times when new plant were acquired. If in addition to depreciating, there were adjustment costs associated with changes in equipment alone, then equipment would be purchased only a countable number of times. It is to be expected that whenever plant were changed, equipment would also; however, it is possible that equipment would sometimes be purchased at a time when plant were not being changed.

5. Continuous versus discrete time

The developments of this chapter are likely to raise the question of whether continuous time models are inherently better or worse than discrete time models. The range of usual answers include (a) superiority of continuous time models, (b) superiority of discrete time models, and (c) use the type which is more tractible. Of course, it is most satisfying when essentially the same results are reached by either route; because that encourages one's wish to use the technique which is easiest for the job at hand.

The argument that time is "truly" continuous; hence continuous time models are better appears contradicted as a general proposition by the results of this chapter. Starting with a continuous time formulation, the model boils down in the end to a discrete time model; however, there are some warnings to be posted. First is that decisions are taken at only discrete intervals because of the presence of adjustment costs associated with some decisions and due to the stationarity of the model. Let demand have a random component or let it grow over time and the firm would surely wish to change price continually. Or remove all adjustment costs, and again, adjustment of both plant and price would be continuous. Perhaps in some applications the assumptions which turn this continuous time model into one of discrete time hold reasonably well, and in others not. There is no substitute for using one's own good judgment.

A second warning is that the discrete time model developed in §2 and §3 is quite unusual, because the time intervals are of an uneven length and they have their length determined within the model. This is, of course, in stark contrast with a usual discrete time model in which the time periods are equal and of a length fixed in advance. Just as a discrete time model might approximate a continuous time model if the periods are sufficiently short, the model of this chapter might also be approximated by a discrete time model of the usual sort if the time periods are made very short.

6. *Noncooperative equilibrium in an n-firm market in which firms can invest*

It is instructive to consider a market of n firms, each of which is like the one described in §2 and §3, and to examine whether such a market has a noncooperative equilibrium. The first step is to reformulate the model for n firms. This is done in §6.1. Next, optimal behavior for one firm, given the choices of the others is considered in §6.2 and, in §6.3, equilibrium for the market is shown.

Intuition may suggest that formulating the n-firm market in the "obvious" way yields a model to which a standard existence theorem for noncooperative equilibrium can be applied. It turns out this is not so, despite the assumption that the firm's own profit rate is a concave function of its own price and capital stock (E3). This is not sufficient to insure concavity, or quasi-concavity of the discounted payoff with respect to price, capital and the timing of investment.[3]

6.1. Reformulating the market model

The demand assumption, E1, is a one firm version of A3–A5 which are introduced in ch. 3 §3.1. It is A3–A5 which are assumed henceforth in place of E1. E2 can very nearly stand as it already is written. The only modification concerns the interpretation of p^+, which, in §2 and §3 is the lowest price at which the firm's sales are zero. Given a demand function for the ith firm of $F_i(p_i, \bar{p}_i)$, the lowest price at which the firm's sales are

[3]Existence theorems are discussed in ch. 7, §2 and §3.

zero depends on $\bar{\mathbf{p}}_i$. For present purposes, the procedure used in A10 (see ch. 5 §2.1) is useful here. Let p_i^* be defined by the conditions $F_i(p_i^*, \mathbf{0}) = 0$ and $F_i(p_i, \mathbf{0}) > 0$ for $p_i < p_i^*$. In the present section (§6), E2 is understood to apply to each firm with p^+ meaning p_i^*, as defined above, for the ith firm.

The counterpart sets to $\mathcal{D}$ are

$$\mathcal{D}_i(\bar{\mathbf{p}}_i) = \{(p_i, K_i^*) | (p_i, K_i^*) \geq 0, p_i \geq C^1(F_i(p_i, \bar{\mathbf{p}}_i), K_i^*)\}. \tag{6.29}$$

E2 assures that, for any $\bar{\mathbf{p}}_i$, $\mathcal{D}_i(\bar{\mathbf{p}}_i)$ is nonempty and has a nonempty interior. E3 is retained, applied individually to the n firms. One final assumption needs to be made to assure concavity of the firm's own discounted payoff with respect to its own choices; however, it is convenient to state this assumption in §6.3.

6.2. *Optimal behavior for one firm*

The main purpose of this section is to show that the results of §2 and §3 carry over, with very little modification, to the present context. Consider the decision problem of firm 1 and suppose that, for each of the other firms, there are only a countable number of points in time when price or plant is changed. Then, for the $n-1$ firms taken together, there are only a countable number of points in time when any one of them changes a price or plant size. From the vantage point of firm 1, only the price changes of the others affect his payoffs; though it is likely that a firm changes its price at any time that it changes its plant. Surely, firm 1 *may* wish to change its price or plant at some of the times that others make changes; however, it is clear from earlier results that it would change either price or plant size on, at most, a countable number of other occasions. Thus

Lemma 6.9. *Under A3–A5, E2 and E3, if the policies of the other firms are such that their prices and/or capital stocks are changed a countable number of times, then an optimal policy for firm 1 requires changes in price and/or capital stock on only a countable number of times.*

One of the results which does not need to hold in quite the form in which it appears is lemma 6.2. The first-order condition, eq. (6.6), which defines the optimal price of the firm given the value for K_{1t}^*, must be modified to take account of $\bar{\mathbf{p}}_{1t}$. In so doing, it is possible that the modified version of eq. (6.6),

$$\pi_{1t}^{1} = F_1(p_{1t}, \bar{p}_{1t}) + (p_{1t} - C_1^1(F_1(p_{1t}, \bar{p}_{1t}), K_{1t}^*))F_1^1(p_{1t}, \bar{p}_{1t}) = 0, \tag{6.30}$$

which is the condition for an interior maximum, does not hold for all t. To see this imagine that $\bar{p}_{1t}$ is relatively large and K_{1t}^* quite small. It is possible that K_{1t}^* is so small that for some values of $\bar{p}_1$, there is no price p_1 such that $(p_1, K_{1t}^*) \in \mathcal{D}_1(\bar{p}_1)$. Were that so, if $\bar{p}_{1t}$ where reduced to such a level then the optimal price for the firm would be one at which its sales would be zero. If the lowest such price were assumed to be used in this instance, the optimal price would still be a continuous function of K_{1t}^*. It would, as well, be a continuous function of $\bar{p}_{1t}$; however, derivatives would not exist for those values of K_{1t}^* and $\bar{p}_{1t}$ such that optimal price implies zero sales; but for which any increase in p_{jt} $(j \neq 1)$ or K_{1t}^* would cause the level of optimal sales to be positive.[4]

Lemma 6.10. *Under A3–A5, E2 and E3, the optimal price at time t for firm 1 is given by the solution to eq. (6.30), where a solution exists, and the second-order condition is satisfied ($\pi_{1t}^{11} < 0$). If these conditions are not met, then the optimal price is taken to be the lowest price at which demand is zero. The function giving the optimal price may be written*

$$p_{1t} = \phi_1(\bar{p}_{1t}, K_{1t}^*); \tag{6.31}$$

ϕ_1 is continuous in all arguments. For values of $\bar{p}_{1t}$ and K_{1t}^* such that $q_{1t} > 0$, $\phi_1^n < 0$ and if $q_{jt} > 0$, then $\phi_1^j > 0$. If $\mathbf{p}_t$ is interior to the region where $q_{jt} = 0$, then $\phi_1^j = 0$.[5]

Lemma 6.10 is the counterpart of lemmas 2 and 3. Lemma 1, which states that, under an optimal plan, the firm acquires a capital stock large enough to allow it to choose prices placing $(p_{1t}, K_{1t}) \in \mathcal{D}_1(\bar{p}_{1t})$, need not hold flatly for the same reasons that eq. (6.30) need not always give the optimal price. However, it remains true that the firm's optimal plan, given the price paths to be followed by the other firms, involves acquiring a strictly positive capital stock, and that the capital stock be large enough to insure that the sum of the total amount of time that price cannot be chosen to exceed marginal cost is finite. As this is not central to the results of the present section, it is not formally stated.

[4]That is, ϕ_i would have *corners* at the $(\bar{p}_i, K_i)$ values at which $\phi_i(\bar{p}_i, K_i)$ just becomes zero.
[5]Recall that ϕ_1^n is the partial derivative with respect to the nth argument, K_{1t}^*. ϕ_1^j is the derivative with respect to $p_{j+1,t}$ for $j < n$.

Lemma 6.4 holds unaltered; for it states that the firm never reduces the number of efficiency units of capital it holds. The causes of this result are unaffected by the expansion of the model to many firms. Theorem 6.5 holds with a minor modification. Given a list of times when the firm is allowed to change its capital stock, the capital stock it chooses at one such time depends on the time of that moment, the time at which the next change in capital occurs and the paths of the prices of the other firms over that interval of time. Finally, the arguments supporting lemmas 6.6 and 6.7 and theorem 6.8 all remain valid, thus these results hold with the modification that the function ψ which gives the optimal plant size, depends on the price paths of other firms (as indicated above) as well as on the date of acquisition and duration that the plant is to be used.

Thus, even with the model expanded to n firms, many characteristics of optimal behavior for a given firm are virtually unchanged from the monopoly model. In particular, optimal behavior for a firm involves purchasing new plant a countably infinite number of times. Further, under an optimal plan, the only times when a firm changes its price is when it acquires new plant or when some other firm changes its price. These considerations lead to the following:

Theorem 6.11. *Under A3–A5, E2 and E3, if all other firms change their prices only at times when at least one firm is acquiring new capital, then, under an optimal policy, the firm acquires new capital a countable number of times, and changes its price only at dates when it or some other firm acquires new capital.*

The importance of theorem 6.11 is that the payoff the firm receives can be written as a sum of pieces for a countable set of time intervals. The beginning and end point of each time interval is when at least one firm changes its capital. At such times, any (and probably all) firms may change prices; however, within a time interval, no price change occurs. Given the rules which have been stated that govern the choice of optimal prices and capital stocks, it is possible to think of the firms' decision problem as only involving the choice of dates at which to invest.

6.3. Noncooperative equilibrium for the market

In this section, the model developed earlier is examined to see whether it has an equilibrium of the Cournot type, suitably generalized. The decision made by, say, firm 1, is $(t_1, \boldsymbol{p}_1, \boldsymbol{K}_1)$ where $t_1 = (0, t_{11}, t_{12}, t_{13}, \ldots)$ is a countable set of dates such that $t_{1k} \leq t_{1,k+1}$ at which the firm changes its price and/or its capital. $\boldsymbol{p}_1 = (p_{10}, p_{11}, p_{12}, \ldots)$ and $\boldsymbol{K}_1 = (K_{10}, K_{11}, K_{12}, \ldots)$, where p_{1k} and K_{1k} are the price and capital put into effect at t_{1k}. It is understood that if

$$e^{\lambda t_{1k}} K_{1k} = e^{\lambda t_{1,k+1}} K_{1,k+1}, \tag{6.32}$$

then no investment has taken place at time $t_{1,k+1}$. The individual p_{1k} are confined to the interval $[0, p_1^+]$, where $\boldsymbol{p}^+$ is the maximal price vector in $\mathscr{A}$ (i.e. for $p_{1t} \geq p_1^+$, $q_{1t} = 0$ no matter what prices are chosen by other firms), and the K_{1k} are confined to the interval $[0, p_1^+ F_1(0, \bar{\boldsymbol{p}}_1^+)/\rho]$. The upper bound on capital is so high that investment at this level could not conceivably be profitable. Using the metric defined in §3.3, let

$$\delta_{1k} = \int_{t_{1k}}^{t_{1,k+1}} e^{-\rho t}\, dt. \tag{6.33}$$

Then the distance between two decisions $(\boldsymbol{t}_1^0, \boldsymbol{p}_1^0, \boldsymbol{K}_1^0)$ and $(\boldsymbol{t}_1^1, \boldsymbol{p}_1^1, \boldsymbol{K}_1^1)$ is given by

$$\sum_{k=0}^{\infty} \delta_{1k}\{|t_{1k}^0 - t_{1k}^1| + |p_{1k}^0 - p_{1k}^1| + |K_{1k}^0 - K_{1k}^1|\}. \tag{6.34}$$

Under this metric, the decision set of each firm is compact and convex. It is already established that the discounted payoff of the firm is bounded and continuous in all variables. Unfortunately, the payoff function need not be quasi-concave with respect to the firm's own strategy. That must be added at this point as another assumption.

E4. *The discounted profits of the firm are a quasi-concave function of the firm's own strategy, $(t_i, \boldsymbol{p}_i, \boldsymbol{K}_i)$.*[6]

Under A3–A5 and E2–E4, the market model may be regarded as a *game* and it is within the class covered by theorem 7.6; hence, the following theorem is stated without proof:

Theorem 6.12. *Under A3–A5 and E2–E4, the n-firm market model has a*

[6]Though it may seem surprising, the concavity already assumed, with respect to prices and capital for fixed investment dates, is not sufficient to make the discounted payoff function concave in $(\boldsymbol{t}, \boldsymbol{p}, \boldsymbol{K})$. The reader may verify this by checking a model in which interior equilibria are assumed, as well as differentiability wherever it is desired.

noncooperative equilibrium. That is, each firm has a decision, (t_i^*, p_i^*, K_i^*), $i = 1, \ldots, n$, such that no single firm can increase its discounted payoff by changing its decision, given the decisions of the other firms.

With some additional information on the model, it is possible that a great deal might be said concerning the nature of an equilibrium. It is now known that in equilibrium, each firm changes its capital a countably infinite number of times, that each time a change is made, the number of efficiency units is raised. Several firms may or may not increase capital simultaneously; however, price changes may behave in a monotone fashion. With the prices of other firms fixed, the optimal price of a firm falls as its own capital rises. If, in addition, ϕ_i were a monotone nondecreasing contraction with respect to $\bar{p}_i$, then, whenever one or more firms increased capital, any price changes would be downward. Thus, prices would only fall when they changed just as capital stocks would only rise.

7. *Concluding comments, with some observations on simultaneous versus nonsimultaneous decision*

In the course of this chapter a model is developed in which the firm, as a monopolist, faces investment and pricing decisions in the presence of a sort of "transactions cost" for changing the configuration of capital. Though time is continuous, given stationary demand and underlying cost conditions, it is seen that changes in capital and in price are made only at discrete intervals. Indeed, if an adjustment cost were introduced for price changes too, then stationarity would not be needed to insure changes at only discrete intervals. For my own part, I am enough convinced that the process of decision-making is, in the real world, costly; hence, decision-making at discrete intervals strikes me as being more "realistic." This does not provide a convincing argument for the usual fixed-period discrete time period models used freely throughout this volume, nor does it argue convincingly against the insightfulness of continuous time models. As usual, the use of either type of model may allow the development of interesting results which may prove unobtainable in a model of the sort developed in this chapter.

The models of this chapter do have their time periods determined endogenously in a way which takes account of the cost of making changes. Such a course is, perhaps, interesting to pursue. After considering monopoly, the next step taken develops an oligopoly version for which conditions are found under which a noncooperative equilibrium exists.

The approach of this chapter, with no prior constraints on who makes decisions when, raises another question of some interest which comes up in the work of Stackelberg (see ch. 4 §5) and elsewhere. One interpretation of the Stackelberg leader–follower model is that, within a given time period, the leader announces first his price for the period, then the follower, hearing what the leader plans to do, announces his price. Thus there is no space of time during which the leader is selling at a price to which the follower has not yet had time to adjust. The rules are that, in effect, the leader announces several minutes before the market opens and the follower announces after him, also before the market opens. This has always seemed to me highly artificial and contrived. There is an empirical question of great importance as to whether real markets exhibit this behavior, or exhibit behavior which this description approximates quite well. If some real markets are, in fact, like this, it is of no consequence that someone's intuition finds it implausible.

As an empirical question, so far as I am aware, there is no real evidence one way or another; however, from the theoretical standpoint, the model in §6 strongly suggests that follower models are not fruitful. This is because the model allows the firms to make changes whenever they please; and it turns out that one thing they do not want to do is make decisions at closely timed points in a way which would put one firm approximately into the position of a leader and others, followers. If one firm changes its price at a particular point in time, it is optimal for others to follow suit at the same point in time. These considerations carry weight even before getting into the question of what should be the leadership–followership order and why.

II Game theory

CHAPTER 7

AN INTRODUCTION TO GAMES AND SOME BASIC RESULTS IN n-PERSON NONCOOPERATIVE GAME THEORY

1. *An overview of game theory*

1.1. An informal description of games

Game theory provides a framework for the study of the interactions of decision-makers whose interests are related, though distinctly different, and whose actions jointly determine all outcomes. There are several different ways of characterizing a game, generally not equivalent to one another; however, they all have certain elements in common. The common elements are the set of decision-makers, called *players*, the rules and regulations concerning the possible decisions that each decision-maker can choose, sometimes called the *set of strategies*, and the rules and regulations governing the way that the players' decisions are related to the rewards or payoffs they receive. With respect to strategies, in some models, they may not provide a full description of the actions players may take, and in other models they may not make an explicit appearance; however, context will make the situation clear.

To make clearer what game theory intends to capture, it is helpful to pursue examples. Imagine a game of checkers played by two people who may be called "red" and "black." To make the rewards explicit assume that when the game is over, the loser will pay the winner $1.00 for each piece the winner has remaining on the board. In this game there is a natural distinction between *move* and *strategy*. A move is the taking of one piece from the square it is on and the placing of it on another square. The game consists of a sequence of moves, first by red, then by black, then by red, etc. Imagine a time in the midst of a game when it is red's move. What he chooses to do depends on what he anticipates black will follow with. Indeed, red is likely to have a plan which looks ahead several moves. The first step of the plan is fixed, and is the immediate move red has decided upon. Next, black moves; however, red cannot be certain which of black's possible moves he will choose. Red's plan takes several, perhaps all, of black's possible moves into account in the sense that for certain of them, his plan includes a decision on how he proposes to respond to various of black's moves. Red may look ahead to black's second move hence and his own response to each of the possible configurations which the board may have at that time, and he may know what he would do in many or all cases.

It is usual for players of checkers and various other parlor games to have plans such as that outlined above. These plans tend to be incomplete in the sense that they may not anticipate all possible combinations of moves. They may exclude those the planner regards as unlikely, and they generally do not extend beyond, say, five or ten moves. They fail to go to the end of the game and to account for every possibility along the way because of the limitations of the human mind. Were red and black able, they would no doubt want to formulate plants which were complete. Plans such that for any possible configuration of the board that, say red, may face, the plan would indicate his move. This plan would start with the first move red would make in the game, take account of every possible move that black might make after it (not only the "likely" or "reasonable" moves) and continue in this full and complete fashion until the game ends. Such a full and complete plan is a *strategy.*

Assume that red and black were each capable of visualizing all possible strategies for both players, and that each would, before the first move of a game, decide which of his strategies to follow. The two strategies would determine the outcome of the game. Indeed, it is possible, at least in principle, to make a chart to show the outcome of the game for every combination of strategies red and black could choose. For red, the strategies could be arbitrarily numbered R1, R2, R3, . . . , and similarly for black. This is illustrated in fig. 7.1 where the rows of the table are labelled with the strategies of red, the columns are labelled with the strategies of black and the entries in the table give the payoff to red for the corresponding strategies of the two players. This game is *zero sum*, meaning that the gain of one player is the loss of the other; hence, the table can also be regarded as a table of losses for black.

It should be clear that incorporated into the choice of strategy are all the considerations the player has taken into account about how he expects the other player to behave at any given point of the game. For the purposes of game theory (and maintaining the assumption that players

PAYOFFS TO RED

	B1	B2	B3	B4	⋯	Bb
R1	6	-3	-2	3	⋯	1
R2	-3	-1	-4	-4	⋯	1
R3	3	3	-2	-5	⋯	-3
R4	2	3	1	7	⋯	-4
	⋮	⋮	⋮	⋮		⋮
Rr	-3	-5	2	2	⋯	-1

Fig. 7.1

choose strategies at the beginning of the game) the game of checkers is completely described by the set of players, red and black, their respective strategy sets ($R1$, $R2$,... and $B1$, $B2$,...) and the payoffs associated with each strategy pair. In checkers the interests of the two players are perfectly opposed. Anything which increases the payoff to red necessarily decreases the payoff to black. This does not depend on the assumption that losses to red, measured by his utility function, equal gains to black, measured by his. It is sufficient that for each player, utility increases with income.[1] The zero sum feature does require that utility be linear in income. A fundamentally interesting aspect of checkers as a game in the von Neumann–Morgenstern sense (i.e. as a game of strategy) is that the strategy which is best for red to play depends on which strategy he thinks black will play, and that the interests of the two players are at least partially opposed.

As a second example, consider a single-period market of the sort studied in ch. 2, with n firms. In such a model there is only one kind of action which players may take – they choose output levels. As the market is single period, a strategy consists of an output choice, making move and strategy the same. n, the number of firms, can be anything. Consider three possibilities, $n = 1$, $2 \leq n < \infty$ and the limiting case, $n \to \infty$. The first is simple monopoly and falls outside the scope of most of game theory because strategic interactions between various rational decision-makers are absent. In the limiting case (where $n \to \infty$) of pure competition, strategic considerations are again absent, making the game, in a sense, degenerate.[2] It is conceivable, and has in fact been the case, that the

[1]A two-person game in which all conceivable outcomes are Pareto optimal is called *strictly competitive*.

[2]The correctness of this statement rests, in part, on the implicit assumption that the players cannot combine into cooperating groups (*coalitions*).

techniques and results of game theory may be useful even in models of competition.

The middle situation, where n is finite and greater than 1, is oligopoly. A new possibility emerges here which is not present in the game model of checkers. Perhaps some of the players may wish to form a *coalition*, which is a group (subset of players) which has an agreement among itself concerning the choices to be made by the members of the group. The rules of the game may either permit or disallow coalitions. In the game of checkers, coalitions would never form even if they were allowed because the two players cannot jointly gain through cooperation. In a general n-person (including 2 person) game it need not be that if one outcome gives more to one player than another outcome, it must give less to some player. Conditions are given in theorem 3.2 under which the Cournot equilibrium (for a single-period price model) is not Pareto optimal, thus there are clearly outcomes in such models which give more to all players than they get at the Cournot equilibrium. Coalition formation, if allowed, might well take place. In ch. 2 §4 a distinction is made between *cooperative* and *noncooperative structure* on the basis of whether the firms are allowed by the rules to make binding agreements. If they are, the structure is called *cooperative*; if they are not, it is called *noncooperative*. This distinction is carried over to the study of games.

Throughout this book, *binding agreements* are taken to mean both (a) binding agreements between two or more players and (b) commitments made unilaterally by one player. Conceptually, these may be separated; and, there may be applications in which it is appropriate to have one present and the other absent. However, there appears to be no compelling reason to give them separate treatment here. Some discussion of commitments and their relationship to binding agreements in general may be found in Aumann (1974, §9). Unless explicitly stated otherwise, the ability to form coalitions is taken as equivalent to the ability to make binding agreements.

Imagine an oligopoly in which coalition formation is allowed. Then, as an alternative to characterizing the game by all the possible strategies and associated payoffs, the game could be described by the various payoff levels which the several coalitions could obtain. For example, if $n = 3$, there are seven possible coalitions: (1), (2), (3), (1, 2), (1, 3), (2, 3), (1, 2, 3). A *coalition structure* is a particular partition of the players into coalitions. For $n = 3$, there are five coalition structures : {(1), (2), (3)}, {(1), (2, 3)}, {(1, 3), (2)}, {(1, 2), (3)} and {(1, 2, 3)}. For each of the seven coalitions, there is a payoff level, or set of them, below which the members of the coalition cannot be forced. For example, it may be that no matter how 2 and 3 behave, there is no way they can prevent 1 from getting at least 3.6. Similarly, say 2 and 3, each alone, can guarantee themselves 5.8 and 2.1. Each of the two-player coalitions also have payoffs they can guarantee themselves. Perhaps (1, 2) can assure itself at least $P_1+3P_2=50$ where P_I is the payoff of the ith player, etc. And the three-player coalition also has a set of payoff vectors which it can guarantee.

In a cooperative game, it is really these guaranteed payoffs which are of interest. It is reasonable that in a cooperative game, an equilibrium should give each player as much as he could guarantee himself and each coalition at least as much as it could guarantee itself. So, for the preceding example, $P_1\geq 3.6$, $P_2\geq 5.8$ and $P_1+3P_2\geq 50$ are all conditions which should be met. These correspond to the shaded area in fig. 7.2. Certain of the conditions noted above are called rationality conditions. In particular, the requirement that each player get at least as large a payoff as he can guarantee himself is called *individual rationality*. The requirement that the coalition consisting of all players in the game receive a payoff vector which is Pareto optimal is called *group rationality*.

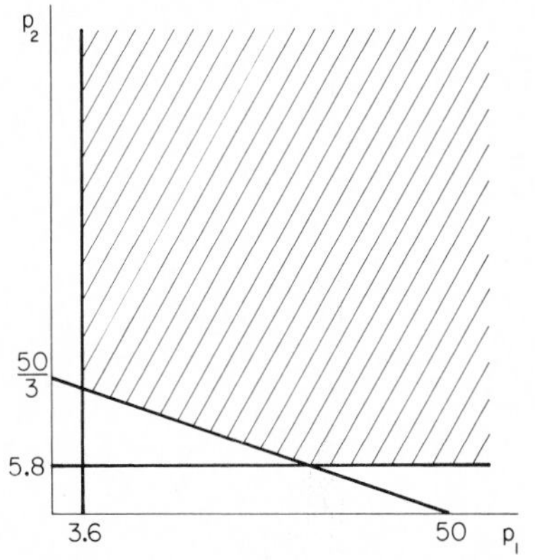

Fig. 7.2

If the equilibrium is to be agreed upon by all the players and is to obey conditions of the sort outlined above, then strategies are of no direct interest. That is, if two strategy vectors lead to the equilibrium payoff vector, it is immaterial to the players which strategy vector is used. Similarly, it is not of interest which strategy of 3 guarantees him 2.1. It only matters that he can guarantee himself that much; because the amount becomes a floor on what he may receive at an equilibrium. As an alternative to deriving these guaranteeable payoffs from a knowledge of payoff functions and possible strategies, a game can be defined from the start without reference to strategies, by specifying the allowable coalitions and what they can guarantee to themselves.

1.2. Formal representations of games

There are three ways to characterize a game: the normal form, the characteristic function form and the extensive form. Roughly speaking, these correspond respectively to describing the game through strategies, through what players and coalitions can guarantee to themselves and through a move by move description. These three approaches are not, in general, equivalent, though each has its uses.

1.2.1. The normal form

There are four parts to the definition of games in normal form: players, strategies, payoff functions and additional rules, if any. The first three may be symbolized by $(\mathcal{N}, \mathcal{S}, \boldsymbol{P})$, where $\mathcal{N} = \{1, 2, \ldots, n\}$ is the set of players, $\mathcal{S}$ is the strategy space of the game and $\boldsymbol{P}$ is the vector payoff functions. The strategy set of the ith player is denoted $\mathcal{S}_i$ and its elements are s_i. The strategy space of the game is the Cartesian product of the individual strategy sets, $\mathcal{S} = \mathcal{S}_1 \times \mathcal{S}_2 \times \cdots \times \mathcal{S}_n$; and a strategy vector, $s \in \mathcal{S}$, is of the form $s = (s_1, \ldots, s_n)$ where $s_i \in \mathcal{S}_i$, $i = 1, \ldots, n$. s is also sometimes written $s = (s_i, \bar{s}_i)$ where $\bar{s}_i$ is the vector of strategies of the $n - 1$ players other than the ith. The payoff function of the ith player is $P_I(s)$ and is scalar valued. A payoff vector is written $\boldsymbol{P}(s) = (P_i(s), \ldots, P_n(s)) = (P_i(s), \bar{\boldsymbol{P}}_i(s))$. Assumptions must be given which fully define both the strategy sets and the payoff functions.

Additional *rules* could take the form of specifying what are the allowable coalitions. To restrict the game to being noncooperative is to say that no coalitions of two or more may form. The normal form is typically associated with noncooperative games and will be used almost exclusively in the present and next three chapters.

1.2.2. The characteristic function form

The characteristic function form is that in which there is a function associated with the coalition structure which gives, for each coalition, a set of attainable payoffs. In the earlier literature, the characteristic function assigns a scalar number to each coalition which represents the payoff which the coalition can guarantee itself.[3] It is clear that this is meaningful where there are transferable utility and side payments. That is, essentially, where utility is measured by money for all players, and the members of a coalition are completely free to divide their joint payoff in any way they wish. In the more general case of nontransferable utility and no side payments, a coalition could usually guarantee to itself any one of a number of payoff vectors, as in the example in §1.1. A game in characteristic function form is completely described by the players, the coalition structure and the specification of what each allowable coalition can guarantee to itself.

There is an important question which, until this point, has been glossed over concerning the meaning of what a coalition can *guarantee itself.* Consider first for a game in normal form what an individual player might guarantee himself. For any s_i that he might choose, the least favorable $\bar{s}_i$ may be found. That is, $\min_{\bar{s}_i} P_i(s_i, \bar{s}_i)$. If the ith player is so pessimistic as to think that whatever strategy he chooses, it will turn out that the others will have chosen so as to minimize his payoff, then he can choose s_i in a way which maximizes this worst outcome. Thus, $\max_{s_i} \min_{\bar{s}_i} P_i(s_i, \bar{s}_i) = P_i(s'_i, \bar{s}'_i) = P_i(s')$. While $P_i(s')$ is literally the largest payoff which the ith player can guarantee himself, it may not be what one really wants to mean or use. In a literal sense, this is the most a player can guarantee himself; because it is possible that the others behave in such a fashion that he can get no more. At the same time, this amount may be unreasonably pessimistic. It is possible that for any choices the others might be expected to make, the minimum payoff to the ith player is much higher.

As an example, consider the game in normal form shown in fig. 7.3. Each player has three strategies, with those of the first player labelling the rows ($S1$, $S2$ and $S3$) and those of the second, the columns ($T1$, $T2$ and $T3$). In the body of the table in fig. 7.3, the first number is the payoff of player 1. So if player 1 chooses $S1$ and player 2 chooses $T3$, the payoffs are $P_1 = -5$, $P_2 = 0$. On the definition in the preceding paragraph of what

[3]See Luce and Raiffa (1957).

		#2		
		T1	T2	T3
#1	S1	5, 2	0, 2	-5, 0
	S2	4, 5	9,-4	-3, 3
	S3	3, 1	8, 0	-2, 0

Fig. 7.3

player 1 can guarantee himself, he can get at least -2; however, to get this minimum he plays $S3$ and his rival plays $T3$.[4] In terms of the notation of the preceding paragraph, $s' = (S3, T3)$. Player 2, however, has no incentive to choose $T3$. For any choice of player 1, P_2 is higher under $T1$ than under $T3$. Taking this into account, player 1 can, in a sense, guarantee himself 4 by choosing $S2$.

The case may be put less starkly. After finding the s' of the preceding paragraph, the next question for player i to consider is whether he has reason to believe the other players, even if they form a coalition, would have any cause to choose $\bar{s}'_i$. In general, there is good reason to think $\bar{s}'_i$ would not be chosen if there is any player j, or subset of players, $j_1, \ldots, j_k$ (not including i) who could choose something else and all obtain higher payoffs than they get using the s'_{j_l}. The point here is to raise, but not solve, a difficulty and to indicate that care must be exercised in using the characteristic function formulation.

1.2.3. *The extensive form*

The extensive form of a game shows the game move by move. Its function is more pedagogical than anything; for it allows one technique for abstractly representing any game which would be easily understood. From the extensive form, the normal form, in particular, is easy to explain and justify. The first three moves of the game tic-tac-toe are represented in fig. 7.4. Figure 7.5 depicts a particular realization or play of a game. The game is played on a 3×3 grid as shown in fig. 7.5. Player 1 marks "x" in any one of the nine boxes of the grid. In fig. 7.5, he chooses a corner, and at this point, all corners are equivalent. Player 2 then places an "o" in one of the empty boxes. In fig. 7.5b, he chooses a corner near to the x. Then player 1 chooses again, and they alternate until one of them has three of his marks lined up in a row (as shown in fig. 7.5g), or until all boxes are

[4]For the sake of simplifying this example, mixed strategies are left out of account.

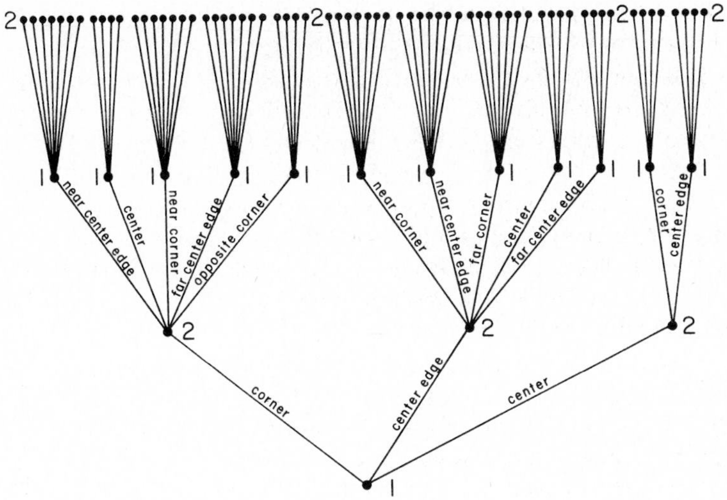
2
2
2
2
1
near center edge
center
near corner
far center edge
opposite corner
near corner
near center edge
far corner
center
far center edge
corner
center edge
2
2
2
corner
center edge
center
1

Fig. 7.4

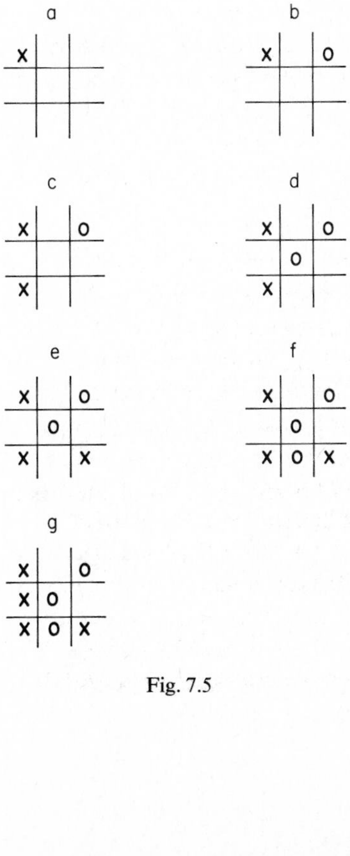
a
b
c
d
e
f
g

Fig. 7.5

filled, but no one has three marks lined up. In the latter case, the game is a draw, and in the former, the player with three lined up wins.

With reference to the *game tree* in fig. 7.4 which is the extensive form, it is possible to describe what is meant by strategy. For example, for player 1, a strategy consists of a specific move for the first move (i.e. corner, center edge or center). Say corner is chosen. Then player 2 has five possible moves; hence the strategy of player 1 must specify what he does for his second move, given each of the five possible configurations which the grid might have, etc. It is clear that the extensive form is entirely impractical except as an illustration; for tic-tac-toe is an extraordinarily simple game with few moves, all of which are easily understood, yet it would be unwieldy if fully represented in extensive form.

1.2.4. Supergames

Supergames may be likened to games in extensive form. The structure is designed so that something analogous to moves is kept explicit, and game strategy remains also. To illustrate, imagine a specific noncooperative game in normal form. Now assume that this game is to be played over and over again, once per time period, by the same set of players and that they understand this. This infinitely repeated play of a game is a supergame. A player's payoff in the supergame consists of the individual payoffs he receives for each of the individual plays of the game. It is necessary to formulate a supergame payoff, which is usually done by summing (discounted) payoffs for the plays. It is clearly possible to investigate supergames in ways which mask the constituent games; however, a point of the supergame formulation is that it is not necessary to do so, and, in certain instances, it is very desirable to keep the individual games in view.

An obvious application of supergames is to oligopoly with the supergame being made from the continual repetition of a single-period model. In such a setting, it is of interest to keep track of the individual single-period choices. Where there is a chance of confusion between supergames and games which are not supergames, the nonsupergames are called *constituent games.*

1.3. Noncooperative games

Noncooperative games and supergames are studied in normal form in chs. 7–10. Although various sets of assumptions are used, as a rule, they have similarities. The central equilibrium concept is the Nash noncooperative equilibrium, which may be defined as follows: s^* is a strategy vector associated with a noncooperative equilibrium if $s^* \in \mathcal{S}$ and $P_i(s^*) = \max_{s_i \in \mathcal{S}_i} P_i(s_i, \bar{s}_i^*)$, $i = 1, \ldots, n$. The equilibrium is characterized by two conditions. The first is that the equilibrium strategies, $s^* = (s_1^*, \ldots, s_n^*)$, are actually in the strategy sets of the players; and the second is that no player could increase his own payoff by deviating from his equilibrium strategy, given that the other players use their equilibrium strategies. What makes the game noncooperative is the same as with the oligopoly models; namely that the players cannot make binding agreements (i.e. cannot form coalitions).

For many models, the noncooperative equilibrium is the only equilibrium which is examined. In some models, notably those of supergames, equilibria are developed which use the noncooperative equilibrium as a basis. These are noncooperative equilibria which satisfy additional conditions as well.

Most proofs of existence of equilibrium follow a similar pattern which it is useful to note. As in chs. 2 and 3, it is possible to define a best reply function. In the present circumstances, the best reply for a player i can be defined for any $\bar{s}_i \in \bar{\mathcal{S}}_i$ which the rivals players may choose. This may be denoted $r_i(\bar{s}_i)$; hence, for any $s \in \mathcal{S}$, there is a best reply vector valued function, $r(s) = (r_1(\bar{s}_1), \ldots, r_n(\bar{s}_n))$. A noncooperative equilibrium is a fixed point of the function r; so, if s^* is a noncooperative equilibrium strategy vector, it satisfies $s^* = r(s^*)$. Existence proofs often follow the pattern of defining the best reply mapping and showing that it must have a fixed point. This, in turn, requires proving that the best reply mapping satisfies the conditions of a fixed point theorem. As there are several somewhat different fixed point theorems, each requiring different assumptions, there are a corresponding variety of models for which noncooperative equilibrium may be proved to exist.

1.4. *Cooperative games*

There is nothing in cooperative game theory which plays a central role comparable to the role of the noncooperative equilibrium in noncooperative games. Two types of equilibrium, studied in chs. 11 and 12, respectively, are the *value* and the *core*. The value approach is directed at finding a vector of payoffs which give a payoff to each player which takes proper account of the threat capabilities of all the players. Nash (1950) and (1953) has a value for two person games without side payments or transferable utility. Shapley (1953b) has a value which applies to n-person games in which payments between the players are possible and in which each player has utility units which are like money. They are the same for everyone and freely transferable. Harsanyi (1963) generalizes Nash and Shapley, dealing with n person games in which utility is not comparable or transferable among players. The value approach is covered in ch. 11.

The core is analogous to the economists' notion of the contract curve. The core consists of those payoff vectors which are, in a specific sense, acceptable to all players. Prior to defining the core more precisely, we should take a closer look at the way cooperative games are often described. Because the players are assumed, in the end, to act in concert, attention is focused on what they can get as final payoffs. Strategies are often ignored, with the game described in characteristic function form.

The characteristic function form of a game is defined by the set of players, the coalition structure and the characteristic function. The coalition structure is the set of coalitions which, under the rules of the game, are allowed to form. At a given moment, one player can be in exactly one coalition. Thus the players may be thought of as being partitioned into coalitions with each player knowing which he is in. In §1.1 it is pointed out that there are seven conceivable coalitions which may be associated with a three-person game. These include the three coalitions consisting of one player. The rules of the game may specify that certain coalitions cannot form. Usually, it is assumed that all conceivable coalitions are able to form, mainly because there is no reason to assume otherwise; however, in particular applications it is possible that certain coalitions may be unable to form. Where that is so, the coalition structure of the game should reflect this. In all subsequent discussion, unless it is explicitly assumed otherwise, it is supposed that all conceivable coalitions may form in cooperative games.

Associated with the coalition $\mathcal{K}$ is a set of payoffs $\mathcal{V}_{\mathcal{K}}$. Take first the *coalition of the whole*, consisting of all players $\mathcal{N}$. $\mathcal{V}_{\mathcal{N}}$ is the set of all payoff vectors $\boldsymbol{P} \in \boldsymbol{R}^n$ which are attainable. To say they are attainable implies that there is, for each $\boldsymbol{P}$, some strategy, s, available to the players which will realize $\boldsymbol{P}$; that is, for which $\boldsymbol{P} = \boldsymbol{P}(s)$.[5] The set $\mathcal{V}_{\mathcal{K}}$, for the

[5]Whenever it is convenient, the notation $s = (s_{\mathcal{K}}, \bar{s}_{\mathcal{K}})$ and $\boldsymbol{P} = (\boldsymbol{P}_{\mathcal{K}}, \bar{\boldsymbol{P}}_{\mathcal{K}})$ is used in a way analogous to $(s_i, \bar{s}_i)$, etc. $s_{\mathcal{K}}$ consists of a vector of strategies $(s_{i_1}, \ldots, s_{i_l})$ the components of which are the individual strategies of the l members of the coalition $\mathcal{K}$. Similarly, $\boldsymbol{P}_{\mathcal{K}}$ is a payoff vector of dimension l which consists of the payoffs to the members of $\mathcal{K}$. $\bar{s}_{\mathcal{K}}$ and $\bar{\boldsymbol{P}}_{\mathcal{K}}$ are the strategies and payoffs of those not in $\mathcal{K}$. It is not assumed that they necessarily form a coalition. Such a coalition would be symbolized by $\mathcal{N}-\mathcal{K}$ or $\bar{\mathcal{K}}$.

coalition $\mathcal{K}$, consists of payoff vectors of dimension equal to the number of members of $\mathcal{K}$. To say that $\boldsymbol{P}_{\mathcal{K}} \in \mathcal{V}_{\mathcal{K}}$ implies that $\boldsymbol{P}_{\mathcal{K}}$ is a vector of payoffs which the members of $\mathcal{K}$ can assure themselves. A simple example is given by a pure trade model of general equilibrium in which each player (economic agent) has an endowment of goods, and players may trade with one another. With externalities ruled out, $\mathcal{V}_{\mathcal{K}}$ consists of all the utility levels which the members of $\mathcal{K}$ can attain by trading only among themselves.

To summarize, the characteristic function, $\mathcal{V}_{\mathcal{K}}$, associates with each coalition, $\mathcal{K}$, the set of payoff vectors which that coalition can get for itself. The core $\mathcal{V}$ consists of payoff vectors in $\mathcal{V}_{\mathcal{N}}$ which cannot be *improved upon* by any coalition. That is, a payoff vector, $\boldsymbol{P} = (\boldsymbol{P}_{\mathcal{K}}, \bar{\boldsymbol{P}}_{\mathcal{K}}) \in \mathcal{V}_{\mathcal{N}}$ is in the core if there is no coalition $\mathcal{K}$ and $\boldsymbol{P}'_{\mathcal{K}} \in \mathcal{V}_{\mathcal{K}}$ such that $\boldsymbol{P}'_{\mathcal{K}} \gg \boldsymbol{P}_{\mathcal{K}}$. To put it in the reverse manner, a payoff vector $\boldsymbol{P}$ can be *improved upon* by a coalition $\mathcal{K}$ if $\mathcal{V}_{\mathcal{K}}$ contains a payoff vector which gives strictly more to each member of $\mathcal{K}$ than he gets under the vector $\boldsymbol{P}$.

Notice that in the very definition of the game in characteristic function form, the question of what is meant by *the coalition $\mathcal{K}$ can get for itself* is sidestepped. If certain payoffs are in $\mathcal{V}_{\mathcal{K}}$, then $\mathcal{K}$ can get them. It is natural to think of the characteristic function as being derived from the information contained in the normal form of the game, in which case, it might not be obvious what should be meant by *the coalition can get for itself.* Intuitively, it is plausible that the best which a coalition can absolutely guarantee itself is often less than the least to which the remaining players can be certain to be able to hold them. This issue will be addressed in chs. 11 and 12.

Chapter 12 is devoted to core theory and follows Scarf (1967, 1971) in proving that the core is not empty for a rather large class of games.

The remainder of this chapter is divided into six sections. In §2 existence of the noncooperative equilibrium is proved for games satisfying one set of axioms. These axioms allow use of the Brouwer fixed point theorem. In §3 other fixed point theorems are presented and the axioms are amended to make use of them, thereby widening the class of models for which equilibrium can be shown to exist. The existence theorems prove just that, and uniqueness is not touched by them. In §4 uniqueness is examined, and it is found that, in a precise and interesting sense, most games have a unique equilibrium. §5 investigates an intriguing notion of "strong" equilibria, while §6 considers the application of the results of the chapter to the single period oligopoly models of chs. 2 and 3. §7 contains concluding comments.

2. Existence of noncooperative equilibrium

2.1. The model

The assumptions to be used in the present section are:

G1. *n, the number of players, is finite.*

G2. *$\mathcal{S}_i$, the strategy set of the ith player, is a compact convex subset of $\mathcal{R}^m$, $i = 1, \ldots, n$.*

G3. *$P_i(s)$, the payoff function of the ith player, is a scalar valued function defined for all $s \in \mathcal{S}$ which is continuous and bounded everywhere, $i = 1, \ldots, n$.*

G4. *$P_i(s)$ is strictly quasi-concave with respect to s_i, $i = 1, \ldots, n$.*

For oligopoly it is no real restriction to be confined to a finite number of players, excepting if entry is to be considered in the model. Even so, it depends on how entry is to be included. Convexity of the strategy set should pose no problems, but compactness, particularly boundedness, might. That the payoff functions are both bounded and continuous are usual and acceptable assumptions. The final assumption, that the payoff of a player is a quasi-concave function of his own strategy, is not something which obviously ought to be true. It is seen in §3 and §6 that it can be weakened somewhat; but it cannot be entirely foregone.

Formal results in game theory began with two-person zero sum games. While such games have not proved useful in economic applications, they are a subset of *n*-person noncooperative games and the noncooperative equilibrium for *n* person noncooperative games is the saddle point equilibrium of von Neumann when the game is restricted to be two-person and zero sum. This is seen in §2.3.

2.2. An existence theorem

The original theorem on existence of noncooperative equilibrium was proved by Nash (1951). The theorem to be proved below is a variant, though not a generalization. His theorem is presented in §3.1 as a special case of another existence theorem. After the theorem, an interesting corollary, due to Debreu (1952), is presented and proved which allows the strategy sets of the players to be somewhat dependent on one another. Note that in the formulation of the present model, the strategy set of the ith player, $\mathcal{S}_i$, is a set of strategies from which he may freely choose, no matter what others decide. It is possible that some actions may be foreclosed to a player, depending on what others decide to do. If, for example, each unit of output of a firm requires one unit of a specific input and the total amount available is no more than Q^* per period, then total output for all firms combined cannot exceed Q^*. One firm could produce Q^* if all others produced nothing; however, any set of decisions resulting in the intent to produce more than Q^* in the aggregate is not meaningful within the model.

Theorem 7.1. *Any game satisfying G1–G4 has a noncooperative equilibrium.*

□ The proof proceeds by first defining the best reply mapping. Then it is seen that the fixed points of the best reply mapping are the same as the noncooperative equilibria of the game. In the remainder of the proof, it is shown that the best reply mapping has a fixed point.

Define $\boldsymbol{r}(s) = (r_1(\bar{s}_1), \ldots, r_n(\bar{s}_n))$ as follows:

$$r_i(\bar{s}'_i) = \{t_i \mid t_i \in \boldsymbol{S}_i,\ P_i(t_i, \bar{s}'_i) = \max_{s_i \in \mathcal{S}_i} P_i(s_i, \bar{s}'_i)\},$$
$$i = 1, \ldots, n,\ \bar{s}'_i \in \bar{\mathcal{S}}_i. \quad (7.1)$$

For any $\bar{s}_i$, $r_i(\bar{s}_i)$ exists because it is defined by the maximization of a continuous function over a compact set. By strict quasi-concavity, $r_i(\bar{s}_i)$ is a function (i.e. takes on only one value for a given $\bar{s}_i$). From the definition of $\boldsymbol{r}(s)$, it is clear that any fixed point of $\boldsymbol{r}$ is a noncooperative equilibrium and vice versa.

Thus if $\boldsymbol{r}(s)$ has a fixed point, the game has a noncooperative equilibrium. $\boldsymbol{r}(s)$ has a fixed point if it satisfies the conditions of the Brouwer fixed point theorem (see Dunford and Schwartz (1957) or Graves (1946)), which are that $\boldsymbol{r}$ be a continuous function which maps a compact, convex set into itself. $\mathcal{S}$ is compact, convex and contains $\boldsymbol{r}(s)$ for all $s \in \mathcal{S}$; therefore it remains to show that $\boldsymbol{r}$ is continuous. This is done by showing that r_i is continuous for any i.

Let $\bar{s}^l_i$ $(l = 1, 2, \ldots)$ be a sequence of strategy vectors in $\bar{\mathcal{S}}_i$ which converge to $\bar{s}^0_i$. Let $s^l_i = r_i(\bar{s}^l_i)$, $l = 1, 2, \ldots$, and $s^0_i = r_i(\bar{s}^0_i)$. r_i is continuous if $\lim_{l\to\infty} s^l_i$ exists and equals s^0_i. Now define a distance between two strategy vectors, s' and s'' in this way. $\|s'_i - s''_i\| = \max_j |s'_{ij} - s''_{ij}|$ and $\|s' - s''\| = \max_i \|s'_i - s''_i\|$. Because all the s^l_i are contained in a compact set, the sequence has cluster points. If r_i is not continuous, then the sequence can have a cluster point other than s^0_i. Assume such a cluster point, s'_i, exists and is the limit of the subsequence $\{s^{l_k}_i\}$. It must be that $P_i(s'_i, \bar{s}^0_i) < P_i(s^0_i, \bar{s}^0_i)$, due to the optimality of s^0_i. Thus $P_i(s^0) - P_i(s'_i, \bar{s}^0_i) = 3\epsilon > 0$. Using continuity of P_i, there is $\delta' > 0$ such that when $\|s - (s'_i, \bar{s}^0_i)\| < \delta'$, then $|P_i(s) - P_i(s'_i, \bar{s}^0_i)| < \epsilon$; and there is $\delta'' > 0$ such that when $\|s - s^0\| < \delta''$, then $|P_i(s) - P_i(s^0)| < \epsilon$. Let δ equal the minimum of δ' and δ''.

There is l^* such that for $l_k > l^*$, $\|s^{l_k}_i - (s'_i, \bar{s}^0_i)\| < \delta$ and $\|(s^0_i, \bar{s}^{l_k}_i) - s^0\| < \delta$. From the former, $P_i(s^{l_k}) < P_i(s'_i, \bar{s}^0_i) + \epsilon$; and from the latter, $P_i(s^0_i, \bar{s}^{l_k}_i) > P_i(s^0) - \epsilon$. Recalling that $P_i(s^0) - P_i(s'_i, \bar{s}^0_i) = 3\epsilon$, it is seen that for $l_k > l^*$, $P_i(s^0_i, \bar{s}^{l_k}_i) - P_i(s^{l_k}) > \epsilon$, which contradicts the optimality of $s^{l_k}_i$. Thus s^l_i cannot have a cluster point other than s^0_i; hence, r_i is continuous, proving that r has a fixed point. Thus a game satisfying G1–G4 has a noncooperative equilibrium. □

Consider now the possibility that the ith player's payoff function is not defined for all $s \in \mathcal{S}_1 \times \cdots \times \mathcal{S}_n$; but that for any $\bar{s} \in \bar{\mathcal{S}}_i$ there is at least one s_i for which the payoff function is defined. As an example of how this could be relevant to an oligopoly market, consider special types of advertising space such as the second page or last page of a newspaper or similar prominent positions in magazines. The various components of the player's strategy may be identified with various decision variables such as price, output level, particular forms of advertising, etc. If one of the decision variables were column inches of advertising space on the last page of the *New York Times*, it would be impossible for the ith firm to use all the space unless the other firms used none.

It is assumed that the payoff function of the ith player is defined on a subset $\mathcal{T}_i$ of $\mathcal{S}$, in such a way that no matter what the others do, there are some strategies for which his payoff is defined.

G5. *$P_i(s)$ is continuous and bounded on $\mathcal{T}_i \subset \mathcal{S}$. For any $\bar{s} \in \bar{\mathcal{S}}_i$ there is at least one $s_i \in \mathcal{S}_i$ such that $(s_i, \bar{s}_i) \in \mathcal{T}_i$. $\mathcal{T}_i$ is compact and convex $(i = 1, \ldots, n)$.*

Note that G5 does not explicitly provide that there are any s for which all the payoff functions are simultaneously defined; however, it is seen below that $\cap_{i=1}^n \mathcal{T}_i$ is not empty. It can happen that if an arbitrary $\bar{s}_i$ is chosen and an s_i is found such that $P_i(s_i, \bar{s}_i)$ is defined, that there are some players j for which $P_j(s_i, \bar{s}_i)$ is not defined.

It is helpful to define a correspondence which provides an alternate way of characterizing $\mathcal{T}_i$,[6]

$$\mathcal{T}_i^*(\bar{s}_i) = \{s_i \mid \bar{s}_i \in \bar{\mathcal{S}}_i, (s_i, \bar{s}_i) \in \mathcal{T}_i\}. \tag{7.2}$$

The object now is to replace G3 with G5 and prove that a game satisfying G1, G2, G4 and G5 has a noncooperative equilibrium. An intermediate result which is needed is:

Lemma 7.2. *If $\mathcal{T}_i$ is compact and convex then $\mathcal{T}_i^*$ is an upper semicontinuous correspondence with compact and convex image sets.*

□ Compactness and convexity of the image sets of $\mathcal{T}_i^*$ follow directly from compactness and convexity of $\mathcal{T}_i$. To see that $\mathcal{T}_i^*$ is upper semicontinuous, let $\bar{s}_i^l$, $l = 1, 2, \ldots$, be a sequence in $\bar{\mathcal{S}}_i$ which converges to $\bar{s}_i^0$ and let $s_i^l \in \mathcal{T}_i^*(\bar{s}_i^l)$ with $\lim_{l\to\infty} s_i^l = s_i^0$. If $s_i^0 \in \mathcal{T}_i^*(\bar{s}_i^0)$ then $\mathcal{T}_i^*$ is upper semicontinuous. Because $\mathcal{T}_i$ is closed, $\lim_{l\to\infty} s^l = s^0 \in \mathcal{T}_i$, which, in turn, means $s_i^0 \in \mathcal{T}_i^*(\bar{s}_i^0)$. Thus $\mathcal{T}_i^*$ is upper semicontinuous. □

Debreu's result may now be proved.[7]

Theorem 7.3. *A game satisfying G1, G2, G4 and G5 has a noncooperative equilibrium. If s^* is a strategy vector corresponding to such an equilibrium, then $s^* \in \cap_{i=1}^n \mathcal{T}_i$.*

□ Best reply functions may be used again, though they cannot be defined by eq. (7.1) because maximization must be restricted to the sets $\mathcal{T}_i$. The best reply functions are

$$r_i(\bar{s}'_i) = \{t_i \mid t_i \in \mathcal{S}_i, P_i(t_i, \bar{s}'_i) = \max_{s_i \in \mathcal{T}_i^*(\bar{s}'_i)} P_i(s_i, \bar{s}'_i)\},$$
$$\bar{s}'_i \in \bar{\mathcal{S}}_i, \quad i = 1, \ldots, n. \tag{7.3}$$

The maximization which takes place in eq. (7.3) maps points of $\mathcal{S}$ into points of $\mathcal{S}$; therefore the argument used in the proof of theorem 7.1 may be used again. The best reply function has a fixed point, such a fixed point

[6] *Correspondence* is defined in §3.1.

[7] In Debreu (1952) a model of greater generality than the present model is presented. Rather than the usual convexity assumption, Debreu proves an existence theorem based upon a fixed point theorem requiring contractibility.

is a noncooperative equilibrium strategy vector for the game, and it is necessarily in $\mathcal{T}_i$, for all i, because $s_i^* = r_i(\bar{s}_i^*) \in \mathcal{T}_i^*(\bar{s}_i^*)$. □

In the proof of theorem 7.3, the compactness and convexity of $\mathcal{T}_i$ is not used directly. That $\mathcal{T}_i^*$ is upper semicontinuous with convex image sets is crucial. Were the image sets not necessarily convex, it would be possible that the image sets of the best reply mapping might not be convex. The required properties of $\mathcal{T}_i^*$ do not imply convexity of $\mathcal{T}_i$. Fig. 7.6 shows a correspondence whose graph is not convex; but whose image sets are convex.

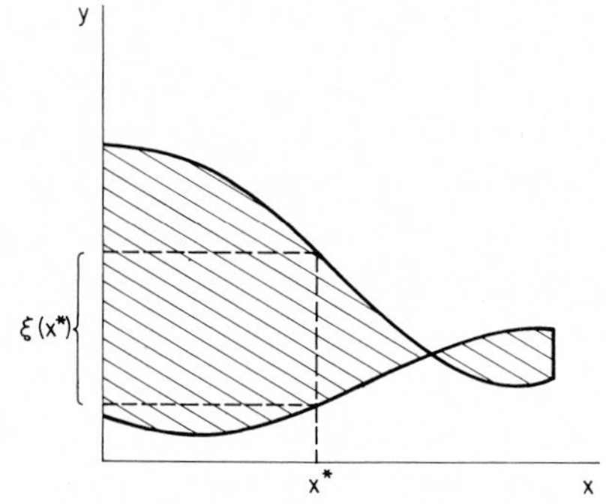

Fig. 7.6

2.3. *A digression on two-person zero sum games*

Although two-person zero sum games have not found application in economics, they have probably been the subject of most interest and results for mathematicians working in game theory. For historical reasons alone it is worthwhile to make use of the background provided by the earlier part of this chapter to state briefly, and without proof, von Neumann's famous saddle point theorem. von Neumann formulated games in a different way from that used in the present chapter. Indeed, a minor variant of the assumptions G1–G4, obtained by weakening G4 to weak, rather than strict, quasi-concavity, provides a model of which von Neumann's is a special case.

First, consider how simple a matter it is to make a game satisfying G1–G4 become two-person and zero sum. It is only necessary to let $n = 2$

and specify that $P_2(s) = -P_1(s)$ for all $s \in \mathscr{S}$. Note that the games dealt with in this chapter are *infinite* in the sense that each player has an infinite number of *pure strategies* from which to choose. von Neumann's games are finite, with m_1 *pure strategies* to the first player and m_2 to the second. Letting the index k run over the integers $1, \ldots, m_1$ for the strategies of player 1 and l, from $1, \ldots, m_2$ for player 2, a_{kl} denotes the payoff to player 1 (and loss to player 2) when player 1 uses his kth strategy and player 2 his lth. These strategies are called *pure strategies* because they have no stochastic elements. von Neumann built upon the pure strategies to construct for each player a larger set of strategies. This is done by supposing that a player, say the first, can assign a probability distribution to his pure strategies and use a probability mechanism with these probabilities to choose the actual pure strategy which is to be played. He may choose any distribution, $p_1, \ldots, p_{m_1}$ for which all the p_k are non-negative and $\Sigma_{k=1}^{m_1} p_k = 1$. Similarly for player 2. In effect, the strategy set for player 1 becomes the unit simplex of dimension $m_1 - 1$ and for player 2, the unit simplex of dimension $m_2 - 1$. A probability distribution, $\boldsymbol{p}$, for player 1 is called a *mixed strategy*. Let $\boldsymbol{q}$ denote a mixed strategy for player 2. Payoffs are defined as the expected payoff under the given pair of mixed strategies:

$$P_1^*(\boldsymbol{p}, \boldsymbol{q}) = \sum_{k=1}^{m_1} \sum_{l=1}^{m_2} p_k q_l a_{kl}. \tag{7.4}$$

In a two-person zero sum game it is reasonable for player 1 to suppose that player 2 wants to minimize player 1's payoff. Doing so is the same as player 2 maximizing his own payoff. For each strategy of his own, $\boldsymbol{p}$, player 1 can determine the worst outcome he could face,

$$\min_{q} P_1^*(\boldsymbol{p}, \boldsymbol{q}). \tag{7.5}$$

A safe strategy for player 1 is one for which the minimum in eq. (7.5) is as large as possible,

$$\max_{p} \min_{q} P_1^*(\boldsymbol{p}, \boldsymbol{q}). \tag{7.6}$$

Player 2, behaving in a parallel fashion, finds

$$\max_{q} \min_{p} P_2^*(\boldsymbol{p}, \boldsymbol{q}) = \min_{q} \max_{p} P_1^*(\boldsymbol{p}, \boldsymbol{q}). \tag{7.7}$$

Eqs. (7.6) and (7.7) have the same solutions, which give equilibrium strategies and payoffs. The equilibrium payoff is called the *value* of the game. The famous minimax theorem which von Neumann proved is that if mixed strategies are allowed, the value of the game to each player is the same:

$$\max_{p} \min_{q} P_1^*(\boldsymbol{p}, \boldsymbol{q}) = \min_{q} \max_{p} P_1^*(\boldsymbol{p}, \boldsymbol{q}). \tag{7.8}$$

Though there may be more than one strategy pair which are equilibrium strategies, all result in the same value. As a final note, the mathematical structure of finite two-person zero sum games and linear programming are very close, in that a game can be cast as a programming problem. The reader interested in pursuing this connection should look at appendices 5 and 6 of Luce and Raiffa (1957) and ch. 13 of Dantzig (1963).

3. *Fixed point theorems and equilibrium in noncooperative games*

The Brouwer theorem states that if $f(\boldsymbol{x})$ is a continuous function which maps the closed m-dimensional unit simplex, $\mathcal{W}$, into itself, then there exists $\boldsymbol{x}^0 \in \mathcal{W}$ such that $\boldsymbol{x}^0 = f(\boldsymbol{x}^0)$. An often used corollary is that the theorem holds if $\mathcal{W}$ is an arbitrary compact convex subset of a finite dimensional vector space. This is the form in which the theorem is used above in §2.2.

If G4 were relaxed to require only weak quasi-concavity then the best reply of player i to $\bar{s}_i$ need not be unique. There may be a set of equally good best replies, making $\boldsymbol{r}_i$ a correspondence, or point-to-set mapping. Kakutani (1941) generalized the Brouwer theorem to correspondences. It is also possible to weaken G2 to do away with the requirement that $\mathcal{S}$ be

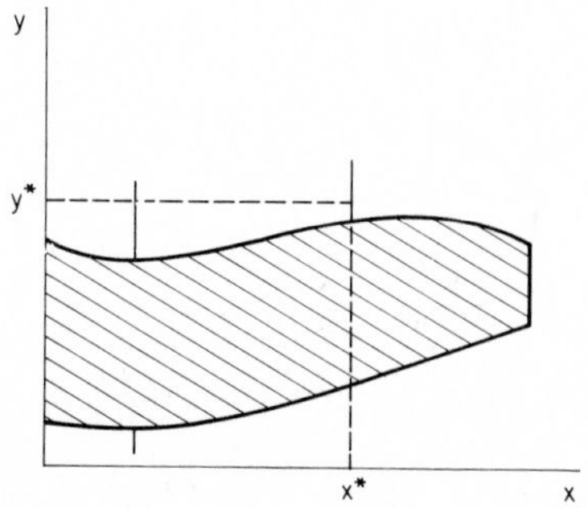

Fig. 7.7

contained in a space of finite dimension. Proof is allowed by a generalization of the Kakutani theorem due to Bohnenblust and Karlin (1950) or Ky Fan (1952). Fan's theorem may also be found in Berge (1963).

§3.1 contains a brief discussion of correspondences. §3.2 contains a generalization of §2.2 based on the Kakutani theorem. §3.3 contains a description of the model of Nash (1951) which is the first for which the existence of noncooperative equilibrium is proved for n-person games. §3.4 has a generalization of the results of §3.2 for a model in which the Ky Fan theorem is used.

3.1. A brief look at correspondences

A correspondence associates with each point in its domain a set of points. A function, associating a single point in the range with a point in the domain, is a special case of correspondence. The correspondences which are of interest for this volume have some sort of continuity property. Three continuity properties are illustrated in figs. 7.6–7.8. Fig. 7.6 shows a correspondence which is continuous, fig. 7.7 one which is upper semicontinuous and fig. 7.8 one which is lower semicontinuous.

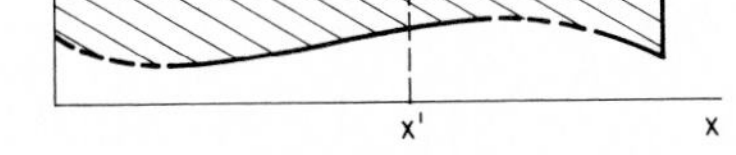

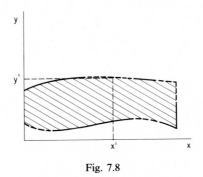

Fig. 7.8

Upper semicontinuity may be defined in either of two equivalent ways. Let $\xi(\boldsymbol{x})$ be a correspondence. $\xi(\boldsymbol{x})$ is *upper semicontinuous* at the point $\boldsymbol{x}^0$ if, when a sequence $\boldsymbol{x}^l$, $l = 1, 2, \ldots$, converges to $\boldsymbol{x}^0$, $\boldsymbol{y}^l \in \xi(\boldsymbol{x}^l)$, and $\boldsymbol{y}^l$, $l = 1, 2, \ldots$, converges to $\boldsymbol{y}^0$ then $\boldsymbol{y}^0 \in \xi(\boldsymbol{x}^0)$. ξ is upper semicontinuous if it is upper semicontinuous at each point of its domain $\mathscr{D}$. The other definition is that the graph of the correspondence is closed. The graph of ξ

is $\{(\boldsymbol{x}, \boldsymbol{y}) | \boldsymbol{x} \in \mathscr{D}, \boldsymbol{y} \in \boldsymbol{\xi}(\boldsymbol{x})\}$. Figs. 6.6 and 6.7 show correspondences which are upper semicontinuous.

Lower semicontinuity is defined by: $\boldsymbol{\xi}(\boldsymbol{x})$ is *lower semicontinuous* at the point $\boldsymbol{x}^0$ if, when $\boldsymbol{x}^l$, $l = 1, 2, \ldots$, converges to $\boldsymbol{x}^0$ and $\boldsymbol{y}^0 \in \boldsymbol{\xi}(\boldsymbol{x}^0)$, there is a sequence $\boldsymbol{y}^l$ such that $\boldsymbol{y}^l \in \boldsymbol{\xi}(\boldsymbol{x}^l)$ and $\lim_{l \to \infty} \boldsymbol{y}^l = \boldsymbol{y}^0$. $\boldsymbol{\xi}$ is lower semicontinuous if it is lower semicontinuous at each point of $\mathscr{D}$. Figs. 7.6 and 7.8 show lower semicontinuous correspondences. The correspondence in fig. 7.8 is not upper semicontinuous. y' is not in the image set of x'; but it would be if the correspondence were upper semicontinuous.

In fig. 7.6, an image set $\xi(x^*)$ is illustrated for a point x^*. The correspondence in this figure is *continuous* because it is both upper and lower semicontinuous. Compare it to fig. 7.7 to see a point such as (x^*, y^*) where lower semicontinuity fails. In fig. 7.8 it is clear that the graph of the correspondence is not closed, for example at (x', y').

3.2. *Weakening strict quasi-concavity*

Kakutani's fixed point theorem, in the form needed for this chapter, may be stated: If $\;(\boldsymbol{x})$ is an upper semicontinuous correspondence which maps a compact convex subset $\mathscr{D}$ of $\mathscr{R}^m$ into closed convex subsets of $\mathscr{D}$, then there exists an $\boldsymbol{x}^0 \in \mathscr{D}$ such that $\boldsymbol{x}^0 \in \boldsymbol{\xi}(\boldsymbol{x}^0)$.

An appropriately weakened version of G4 is

G6. *$P_i(s)$ is quasi-concave with respect to s_i, $i = 1, \ldots, n$.*

The following existence theorem may be proved.

Theorem 7.4. *Any game satisfying G1–G3 and G6 has a noncooperative equilibrium.*

□ The proof is largely parallel to that of theorem 7.1. First the best reply correspondences must be defined:

$$\mathfrak{z}_i(\bar{s}'_i) = \{t_i | t_i \in \mathscr{S}_i, P_i(t_i, \bar{s}'_i) = \max_{s_i \in \mathscr{S}_i} P_i(s_i, \bar{s}'_i)\}, \qquad \bar{s}'_i \in \bar{\mathscr{S}}_i,$$
$$i = 1, \ldots, n. \tag{7.9}$$

$\mathfrak{z}$ is then given by

$$\mathfrak{z}(s) = \mathfrak{z}_1(\bar{s}_1) \times \mathfrak{z}_2(\bar{s}_2) \times \cdots \times \mathfrak{z}_n(\bar{s}_n); \tag{7.10}$$

that is, $t \in \boldsymbol{\imath}(s)$ if $t \in \boldsymbol{\imath}_i(\bar{s}_i)$, $i = 1, \ldots, n$. t is a best reply to $\bar{s}_i$. Again, clearly the fixed points of $\boldsymbol{\imath}$ must coincide with the strategies which are associated with noncooperative equilibria. $\boldsymbol{\imath}$ must be shown to be an upper semicontinuous correspondence from $\mathcal{S}$ to $\mathcal{S}$ with closed and convex image sets. Quasi-concavity assures that $\boldsymbol{\imath}(s)$ is convex, and continuity of the P_i assures that it is closed. Upper semicontinuity follows from precisely the same argument used in the proof of theorem 7.1 to show continuity of $\boldsymbol{r}$. □

We also have, without further proof, a parallel result to theorem 7.3:

Corollary 7.5. *A game satisfying G1, G2, G5 and G6 has a noncooperative equilibrium. If s^* is a strategy vector corresponding to such an equilibrium then $s^* \in \bigcap_{i=1}^n \mathcal{T}_i$.*

3.3. The original result of Nash, and the relationship of finite to infinite games

The model used by Nash (1951) is finite. The ith player has k_i pure strategies. A pure strategy for the ith player may be denoted j_i, which can take on values $1, 2, \ldots, k_i$. His payoff is $e^i_{j_1,\ldots,j_n}$ when the first player uses strategy j_1, the second uses j_2, etc. Thus each player has a payoff matrix whose size is $k_1 \times k_2 \times \cdots \times k_n$. Following von Neumann and Morgenstern, Nash introduces mixed strategies. A mixed strategy for the ith player is a probability distribution $\boldsymbol{v}_i = (v_{i1}, \ldots, v_{ik_i})$ over his pure strategies with $v_{ij} \geq 0$, $j = 1, \ldots, k_i$ and $\sum_{j=1}^{k_i} v_{ij} = 1$. The pure strategies are, of course a special case of the mixed. Letting $\boldsymbol{v} = (\boldsymbol{v}_1, \ldots, \boldsymbol{v}_n)$, a payoff function in terms of mixed strategies may be given for each player,

$$P_i^*(\boldsymbol{v}) = \sum_{j_1=1}^{k_1} \sum_{j_2=1}^{k_2} \cdots \sum_{j_n=1}^{k_n} v_{1,j_1} v_{2,j_2} \cdots v_{n,j_n} e^i_{j_1,\ldots,j_n}, \qquad i = 1, \ldots, n. \tag{7.11}$$

This model satisfies the assumptions G1–G3 and G6. The number of players is finite. $\boldsymbol{v}_i$ is restricted to the $k_i - 1$ dimensional unit simplex $\mathcal{V}_i$, which is compact and convex. The P_i^* are continuous in all $\boldsymbol{v}_j$ and concave (hence quasi-concave) in $\boldsymbol{v}_i$.[8] As a result it is unnecessary to

[8]In fact $P_i^*(\boldsymbol{v})$ is concave in $\boldsymbol{v}$, not merely in $\boldsymbol{v}_i$.

prove Nash's theorem, as it is a special case of theorem 7.4. In Nash (1951), where his theorem is proved, the proof does not use the Kakutani theorem, nor does it proceed along the lines of the proofs to theorems 7.1 and 7.4. Instead, an ingenious proof is constructed using the Brouwer theorem.

3.4. Further relaxation of the assumptions

The assumption G2 under which the strategy sets of the players are assumed to be subsets of $\mathcal{R}^m$ may be relaxed. Let $\mathscr{E}$ be a Banach space – that is, a normed space which is complete (see Dieudonné, 1960, ch. V §1). An example is the space obtained by taking a countable product of the real line.[9] Thus, G2 is relaxed by replacing $\mathcal{R}^m$ with $\mathscr{E}$. Consequently, a generalization of the Kakutani theorem is needed which extends it to Banach spaces. This is done by the fixed point theorem of Bohnenblust and Karlin (1950), which requires that the set which is mapped into itself be sequentially compact (i.e. that every sequence in the set have a convergent subsequence). The further generalization of Fan (1952) extends the Bohnenblust and Karlin result to locally convex spaces. Their theorem is:

Fixed point theorem. *Let $\mathscr{D}$ be a convex, closed and sequentially compact subset of a Banach space; and let $\xi(\boldsymbol{x})$ be an upper semicontinuous correspondence on $\mathscr{D}$. If $\xi(\boldsymbol{x})$ is convex for all $\boldsymbol{x} \in \mathscr{D}$, then there exists a point $\boldsymbol{x}^0 \in \mathscr{D}$ such that $\boldsymbol{x}^0 \in \xi(\boldsymbol{x}^0)$.*

In modifying theorem 7.4 to allow strategy sets to be subsets of a Banach space, it is necessary to be certain that sequential compactness holds. If it does, it is immediate that the best reply mapping defined as in the proof of theorem 7.4 satisfies all required conditions – it is upper semicontinuous, has convex image sets and maps $\mathscr{S}$ into $\mathscr{S}$. The norm of $\mathscr{E}$ is given by $\|\boldsymbol{x}\| = \sum_{j=1}^{\infty} |x_j|$ for $\boldsymbol{x} \in \mathscr{E}$.

G7. *$\mathscr{S}_i$, the strategy set of the ith player, is a closed, convex and bounded subset of a Banach space $\mathscr{E}$ (for $i = 1, \ldots, n$).*

G8. *There is $s^+ \in \mathscr{E}$ such that for any s, $s' \in \mathscr{S}_i$, $|s_j - s'_j| \leq s^+_j$ and $\|s^+\| < \infty$, $i = 1, \ldots, n$.*

[9] Loosely speaking, a suitable $\mathscr{E}$ is obtained by letting $\mathscr{E} = \lim_{m \to \infty} \mathcal{R}^m$.

Theorem 7.6. *A game satisfying G1, G3, G6–G8 has a noncooperative equilibrium.*

□ Proof of the theorem is achieved by showing that the sets $\mathscr{S}_i$ are sequentially compact. To do this, let s^l, $l = 1, 2, 3, \ldots$, be a sequence of elements of $\mathscr{S}_i$. Using the Cantor diagonal process, a subsequence of the original sequence may be found such that the subsequence converges in the first component, then a subsequence of the first subsequence may be found which converges in the second component, etc. Two members of the first subsequence can be no farther apart than $\Sigma_{j=1}^{\infty} s_j^+$, and, in general, two members of the kth subsequence can be no farther apart than $\Sigma_{j=k}^{\infty} s_j^+ = h_k$. By G8, $h_k \to 0$ as $k \to \infty$. Therefore, the original sequence has a convergent subsequence. □

4. Solutions, equilibria and uniqueness

4.1. Solutions and equilibria

To this point the concept *solution* has not been used. A solution may be thought of as a "natural" or "expected" equilibrium. Imagine a game in which the noncooperative equilibrium is unique. It is reasonable to call such an equilibrium the *solution* of the game, meaning that rational players may be expected to choose the particular strategies which lead to the (unique) equilibrium. Now imagine a game in which there are several noncooperative equilibria, as illustrated in fig. 7.9. On the face of it, there need be no particular one of the equilibria which is special – which the players might be expected to choose in preference to the others. There are in fig. 7.9 three equilibria which rational players might rule out. These are *D*, *E* and *F*. They are each dominated. That is, for each there is another equilibrium which offers a higher payoff to each player. *D* is dominated by *A*, *E* is dominated by both *A* and *B*; while *F* is dominated by *C*. As between *A*, *B* and *C* there is no obvious way to choose one as a solution.

The difficulty which arises when a game has no solution, though it may have several equilibria, is that there is no way to make an informed prediction of how sensible players would choose to play. It cannot even be said that one of the equilibria should be chosen; for, to do that, the players would have to coordinate. In a noncooperative game which also lacks communication between players, coordination is not possible; however, if communication is admitted, agreement upon a particular

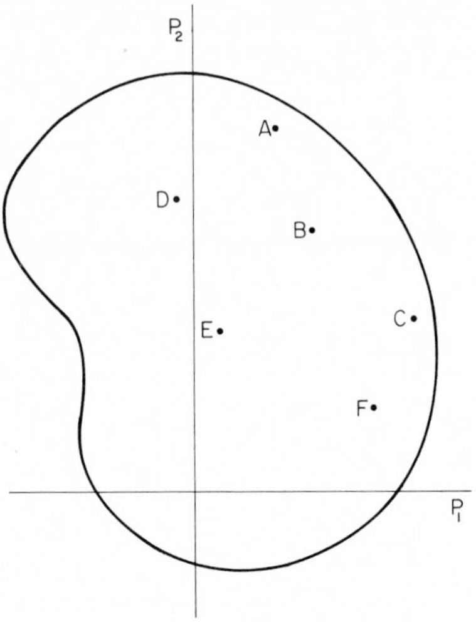

Fig. 7.9

noncooperative equilibrium is possible. Indeed, in such a game, it is quite reasonable to say that the solution consists of all the undominated noncooperative equilibria; because verbal agreement on a particular noncooperative equilibrium would be self-enforcing. Returning to games lacking communication, should one player choose a strategy corresponding to one of the equilibria, where is any assurance that other players will choose strategies corresponding to the same equilibrium?

There are two possible ways out of this problem. One is to try to determine which classes of games have a unique noncooperative equilibrium. The other is to find some way to naturally select one noncooperative equilibrium from the set of them. Both avenues are explored, with limited success for each, in the following two subsections.

4.2. A sufficient condition for uniqueness of the noncooperative equilibrium

It is well known that when a contraction mapping has a fixed point, the fixed point is unique (see Bartle (1964, theorem 16.14)). Thus, if the best reply mappings, ***r***, were contractions, they would have exactly one fixed point and it would be the unique equilibrium of the game. To say that ***r*** is a contraction means that, for each player, $|r_i(\bar{s}'_i) - r_i(\bar{s}''_i)| < \|\bar{s}'_i - \bar{s}''_i\|$. Intuitively, if the other players change their strategies by some amount, the best reply of the player is altered by a lesser amount. The result may be stated formally:

Theorem 7.7. *In a game having a noncooperative equilibrium, if the best reply mapping,* $\mathbf{r}$*, is a contraction, then the equilibrium is unique.*

The requirement that $\boldsymbol{r}$ be a contraction may be undesirable in some applications. Furthermore, it may be difficult to impose restrictions on the payoff functions to achieve this end.

4.3. Harsanyi's tracing method for finding a solution

In Harsanyi (1974) a method is proposed for choosing a particular noncooperative equilibrium as a solution. The method works for virtually all games of the sort discussed in §3.3. In the present section, the method is described and results given but there are no proofs.

Recall that the class of games being dealt with are games in which each player has a finite number of pure strategies, and that each player can also choose mixed strategies, $\boldsymbol{v}_i = (v_{i1}, \ldots, v_{i,k_i})$, which are probability distributions over the player's own pure strategies. Player i's payoff function is $P_i^*(\boldsymbol{v}) = P_i^*(\boldsymbol{v}_i, \bar{\boldsymbol{v}}_i)$. Finding the solution is accomplished by assuming the players go through a fictional thought process in which they begin with an assumption about how other players choose, then each associates the same particular noncooperative equilibrium with the initial assumptions. The initial assumptions on behavior take the form of probability distributions, $\boldsymbol{q}_i = (q_{i1}, \ldots, q_{i,k_i})$, $i = 1, \ldots, n$. $\boldsymbol{q}_i$ is the initial probability distribution which all players other than player i assume player i uses. Using $\boldsymbol{q} = (\boldsymbol{q}_1, \ldots, \boldsymbol{q}_n)$, a family of games can be defined which are parameterized by t, which ranges from 0 to 1. A member of the family is denoted $\boldsymbol{\Gamma}^t$. The strategies and strategy sets for all members of $\boldsymbol{\Gamma}^t$ are the same and are precisely those of the original game. The payoff functions are:

$$tP_i^*(\boldsymbol{v}_i, \bar{\boldsymbol{v}}_i) + (1-t)P_i^*(\boldsymbol{v}_i, \bar{\boldsymbol{q}}_i), \qquad i = 1, \ldots, n. \tag{7.12}$$

When $t = 0$, the game is one in which the payoff of the ith player is completely independent of the actual behavior of the others (i.e. of $\bar{\boldsymbol{v}}_i$); and when $t = 1$, the game is the original game.

For each value of t, it is possible to characterize the (clearly nonempty) set of noncooperative equilibria. Denote them $\mathscr{E}(t)$. $\mathscr{E}(t)$ is a correspondence whose domain is $t \in [0, 1]$. The members of $\mathscr{E}(1)$ are the noncooperative equilibria of the original game. The *linear tracing procedure* is a connected path from a member of $\mathscr{E}(0)$ to a member of $\mathscr{E}(1)$. The procedure is said to be *feasible* if such a path exists and *well defined* if there is exactly one such path. For a large class of games, the procedure is feasible; but not necessarily well defined. Note that for $t < 1$, the set $\mathscr{E}(t)$ depends upon the particular prior distributions $\boldsymbol{q}$. With different prior distributions, $\boldsymbol{q}'$, new sets $\mathscr{E}'(t)$ arise; however, $\mathscr{E}'(1) = \mathscr{E}(1)$. The particular equilibrium in $\mathscr{E}(1)$ which the tracing procedure singles out is likely to be sensitive to the choice of $\boldsymbol{q}$.

To define the class of games for which the procedure is feasible, it is first necessary to define *stability sets* and *inferior strategies*. For the original game (i.e. the game with $t = 1$), and the ith player, it is possible to take one pure strategy, say j_i, and look at the set of $\bar{\boldsymbol{v}}_i$ for which j_i is the best reply. That is the *stability set* of j_i. Denote this set $\mathscr{S}_i^*(j_i)$. Clearly $\bigcup_{j_i=1}^{k_i} \mathscr{S}_i^*(j_i) = \bar{\mathscr{V}}_i$, which means that some of the $\mathscr{S}_i^*(j_i)$ have nonempty interiors relative to $\bar{\mathscr{V}}$; although it is possible that some $\mathscr{S}_i^*(j_i)$ are empty. The corresponding pure strategies may be entirely removed from the game on the ground that they will never be used, being best replies to nothing. Other of the $\mathscr{S}_i^*(j_i)$ may be sets which are nonempty but which have no interior relative to $\bar{\mathscr{V}}_i$. These are called *inferior*, and it may be argued that, though they are best replies to some choices of the other players, they will not actually be used. The reason is that any $\bar{\boldsymbol{v}}'_i$ has a best reply which is not inferior; hence it is not necessary to use inferior strategies. And using a noninferior strategy has the advantage that it remains a best reply for a much larger set of $\bar{\boldsymbol{v}}_i$ in the neighborhood of $\bar{\boldsymbol{v}}'_i$. Harsanyi restricts himself to games having no inferior strategies. Anyone willing to assume inferior strategies will never be used may remove them from an arbitrary game, leaving only noninferior strategies.

One main result is that for any fixed $\boldsymbol{q}$ and almost all games, the linear tracing procedure is well defined. That is, associated with a particular game and prior distribution is a unique noncooperative equilibrium. Further results are obtained by something called the *logarithmic tracing procedure*, which involves a perturbation of the model already described. Strategy sets remain the same, and, as before, a family of games is defined which are parameterized by t. Their payoffs are:

$$tP_i^*(\boldsymbol{v}_i, \bar{\boldsymbol{v}}_i) + (1-t)P_i^*(\boldsymbol{v}_i, \bar{\boldsymbol{q}}_i) + \epsilon(1-t)\sum_{k=1}^{k_i} \ln v_{ik} \tag{7.13}$$

for an arbitrary $\epsilon > 0$. As t goes from 0 to 1 a set of equilibria is generated from which a connected path may be selected, as before. Denote by $\boldsymbol{v}^*(\epsilon)$ the equilibrium obtained for $t = 1$ and the given ϵ; and denote by $\boldsymbol{v}^*$ the equilibrium obtained under the linear tracing procedure for $t = 1$. It need not be that $\boldsymbol{v}^* = \boldsymbol{v}^*(\epsilon)$; however, $\boldsymbol{v}^* = \lim_{\epsilon \to 0} \boldsymbol{v}^*(\epsilon)$. A second way to take limits is to find a limit path by allowing ϵ to go to zero for each value of t. This also results in a path whose end point, for $t = 1$, is $\boldsymbol{v}^*$. The logarithmic tracing procedure, which has been characterized in two ways, is a way to start from a prior distribution $\boldsymbol{q}$ with its associated best replies and go to a noncooperative equilibrium for the game, $\boldsymbol{v}^*$. In contrast to the linear tracing procedure, the logarithmic procedure always exists and is always unique; hence, it is always well defined. In addition, when the linear tracing procedure is well defined, its solution is the same as the solution of the logarithmic procedure.

There remains the question of determining q and of assuring that it is reasonable for all players (other than the ith) to have precisely the same expectations of the ith. While this is not addressed in Harsanyi (1974), he promises a solution.

5. *Strong and weak equilibria*

To this point, cooperative and noncooperative games have been kept separated by the greatest possible distance. The latter are characterized by the ability of the players to make binding, unbreakable agreements; while the former have been characterized by the absence of that ability, together with the stipulation that the players cannot even have a conversation or send letters to one another before committing themselves to strategies, i.e. no preplay communication is allowed. In §4 it is pointed out that in a game having several undominated noncooperative equilibria, if preplay communication is ruled out, it cannot be supposed that the players choose strategies which place them at an equilibrium point. With preplay communication admitted, it is surely reasonable to expect players to select an equilibrium point because they can easily coordinate to that extent; and any point which is not an equilibrium point has strikes against it which have been much discussed already.

But perhaps just being an undominated equilibrium point is not enough? For a game satisfying, say, G1–G4, a *strong equilibrium point* may be defined. s^* is the strategy vector associated with a strong equilibrium point if $s^* \in \mathcal{S}$ and $\boldsymbol{P}_{\mathcal{K}}(s^*) \geq \boldsymbol{P}_{\mathcal{K}}(s_{\mathcal{K}}, \bar{s}^*_{\mathcal{K}})$ for all $s_{\mathcal{K}} \in \boldsymbol{S}_{\mathcal{K}}$ for all subsets of players, $\mathcal{K}$. Thus, to qualify as a strong equilibrium, it must be impossible for one player to change his strategy (all others' remaining the same) and increase his payoff, it must be impossible for a subset of players to jointly change their strategies (all others' remaining the same) and increase their payoffs, and it must be impossible for all players, jointly, to change their strategies and increase their payoffs. Thus a strong equilibrium yields payoffs on the payoff frontier. Even casual thought suggests that many games having noncooperative equilibria lack strong equilibrium points. Clearly this is so for any game in which all noncooperative equilibria are interior to the payoff space. For example the single period oligopolies which are the subject of theorems 2.2 and 3.2 have no strong equilibria for just this reason (see also Aumann (1960)).

6. Noncooperative games and oligopoly theory

Game theory can make both formal and informal contributions to oligopoly theory, either of which can be illustrated with reference to material in chs. 2 and 3. On the formal side is the direct application of the noncooperative equilibrium to the oligopoly models of chs. 2–6. When that application is made in §6.1 and §6.2 it is seen that the Cournot equilibrium is the same as the noncooperative equilibrium for the single-period oligopoly models, and it is possible to use the existence theorems of the present chapter to prove existence of the Cournot equilibrium. One step which must be taken is to interpret each oligopoly model as a game. That means identifying in the model what is to be meant by *player*, *strategy*, and *payoff*. Then the next step is to see whether the oligopoly model, thus interpreted as a game, meets the assumptions of theorems which could yield interesting results. It is also possible to see whether an equilibrium from the oligopoly literature, such as the Cournot equilibrium, turns out to be a special case of an equilibrium in the game theory literature.

Game theory focuses attention on the actions which players might take, and encourages a thorough consideration of all the possible actions which might be open to a player, as well as suggesting that equilibria should have some reasonable consistency. Thus, for example, in pondering the Bowley or Stackelberg models in the light of game theory, it becomes clear that their equilibria are not noncooperative equilibria. Because of this, they do not seem reasonable precisely because at least one player is failing to use an available strategy which is a best response to the strategy of the other player. Furthermore, in the Stackelberg model particularly, attention is confined to only two strategies (leader and follower), while it is possible to imagine many other reaction functions than those. It may turn out that only those two are interesting, though that appears unlikely in view of the results of ch. 5; however, training in game theory makes it almost automatic to want to specify all the possible reaction functions which are conceivable within the model and start with them as strategy sets.

6.1. Noncooperative equilibrium for single-period oligopoly models

For both the quantity and price models, it is obvious how to interpret them as games. Taking the quantity model first, the players are, of course, the firms. The strategy of the *i*th firm is its quantity choice, q_i and the strategy set is the interval $[0, \bar{Q}]$. The payoff function of the *i*th firm is its profit function, $\pi_i(\boldsymbol{q})$. It is easy to show that a game satisfying A1, A2 and A6 has a Cournot equilibrium by showing that G1–G4 are satisfied, and that the Cournot equilibrium is a noncooperative equilibrium.

Theorem 7.8. *Strategy sets of $[0, \bar{Q}]$ for each player together with A1, A2 and A6 imply G1–G4, and a Cournot equilibrium is a noncooperative equilibrium; therefore, a game having $[0, \bar{Q}]$ for strategy sets and satisfying A1, A2 and A6 has a Cournot equilibrium.*

□ G1 is satisfied, as the number of firms, *n*, is finite. G2 is satisfied; for the strategy sets are all of the form $\mathcal{S}_i = [0, \bar{Q}]$, where $\bar{Q}$ is finite making $[0, \bar{Q}]$ a compact interval. From A1 and A2 the demand and cost functions are both continuous and bounded; hence, so are the profit functions. Finally, A6 is strict concavity of π_i with respect to q_i, satisfying G4. This establishes that the single period Cournot quantity model has a noncooperative equilibrium. It is immediate from the definitions of the Cournot equilibrium and of the noncooperative equilibrium that, in the single-period quantity model they are the same. □

Interpreting the single-period price model as a game is similar to doing so for the quantity model. Each firm is a player and the strategy set of the *i*th firm is the interval $[0, p_i^+]$. As before, the profit functions are the payoff functions. A parallel theorem to theorem 7.8 may be stated and proved.

Theorem 7.9. *Strategy sets of $[0, p_i^+]$, $i = 1, \ldots, n$, A2–A5, and A8 imply G1–G3 and G6. A Cournot equilibrium is a noncooperative equilibrium; therefore, a game satisfying the preceding conditions has a Cournot equilibrium.*

□ That G1, G2 and G3 hold is immediate. To see that G6 holds, the interval $[0, p_i^+]$ must, for a fixed $\bar{p}_i$, be broken into three subintervals. These are $[0, p_i')$, $[p_i', p_i'')$ and $[p_i'', p_i^+]$. The first interval is the set of prices for which the firm's marginal cost exceeds its price, the second is the set of prices for which price is at least as great as marginal cost and demand is positive, and the third is the set of prices for which demand is zero. Quasi-concavity of π_i with respect to p_i means that if, at any point, π_i ceases falling as p_i increases, then for all larger values of p_i, π_i is nonincreasing as p_i rises. Look at π_i^i:

$$\pi_i^i(p) = F_i(p) + (p_i - C_i'(F_i(p)))F_i^i(p). \tag{7.14}$$

On the first interval, $\pi_i^i > 0$; on the second it is strictly decreasing and is positive at p_i' and negative at p_i''; and on the third interval, it is zero. Thus, if all three intervals are nonempty, quasi-concavity is satisfied. On the other hand, if one or two of the intervals is empty, or consists of only one point, it is clear that quasi-concavity still is satisfied. Where the middle interval is nonempty, it always contains the best reply. Again, it is clear that the Cournot equilibrium is the noncooperative equilibrium for the single-period price model. □

The assumption A10 assures that the middle price interval $[p_i', p_i'')$ is nonempty for any $\bar{p}_i$, which assures that Cournot equilibria occur at price vectors for which all firms have strictly positive demand as well as strictly positive prices.

Corollary 7.10. *Under the assumptions A2–A5, A8 and A10, and with compact strategy sets $[0, p_i^+]$, $i = 1, \ldots, n$, any Cournot equilibrium is associated with strictly positive prices and demand for all firms.*

With respect to uniqueness of the Cournot equilibrium, adding A7 to the assumptions of theorem 7.8 and A9 to the assumptions of corollary 7.10 guarantees uniqueness for the quantity and price models, respectively. Each is accomplished in the same way. They imply that the best reply functions are contractions. A contraction has at most one fixed point; and, as it is already known these best reply functions have at least one fixed point, there must be precisely one.

Theorem 7.11. *A1, A2, A7 and compact strategy sets imply uniqueness of the Cournot equilibrium for the quantity model.*

Theorem 7.12. *A2–A5, A9, A10 and compact strategy sets imply uniqueness of the Cournot equilibrium for the price model.*

6.2. Noncooperative equilibrium for reaction function models

This section is far more concerned with why results are not obtainable than it is with developing new results. All comments are directed toward the n-firm price model with simultaneous decision which is studied in ch. 5. First, the reaction function model should be formulated as a game. Again, the players are clearly the firms. A strategy for a firm is most appropriately a reaction function; hence, the strategy set of a player is the set from which he may choose his reaction function. Finally, the payoff function of a player is the sum of discounted profits of the firm.

In ch. 5 §2 a maximization process is described for a firm which results in the reaction function which is the firm's best reply to the reaction functions of the other firms. In the notation of §2.4, $\phi_i = W_i(\bar{\boldsymbol{\psi}}_i)$ where $\bar{\boldsymbol{\psi}}_i$ are the reaction functions of the other firms and ϕ_i is the ith firm's best reply to $\bar{\boldsymbol{\psi}}_i$. For all firms, $\boldsymbol{\phi} = \boldsymbol{W}(\boldsymbol{\psi})$ is the counterpart to the best reply mapping, $\boldsymbol{r}$, used earlier in the present chapter. Note that $\boldsymbol{W}$ carries one function ($\boldsymbol{\psi}$) into another ($\boldsymbol{\phi}$), rather than carrying a vector into another; however, it remains true that a fixed point of $\boldsymbol{W}$ is a noncooperative equilibrium for the reaction function model. Before even trying to see whether a fixed point theorem can be made to apply, there is a known fixed point for the mapping $\boldsymbol{W}$. It is $\psi_i(\boldsymbol{p}) = p_i^c$, $i = 1, \ldots, n$, where it will be recalled $\boldsymbol{p}^c$ is the Cournot equilibrium price vector for the single-period model which forms the basis of the reaction function model. What is of great interest is whether there is a noncooperative reaction function equilibrium *other than* the degenerate one corresponding to constant use of the single-period Cournot equilibrium prices. Thus, a fixed point theorem would be useless at this stage. It would, if successfully applied, state that $\boldsymbol{W}$ has a fixed point; but, as it is already known that $\boldsymbol{W}$ has a particular fixed point and it is of interest to know if there is a different one, such a result is no help.

An alternative might be to restrict $\boldsymbol{\psi}$ to a set which does not include the constant functions or to force the domain of ψ to exclude p^c. Neither of these devices work, as they result in sets for ψ which do not map into themselves. For example, if ψ were restricted so that its first derivatives were bounded strictly away from zero, it would not prove possible to force $W(\psi)$ to obey the same lower bound.

Taking the content of this section altogether, it is clear that game theory provides some substantive results for some of the models of oligopoly considered earlier, the single period models. With respect to existence of the Cournot equilibrium there is a proof to be found in Frank and Quandt (1963) for a quantity model. The proof is similar to proofs of the kind used earlier in this chapter for game models. For the reaction function models, it is not possible to claim substantive results stemming from the application of game theory; however, the attempt to apply the noncooperative equilibrium results in pinpointing clearly the strengths and weaknesses of the reaction function equilibrium developed in ch. 5. Indeed it is quite possible that the sort of work undertaken by Friedman and Cyert and de Groot would not have been undertaken at all without the general guidance which comes from a knowledge of the game theoretic approach to oligopoly.

It is left as an exercise to the reader to verify that theorem 6.12 is a special case of theorem 7.6.

7. Concluding comments

The central focus of this chapter is the noncooperative equilibrium in various forms and its application to the oligopoly models presented in earlier chapters. Looking ahead to later chapters, the noncooperative equilibrium and the best reply principle are worked quite hard. The particular existence theorems presented in §3 find application and equilibria are also developed which obey conditions additional to those which define the noncooperative equilibrium.

CHAPTER 8

NONCOOPERATIVE EQUILIBRIA FOR SUPERGAMES LACKING TIME DEPENDENCE

1. *Introduction*

The term *supergame* was apparently coined by Luce and Raiffa (1957) who discuss temporal repetitions of the prisoner's dilemma and of two person zero sum games.[1] When referring to supergames in the present volume, I always mean a situation in which a fixed set of players $\mathcal{N}$ play a countably infinite sequence of games. Each game in the sequence is called a *constituent game.* Supergames can be categorized according to whether they are stationary and whether they have time dependence. The supergame is stationary if all the constituent games are identical; and it is time dependent if the payoff to a constituent game is a function of past, as well as current, actions. The study of games with time dependence is deferred to ch. 9.

Aumann (1959, 1960) has studied cooperative supergames with some results relevant to noncooperative games. His work is briefly examined in ch. 12. §2 and §3 of this chapter are based on Friedman (1971a, 1972a). §2 deals with stationary supergames. Noncooperative equilibria are examined and existence is proven for a particular noncooperative equilibrium called the *balanced temptation equilibrium.* In §3 the results of §2 are extended to nonstationary supergames. §4 takes up the application of the supergame results to oligopoly and §5 contains concluding comments.

2. *Balanced temptation equilibria for stationary supergames*

As a starting point, consider a game satisfying G1–G3 and G6. Certain payoff vectors for the game are associated with noncooperative equilib-

[1]See also Telser (1972, pp. 144–145) and Owen (1968, p. 153) where temporal repetition of the prisoner's dilemma is discussed.

ria. It would not be surprising if none of the equilibrium payoff vectors gave Pareto optimal payoffs to the players; however, circumstances change drastically when a given game is made the basis of a supergame (that is, when the particular game is to be repeated indefinitely). It is then possible that many other payoff vectors, including many Pareto optimal payoff vectors, are associated with noncooperative equilibria in the supergame.

To see the mechanism which is in operation, imagine that the game which the players repeatedly play has a unique noncooperative equilibrium having payoffs interior to the payoff space (i.e. which are not Pareto optimal). Now select a point on the payoff frontier which gives a larger payoff to each of the players. It is possible that this payoff vector is associated with a noncooperative equilibrium in the supergame if each of the players has a supergame strategy under which he chooses the strategy corresponding to the frontier point as long as all the other players do the same; but, if some player fails to choose his strategy corresponding to the frontier point in some time period, then, from the next period onward, all the other players choose strategies corresponding to the constituent game noncooperative equilibrium. The contingent plan to move to the constituent game noncooperative equilibrium from the frontier point acts like a threat which may hold the players to choosing the frontier point. These equilibria are examined in §2.2, following §2.1 in which the remaining assumptions of the underlying model are presented. §2.3 deals with the balanced temptation equilibrium, which is an equilibrium satisfying additional conditions to those satisfied by the equilibria of §2.2. Throughout §2 attention is restricted to stationary supergames. These are supergames in which exactly the same constituent game is repeated over and over, and in which the discounted payoff function is also stationary. That is, the discount parameter of the player is a constant, α_i.

2.1. The stationary supergame model

In addition to G1–G3 and G6, several assumptions are made about the nature of the attainable payoff space in the constituent game. Let $\mathscr{H}$ denote the set of attainable payoffs. Then $\mathscr{H} = \{\boldsymbol{P}(s) | s \in \mathscr{S}\}$. The subset of $\mathscr{H}$ containing Pareto optimal payoff points is denoted $\mathscr{H}^*$ and is defined as $\mathscr{H}^* = \{\boldsymbol{P} | \boldsymbol{P} \in \mathscr{H}, \boldsymbol{P}' \in \mathscr{H} \text{ and } \boldsymbol{P}' \neq \boldsymbol{P} \Rightarrow \boldsymbol{P}' \not> \boldsymbol{P}\}$. A payoff vector is said to be Pareto optimal if there is no other payoff vector in the attainable set which gives strictly more to each player. This definition of Pareto optimality allows a payoff vector to be Pareto optimal when there are other payoff vectors giving larger payoffs to some, but not all, of the players and equal payoffs to the rest. Two of the additional assumptions are:

G9. *If* $\mathbf{P}' \ll \mathbf{P}''$ *and* $\mathbf{P}', \mathbf{P}'' \in \mathcal{H}$, *then for* $\mathbf{P}' \leq \mathbf{P} \leq \mathbf{P}''$, $\mathbf{P} \in \mathcal{H}$.

G10. $\mathcal{H}^*$ *is a concave surface. That is, if* $\mathbf{P}', \mathbf{P}'' \in \mathcal{H}^*$ *and* $\mathbf{P}^\lambda = \lambda \mathbf{P}' + (1 - \lambda)\mathbf{P}''$, *there is* $\mathbf{P}^0 \in \mathcal{H}^*$ *such that* $\mathbf{P}^0 \geq \mathbf{P}^\lambda$.

G9 assures that a ray of nonnegative slope through an arbitrary interior point in $\mathcal{H}$ intersects a point in $\mathcal{H}^*$. Figure 8.1 illustrates the effect of both G9 and G10. The cross hatched region, *A*, in the figure is a possible payoff space under G1–G3 and G6. It is a compact and connected set. The effect of G9 is to add the shaded regions, *B*. G10 causes the region *C* to be added to the payoff space. The Pareto optimal set, $\mathcal{H}^*$, consists of the border of the payoff space *abcd*.

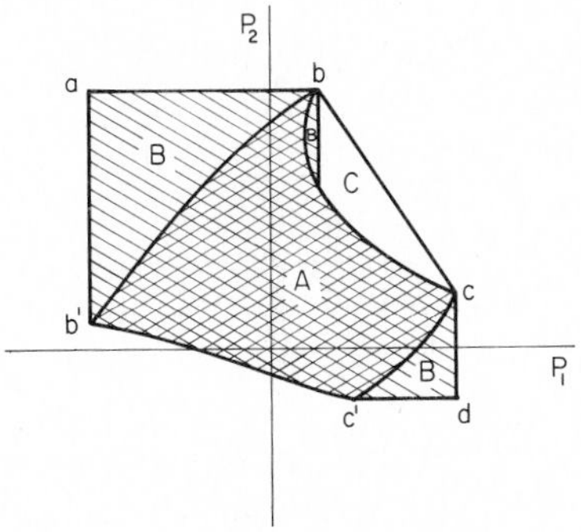

Fig. 8.1

An interesting aside is that if the constituent games were finite, with mixed strategies, the payoff space would be convex, which would imply G10. The roles played by these assumptions should be made apparent below.

There is a final assumption to be made which concerns the relationship between $\mathcal{H}^*$ and the strategies which generate the points in $\mathcal{H}^*$. The inverse mapping of the payoff function $\mathcal{P}^{-1}$ may be defined as

$$\mathcal{P}^{-1}(\mathbf{P}) = \{\mathbf{s} | \mathbf{s} \in \mathcal{S} \text{ and } \mathbf{P}(\mathbf{s}) = \mathbf{P}\}. \tag{8.1}$$

$\mathscr{P}^{-1}(\boldsymbol{P})$ consists of any and all strategies which generate the payoff vector $\boldsymbol{P}$. This correspondence is required to be lower semicontinuous, with convex image sets, for the domain $\mathscr{H}^*$.

G11. *The correspondence $\mathscr{P}^{-1}$ from $\mathscr{H}^*$ into $\mathscr{S}$ is lower semicontinuous with convex image sets.*

Denoting the discount parameter of the ith player by α_i, the supergame payoff functions are

$$\sum_{t=1}^{\infty} \alpha_i^{t-1} P_i(s_t), \qquad i = 1, \ldots, n. \tag{8.2}$$

As with the reaction function models of chs. 4 and 5, it is assumed that at time t, before the tth action is taken, each player knows all past constituent game strategies chosen by all players. Thus choice is simultaneous in each period with full knowledge of the past. Regarding the discount parameters, α_i, two interpretations are possible. One is that the players simply discount future payoffs as indicated. The second is that payoffs are not discounted; but the supergame has a positive probability of ending at any given period. In particular, $1 - \alpha_i$ would be interpreted as the probability estimate of the ith player that the current play is the last one. If all players have the same beliefs or if the probability is given from outside, then $\alpha_i = \alpha_j$ for all i and j. With respect to economic applications, the discount parameter interpretation is usually the more useful.

Taking into account the supergame payoff function and the information conditions which are assumed there is a natural set of supergame strategies to consider. This set turns out to be too large to be of practical use; however, it is well to understand what it is. Let s_{it} be the constituent game strategy choice of the ith player in period t. A typical member of the supergame strategy set consists of an initial action, s_{i1}, a function which chooses s_{i2} as a function of all actions of the first time period, $\boldsymbol{u}_{i2}(s_1)$, a function which chooses s_{i3} as a function of s_1 and s_2, the past actions of both earlier periods, $\boldsymbol{u}_{i3}(s_1, s_2)$, etc. Denoting a supergame strategy for the ith player by $\boldsymbol{\sigma}_i$, this may be written as

$$\boldsymbol{\sigma}_i = (s_{i1}, \boldsymbol{u}_{i2}(s_1), \boldsymbol{u}_{i2}(s_1, s_2), \ldots, \boldsymbol{u}_{it}(s_1, \ldots, s_{t-1}), \ldots). \tag{8.3}$$

The weakest restrictions which might be put on the u_{it} is that their domain is the $t-1$ fold product of $\mathscr{S}$ and, for any $(s_1, \ldots, s_{t-1})$ such that $s_\tau \in \mathscr{S}$, $\tau = 1, \ldots, t-1$, $\boldsymbol{u}_{it}(s_1, \ldots, s_{t-1})$ is in $\mathscr{S}_i$.

2.2. *A large class of noncooperative equilibria for stationary supergames*

The existence of one noncooperative equilibrium for the supergame is immediate. Let s^c denote a strategy vector corresponding to a noncooperative equilibrium for the constituent game. A vector of supergame strategies consisting of the repeated play of s^c; that is, where $u_{it}(s_1, \ldots, s_{t-1}) = s_i^c$, $i = 1, \ldots, n$, $t = 1, 2, \ldots$, is obviously a noncooperative equilibrium for the supergame. The reason is that the action of any player in any time period is entirely independent of all past actions. Thus what a player chooses to do in time t has no effect whatever on the later choices of any other players, and no effect on the player's own payoffs in later periods. The constituent games are, in this case, strategically independent, as well as being structurally independent. Thus a supergame strategy consisting of a sequence of plays of constituent game noncooperative equilibria is a noncooperative equilibrium for the supergame as long as the actions in each time period are independent of actions taken in the past. This result remains true even if the constituent game has many noncooperative equilibria and it is not the same one which is played in each period. Let $\mathcal{S}^c \subset \mathcal{S}$ denote the set of noncooperative equilibria for the constituent game. This proves

Lemma 8.1. *For a stationary supergame satisfying G1–G3 and G6, let supergame strategies* $\boldsymbol{\sigma}_i$ *be given by* $s_{i1} = s'_{i1}$, *and* $s_{it} = u_{it}(s_1, \ldots, s_{t-1}) = s'_{it}$, $t = 2, 3, \ldots$, $i = 1, \ldots, n$. *If* $s'_t \in \mathcal{S}^c$ *for* $t = 1, 2, \ldots$, *then* $\boldsymbol{\sigma}$ *is a noncooperative equilibrium strategy vector for the supergame.*

Now let s^c be a particular member of $\mathcal{S}^c$ such that $\boldsymbol{P}(s^c)$ is not in the Pareto optimal set $\mathcal{H}^*$. There is, as illustrated in fig. 8.2, a subset of the payoff space for the constituent game containing payoff vectors which dominate $\boldsymbol{P}(s^c)$. Members of this subset give at least as much to all players and more to some as compared with $\boldsymbol{P}(s^c)$. Indeed, the interior of this set plus the part of it in $\mathcal{H}^*$, which is not empty if $\boldsymbol{P}(s^c) \notin \mathcal{H}^*$, contains only payoff vectors which give strictly more to each player than he gets at s^c. The shaded area in fig. 8.2 is the set of points which weakly dominate $\boldsymbol{P}(s^c)$.[2] That is the set for which $\boldsymbol{P}(s) \geq \boldsymbol{P}(s^c)$, $s \in \mathcal{S}$. Denote the

[2] A payoff vector $\boldsymbol{P}'$ is said to *dominate* another payoff vector $\boldsymbol{P}''$ if $\boldsymbol{P}' \geq \boldsymbol{P}''$. Domination is *strong* if $\boldsymbol{P}' \gg \boldsymbol{P}''$.

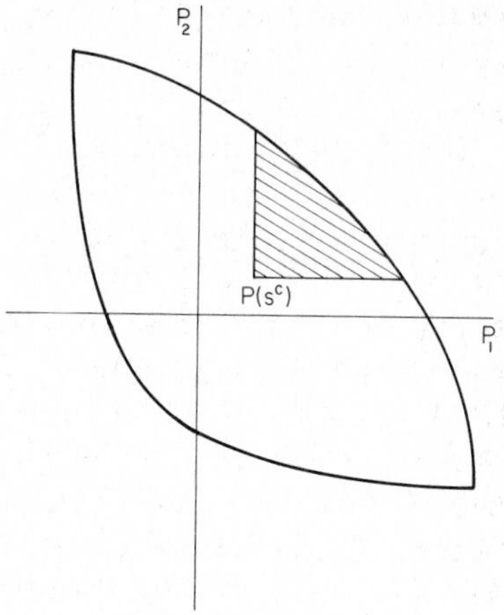

Fig. 8.2

set of corresponding strategies $\mathscr{B}(s^c)$,

$$\mathscr{B}(s^c) = \{s \mid s \in \mathscr{S}, \boldsymbol{P}(s) \geq \boldsymbol{P}(s^c)\}. \tag{8.4}$$

Let $\mathscr{B}'(s^c)$ denote the subset of $\mathscr{B}(s^c)$ for which corresponding payoffs strongly dominate $\boldsymbol{P}(s^c)$,

$$\mathscr{B}'(s^c) = \{s \mid s \in \mathscr{S}, \boldsymbol{P}(s) \gg \boldsymbol{P}(s^c)\}. \tag{8.5}$$

Now a supergame strategy may be defined which, depending on the size of discount parameters, may be a noncooperative equilibrium. Choose $s' \in \mathscr{B}'(s^c)$ and define a strategy, $\boldsymbol{\sigma}'_i$, for the ith player as follows:

$$\begin{aligned} s_{i1} &= s'_i, \\ s_{it} &= s'_i, \quad \text{if } s_{j\tau} = s'_j, \quad j \neq i, \quad \tau = 1, \ldots, t-1, \quad t = 2, 3, \ldots, \\ s_{it} &= s^c_i, \quad \text{otherwise.} \end{aligned} \tag{8.6}$$

The gist of the strategy is this: the player begins by choosing s'_i and continues to choose s'_i in subsequent periods as long as all other players choose s'_j ($j \neq i$). If ever any one player should fail to choose s'_j in, say, period t, then, from period $t+1$ onward, player i always chooses s^c_i.

Theorem 8.2. *Assume a stationary supergame in which G1–G3 and G6 hold. Let s^c be a noncooperative equilibrium strategy vector for the* *constituent game such that* $s^c \notin \mathcal{H}^*$. *Then for an arbitrary* $s' \in \mathcal{B}'(s^c)$, *the corresponding supergame strategy* $\boldsymbol{\sigma}'$, *defined by eqs. (8.6), is a noncooperative equilibrium for the supergame for values of the discount parameters between* $\alpha'_i(s')$ *and one, where* $\alpha'_i(s') < 1$, $i = 1, \ldots, n$.

□ To prove the theorem, values for the discount parameters must be found for which $\boldsymbol{\sigma}'_i$ is a best reply against $\bar{\boldsymbol{\sigma}}'_i$ $(i = 1, \ldots, n)$. Before addressing directly what values of the α_i might suffice, attention is first turned to the best reply of the ith player to $\bar{\boldsymbol{\sigma}}'_i$. There is a simplicity to $\bar{\boldsymbol{\sigma}}'_i$ which implies that there are only two candidates for best reply. When $\bar{\boldsymbol{\sigma}}'_i$ is used by the other players, the ith player receives a payoff of $P_i(s')$ per period as long as he chooses, and continues to choose, s'_i. If he should fail to choose s'_i, say in period t, then from period $t+1$ onward the others choose $\bar{s}^c_i$ irrespective of what player i elects to do. Thus from $t+1$ onward, the ith player can do no better than to choose s^c_i and receive $P_i(s^c)$ per period.

Certain characteristics of a best reply are now clear. It may consist of playing s'_i in every period. If it does not, and at some time $t s'_i$ is not chosen, the best reply from $t+1$ onward is s^c_i. This latter alternative is of the form $\boldsymbol{\sigma}''_i(t) = (s'_i, \ldots, s'_i, s''_i, s^c_i, s^c_i, \ldots)$ where s''_{it} and t are yet to be determined. Note that as long as $s''_i \neq s'_i$, payoffs in all periods other than period t are totally independent of s''_i. The supergame payoff is

$$\sum_{\tau=1}^{t-1} \alpha_i^{\tau-1} P_i(s') + \alpha_i^t P_i(s''_i, \bar{s}'_i) + \sum_{\tau=t+1}^{\infty} \alpha_i^{\tau-1} P_i(s^c). \tag{8.7}$$

Equation (8.7) is maximized, for given t, by choosing s''_i so that

$$P_i(s''_i, \bar{s}'_i) = \max_{s_i \in \mathcal{S}_i} P_i(s_i, \bar{s}'_i). \tag{8.8}$$

Assume now that s''_i is chosen according to eq. (8.8). The best response to $\bar{\boldsymbol{\sigma}}'_i$ is either $\boldsymbol{\sigma}''_i(t)$ or $\boldsymbol{\sigma}'_i$. Due to the stationarity of the model, if $\boldsymbol{\sigma}''_i(t)$ is better, it is best for $t = 1$. The payoffs under $\boldsymbol{\sigma}'_i$ and $\boldsymbol{\sigma}''_i(1)$ are, respectively, $P_i(s')/(1-\alpha_i)$ and $P_i(s''_i, \bar{s}'_i) + \alpha_i P_i(s^c)/(1-\alpha_i)$. The payoff using $\boldsymbol{\sigma}'_i$ is larger if and only if

$$\alpha_i(P_i(s') - P_i(s^c))/(1-\alpha_i) > P_i(s''_i, \bar{s}'_i) - P_i(s'), \tag{8.9}$$

which is equivalent to

$$\alpha_i > (P_i(s''_i, \bar{s}'_i) - P_i(s'))/(P_i(s''_i, \bar{s}'_i) + P_i(s^c)). \tag{8.10}$$

The right-hand expression in eq. (8.10) is clearly nonnegative and less than one because $P_i(s''_i, \bar{s}'_i)$ cannot be less than $P_i(s')$ and $P_i(s')$ is larger than $P_i(s^c)$. Thus for

$$\alpha_i > \alpha'_i(s') = (P_i(s''_i, \bar{s}'_i) - P_i(s'))/(P_i(s''_i, \bar{s}'_i) - P_i(s^c)), \qquad i = 1, \ldots, n, \qquad (8.11)$$

$\boldsymbol{\sigma}'$ is a supergame strategy vector corresponding to noncooperative equilibrium for the supergame. □

What is crucial with respect to the strategies $\boldsymbol{\sigma}''_i$ and $\boldsymbol{\sigma}'_i$ is clear and is illustrated in fig. 8.3. To use $\boldsymbol{\sigma}''_i$ gives the player, in one period only, an extra payoff of $P_i(s''_i, \bar{s}'_i) - P_i(s')$ as compared with using $\boldsymbol{\sigma}'_i$. That one period gain is followed by a loss of $P_i(s') - P_i(s^c)$ in each succeeding period. $\boldsymbol{\sigma}'_i$ is superior when the losses, discounted back to the period when s''_i is used, are larger than the one time gain. In fig. 8.3 the one period gain for player 2 is ab and the per period loss afterward is bc. Thus for suitably high values for the discount parameters, any $s \in \mathcal{B}'(s^c)$ may be associated with a supergame noncooperative equilibrium in which the payoffs are at the rate of $\boldsymbol{P}(s)$ per period. Definitely included are payoffs which are Pareto optimal, even when the constituent game has no noncooperative equilibrium associated with Pareto optimal payoffs.

2.3. *The balanced temptation equilibrium*

The number of potential equilibria associated with points in $\mathcal{B}'(s^c)$ is very large; however, there are natural ways to narrow them down. Consider the set $\mathcal{B}^*(s^c)$ consisting of constituent game strategies in $\mathcal{B}(s^c)$ with associated payoffs which are Pareto optimal. $\mathcal{B}^*(s^c) = \{s | s \in \mathcal{B}(s^c), \boldsymbol{P}(s) \in \mathcal{H}^*\}$. I want to go one step further and propose an equilibrium called the *balanced temptation equilibrium*. Let $s' \in \mathcal{B}(s^c)$ and define s'' by eq. (8.8). s' is said to have the *balanced temptation property* if $\alpha'_i(s') = \alpha'_j(s')$ for $i, j = 1, \ldots, n$. Figure 8.4 shows a point $\boldsymbol{P}(s')$ which has the balanced temptation property. One way to view the property is that it is characterized by all players having the same "break even" discount parameter. (i.e. the $\sigma'_i(s')$ are the same for all i.) Another, equivalent, property is that

$$(P_i(s''_i, \bar{s}'_i) - P_i(s'))/(P_i(s') - P_i(s^c))$$
$$= (P_j(s''_j, \bar{s}'_j) - P_j(s'))/(P_j(s') - P_j(s^c)), \qquad i, j = 1, \ldots, n. \quad (8.12)$$

That is, the ratio of the one-period gain to the per period later loss is the

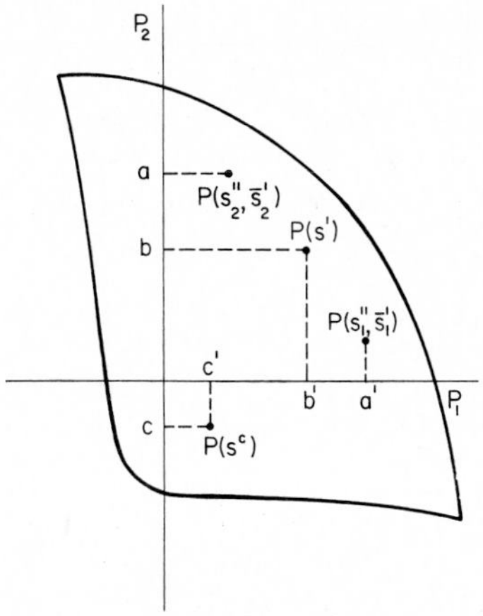

Fig. 8.3

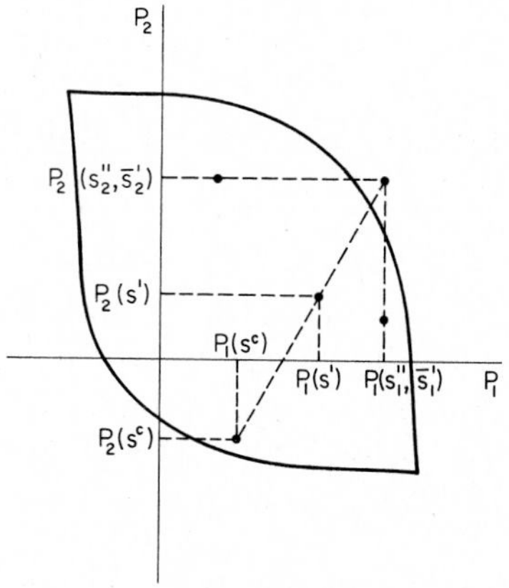

Fig. 8.4

same for all players. Eq. (8.12) could be regarded as a measure of the temptation of a player to choose the $\boldsymbol{\sigma}_i''$ strategy.

The *balanced temptation equilibrium* is a supergame noncooperative equilibrium of the type outlined in theorem 8.2 which satisfies the additional conditions that $s' \in \mathscr{B}^*(s^c)$ and s' has the balanced temptation property. Under the assumptions used in theorem 8.2 it is not possible to prove the existence of a balanced temptation equilibrium because the players' discount parameters may be too low. It is possible to prove the existence of a point $\mathscr{B}^*(s^c)$ which has the balanced temptation property. Call such a point s^*. Then s^*, playing the role of s' in theorem 8.2, is associated with a balanced temptation equilibrium if the players' discount parameters are sufficiently high.

Prior to proving the main theorem, a needed intermediate result is established. Let

$$\phi_i(\bar{s}_i) = \max_{s_i \in \mathscr{S}_i} P_i(s_i, \bar{s}_i), \qquad \text{for } \bar{s}_i \in \bar{\mathscr{S}}_i \text{ and } i = 1, \ldots, n. \tag{8.13}$$

ϕ_i might be called the best reply payoff of the ith player.

Lemma 8.3. *Given G1–G3, the functions ϕ_i, $i = 1, \ldots, n$, are continuous.*

□ Proof follows quite directly from continuity of the P_i. Let $\bar{s}_i^0$ be an arbitrary point in $\bar{\mathscr{S}}_i$, s_i^0 be chosen so that $P_i(s_i^0, \bar{s}_i^0) = \phi_i(\bar{s}_i^0)$ and $\bar{s}_i^l$, $l = 1, 2, \ldots$, be a sequence of points in $\bar{\mathscr{S}}_i$ such that $\bar{s}_i^l \to \bar{s}_i^0$ as $l \to \infty$. By definition of ϕ_i, there is a s_i^l associated with $\bar{s}_i^l$ such that $P_i(s_i^l, \bar{s}_i^l) = \phi_i(\bar{s}_i^l)$, $l = 1, 2, \ldots$. To say that ϕ_i is continuous is to say that $\lim_{l\to\infty} \phi_i(\bar{s}_i^l) = \phi_i(\bar{s}_i^0)$. From continuity of P_i, $\lim_{l\to\infty} P_i(s_i^0, \bar{s}_i^l) = P_i(s_i^0, \bar{s}_i^0) = \phi_i(\bar{s}_i^0)$; therefore, $\lim_{l\to\infty} \phi_i(\bar{s}_i^l) \geq \phi_i(\bar{s}_i^0)$. But if strict inequality held, $\lim_{l\to\infty} \phi_i(\bar{s}_i^l) = P_i(s_i', \bar{s}_i^0) > \phi_i(\bar{s}_i^0)$ where s_i' is a cluster point of the sequence s_i^l, $l = 1, 2, \ldots$, which contradicts the definition of $\phi_i(\bar{s}_i^0)$. □

Note that if a point s^* has the balanced temptation property, it has it relative to a particular constituent game noncooperative equilibrium. In the discussion following lemma 8.1, s^c is taken to be a particular element of $\mathscr{S}^c$, and the equilibria of theorem 8.2 as well as the definition of the balanced temptation property are relative to s^c.

Theorem 8.4. *Assume a stationary supergame satisfying G1–G3, G6, G9, and G11. Let s^c be a noncooperative equilibrium strategy vector for the constituent game such that $s^c \notin \mathscr{H}^*$. There is $s^* \in \mathscr{B}^*(s^c)$ which has the balanced temptation property.*

□ The method of proof to be employed is to construct a function which carries each Pareto optimal point in $\mathscr{H}$ which dominates $\boldsymbol{P}(s^c)$ into another such point. The function is chosen so that any fixed point has the balanced temptation property.

Let $\mathscr{H}^{**}(s^c)$ denote the subset of $\mathscr{H}^*$ which weakly dominates $\boldsymbol{P}(s^c)$. Then each element of $\mathscr{B}^*(s^c)$ generates an element of $\mathscr{H}^{**}(s^c)$, all the strategies which generate members of $\mathscr{H}^{**}(s^c)$ are in $\mathscr{B}^*(s^c)$ and all elements of $\mathcal{H}^{**}(s^c)$ are generated by members of $\mathcal{B}^*(s^c)$. By G11 the mapping $\mathcal{P}^{-1}$ from $\mathcal{H}^{**}(s^c)$ onto $\mathcal{B}^*(s^c)$ is lower semicontinuous and each set $\mathcal{P}^{-1}(\boldsymbol{P})$ is convex. Therefore $\mathcal{P}^{-1}$ admits a continuous selection. That is, there is a function $\boldsymbol{h}(\boldsymbol{P})$ from $\mathcal{H}^{**}(s^c)$ into $\mathcal{B}^*(s^c)$ which is continuous and such that for any $\boldsymbol{P} \in \mathcal{H}^{**}(s^c)$, $\boldsymbol{h}(\boldsymbol{P}) \in \mathcal{P}^{-1}(\boldsymbol{P})$. See lemma 1.1 of Parthasarathy (1972) or Michael (1956).[3]

G9 insures for any $\boldsymbol{\rho} = (\rho_1, \rho_2, \ldots, \rho_n) > 0$ with $\Sigma_{i=1}^n \rho_i = 1$, there is exactly one Pareto optimal point, $\boldsymbol{P} \in \mathcal{H}^{**}(s^c)$, such that

$$(P_i - P_i(s^c))/\left(\sum_{j=1}^{n} [P_j - P_j(s^c)]\right) = \rho_i, \qquad i = 1, \ldots, n. \tag{8.14}$$

Equation (8.14) describes a function $g(\boldsymbol{\rho})$ which is continuous and one to one from the $n-1$ dimensional unit simplex onto $\mathcal{H}^{**}(s^c)$.

It is now possible to use the functions g, $\boldsymbol{h}$, and $\boldsymbol{\phi}$ to define a function from the unit simplex to itself. Let

$$\Omega_i(\boldsymbol{\phi}) = (\phi_i - P_i(s^c))/\left[\sum_{j=1}^{n} (\phi_j - P_j(s_j^c))\right], \qquad i = 1, \ldots, n. \tag{8.15}$$

$\boldsymbol{\Omega}$ is a member of the $n-1$ dimensional unit simplex; hence, g takes $\boldsymbol{\rho}$ into $\boldsymbol{P}$, $\boldsymbol{h}$ takes $\boldsymbol{P}$ into s, $\boldsymbol{\phi}$ takes s into $\boldsymbol{\phi}$ and $\boldsymbol{\Omega}$ takes $\boldsymbol{\phi}$ into $\boldsymbol{\rho}$. That is, the function $\boldsymbol{V}(\boldsymbol{\rho}) = \boldsymbol{\Omega} \circ \boldsymbol{\phi} \circ \boldsymbol{h} \circ g(\boldsymbol{\rho})$ is from the simplex to itself. It is continuous if $\boldsymbol{\Omega}$, $\boldsymbol{\phi}$, $\boldsymbol{h}$ and g are continuous. All except $\boldsymbol{\Omega}$ are already known to be continuous, and the continuity of $\boldsymbol{\Omega}$ is assured because $\boldsymbol{\phi}$ is continuous and the denominator of eq. (8.15) is bounded strictly away from zero. By the Brouwer fixed point theorem, the function $\boldsymbol{V}$ has a fixed point.

It remains to show that a fixed point of $\boldsymbol{V}$ is associated with a point having the balanced temptation property. Let $\boldsymbol{\rho}^*$ be a fixed point of $\boldsymbol{V}$. $\boldsymbol{P}^* = g(\boldsymbol{\rho}^*)$ is a vector of Pareto optimal payoffs in $\mathcal{H}^{**}(s^c)$ which are uniquely associated with $\boldsymbol{\rho}^*$. $\boldsymbol{P}^*$ can be generated by the strategy vector $s^* = \boldsymbol{h}(\boldsymbol{P}^*)$. $\boldsymbol{\phi}$ is the vector of best response payoffs to s^*, and, because $\boldsymbol{\rho}^* = \boldsymbol{\Omega}(\boldsymbol{\phi}(s^*))$ we have

$$\begin{aligned}(\phi_i(\bar{s}_i^*) - P_i(s^c))/&\left[\sum_{j=1}^{n} (\phi_j(\bar{s}_j^*) - P_j(s^c))\right] \\ &= (P_i(s^*) - P_i(s^c))/\left[\sum_{j=1}^{n} (P_j(s^*) - P_j(s^c))\right] = \rho_i, \qquad i = 1, \ldots, n.\end{aligned} \tag{8.16}$$

[3]Rather than assume G11 which implies the existence of a continuous selection, G11 could instead have stated that the mapping $\mathcal{P}^{-1}$ from $\mathcal{H}^{**}$ to $\mathcal{B}^*$ is a continuous function.

Equation (8.16) follows from the fixed point property, and it may be written as

$$(\phi_i(\bar{s}_i^*) - P_i(s^c))/(P_i(s^*) - P_i(s^c))$$
$$= \left[\sum_{j=1}^{n} (P_j(s^*) - P_j(s^c))\right] \Big/ \left[\sum_{j=1}^{n} (\phi_j(\bar{s}_j^*) - P_j(s^c))\right], \quad i = 1, \ldots, n. \tag{8.17}$$

Note that the right-hand side of eq. (8.17) is the same for all i; hence, the left-hand sides are equal for all i. This equality is what defines the balanced temptation property. Thus a fixed point of $\boldsymbol{V}$ is associated with a strategy vector s and Pareto optimal payoffs $\boldsymbol{P}(s)$ which have the balanced temptation property. As $\boldsymbol{V}$ has a fixed point, there is a payoff vector in $\mathcal{H}^{**}(s^c)$ which has the balanced temptation property. □

The following theorem on existence of a balanced temptation equilibrium is obvious, given theorem 8.4.

Theorem 8.5. *Given a supergame satisfying the assumptions of theorem 8.4 and $s^* \in \mathcal{H}^{**}(s^c)$ having the balanced temptation property, s^* is associated with a balanced temptation equilibrium if*

$$\alpha_i > (\phi_i(\bar{s}_i^*) - \boldsymbol{P}_i(s^*))/(\phi_i(\bar{s}_i^*) - \boldsymbol{P}_i(s^c)), \qquad i = 1, \ldots, n. \tag{8.18}$$

Furthermore, if G10 also holds, the supergame payoffs associated with the balanced temptation equilibrium are Pareto optimal.

Of course, the right-hand side of eq. (8.18) has the same value for all players because $\boldsymbol{P}(s^*)$ has the balanced temptation property.

Other balanced temptation equilibria are possible within the present model if the constituent game noncooperative equilibrium is not unique. The balanced temptation equilibrium shown in theorem 8.5 is defined in relation to a particular noncooperative equilibrium, s^c. Now let $s_t^c \in \mathcal{S}^c$, $t = 1, 2, \ldots$, be a sequence of constituent game noncooperative equilibria. It need not be, of course, that all the s_t^c are distinct from one another. Let $s_t^* \in \mathcal{H}^{**}(s_t^c)$ $t = 1, 2, \ldots$, be chosen so that s_t^* has the balanced temptation property in relation to s_t^c. Then the following theorem is obvious.

Theorem 8.6. *Given a supergame satisfying the assumptions of theorem 8.4, a sequence of constituent game noncooperative equilibria, $s_t^c \in \mathcal{S}^c$, $t = 1, 2, \ldots$, and a sequence of points $s_t^* \in \mathcal{H}^{**}(s_t^c)$, $t = 1, 2, \ldots$, with the*

s_t^* *having the balanced temptation property in relation to* s_t^c, *the* s_t^*, $t = 1, 2, \ldots$, *are associated with a balanced temptation equilibrium if*

$$\sum_{\tau=t}^{\infty} \alpha^{\tau-t} P_i(s_\tau^*) > \phi_i(\bar{s}_{it}^*) + \sum_{\tau=t+1}^{\infty} \alpha_i^{\tau-t} P_i(s_\tau^c),$$

$$i = 1, \ldots, n, \quad t = 1, 2, \ldots. \tag{8.19}$$

The condition given in eq. (8.19) is that there is no time period for which the one-period gain obtained by deviating from s_{it}^* is at least as large as the discounted stream of foregone payoffs which would result from obtaining the one-period gain.

A balanced temptation equilibrium under the conditions of theorem 8.5 gives the players supergame payoffs which are Pareto optimal; however, under the conditions of theorem 8.6, supergame payoffs need not be Pareto optimal even though each constituent game payoff vector is. This is illustrated in fig. 8.5 where the payoff $\boldsymbol{P}(s^*)$ may be associated with even numbered periods and $\boldsymbol{P}(s^{**})$ with odd numbered. The payoff $\boldsymbol{P}(s')$ is larger in both components than the mean of $\boldsymbol{P}(s^*)$ and $\boldsymbol{P}(s^{**})$. This example is slightly muddied by the presence of discounting; however, it is clear that discount parameters and payoff functions could easily be chosen in a way which would ensure the absence of Pareto optimality for the supergame payoffs.

Turn now to the assumptions G9–G11 to see what role they play and whether they may be dispensed with. G11 assures that the function g

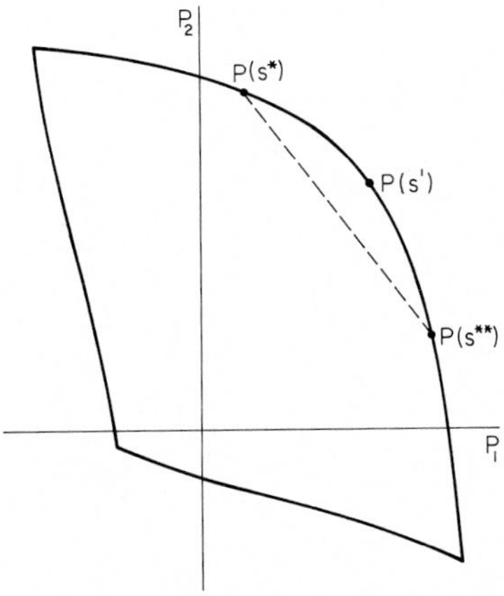

Fig. 8.5

exists and is continuous; hence, to dispense with it would mean being unable to prove the existence of points having the balanced temptation property. G9 and G10 could be weakened with the only casualty being the Pareto optimality of the balanced temptation equilibrium of theorem 8.5. G10 could be dropped and G9 weakened to only require that a nonnegatively sloped ray through a point in $\mathcal{S}^c$ must, in the "upward" direction, intersect the boundary of the payoff set in only one place. Multiple intersections, which must be ruled out, are illustrated in fig. 8.6. Again, multiple intersections make it impossible to have the continuous function g.

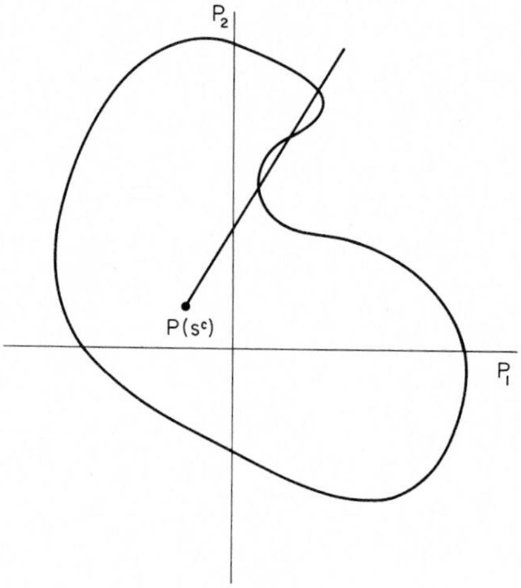

Fig. 8.6

3. Balanced temptation equilibria for nonstationary supergames

It is clear from the discussion of Pareto optimality which follows theorem 8.6 that balanced temptation equilibria for nonstationary supergames do not, in general, yield Pareto optimal supergame payoffs. Thus G10 may be costlessly dropped from the list of assumptions to be used in this section. The nonstationary supergame is based upon a sequence of constituent games $(\mathcal{N}, \mathcal{S}_t, \mathbf{S}_t)$, $t = 1, 2, \ldots$, each of which is assumed to satisfy G1–G3, G6, G9 and G11. It is no longer necessary to assume the players' discount parameters to be constant over time. α_{it} denotes the value placed in period 1 on a unit of payoff to be received in period t. The supergame payoff is now

$$\sum_{t=1}^{\infty} \alpha_{it} P_{it}(s_t). \tag{8.20}$$

The supergame is now no longer characterized by a set of players who engage in a (countable) temporal repetition of a given game. Instead, the players engage in a sequence of games, no two of which need be the same. As before, it is an unchanging set of players throughout, and there are a countably infinite number of plays. Two additional assumptions, needed to assure the boundedness of supergame payoffs are:

G12. *The constituent game payoff functions, P_{it}, are bounded on $\mathcal{S}_t$ independently of t, $i = 1, \ldots, n$.*

G13. *The discount parameters are declining over time and have a bounded sum. $\alpha_{i1} = 1$, $\alpha_{it} \geq \alpha_{i,t+1} > 0$, $t = 1, 2, \ldots$, and $\sum_{t=1}^{\infty} \alpha_{it} < \infty$.*

While G12 and G13 do guarantee that supergame payoffs are bounded, it would be possible to accomplish this in other ways. One example is merely to assume the α_{it} are positive and that $\alpha_{it}P_{it}$ is bounded independently of t. That the α_{it} are declining with time plays no role in creating bounds; however, it is natural to assume in view of the interpretation of the α_{it} as discount parameters.

All the various sets defined in §2 such as $\mathcal{H}$, $\mathcal{H}^*$, $\mathcal{S}^c$, $\mathcal{B}(s^c)$, $\mathcal{B}'(s^c)$, $\mathcal{B}^*(s^c)$ and $\mathcal{H}^{**}(s^c)$ are considered defined for the constituent games in the present section. Thus $\mathcal{H}_t$ denotes the set of attainable payoffs in the tth game, $\mathcal{H}_t^*$, the Pareto optimal payoffs, etc. Also, the functions $\boldsymbol{\phi}_t$ may be defined analogously to $\boldsymbol{\phi}$ of §2. It is an easy matter to give a theorem on existence of a balanced temptation equilibrium for the nonstationary supergame.

Theorem 8.7. *Given a supergame whose constituent games satisfy G1–G3, G6, G9, G11–G13, a sequence of constituent game noncooperative equilibria, $s_t^c \in \mathcal{S}_t^c$, $t = 1, 2, \ldots$, and a sequence of points $s_t^* \in \mathcal{H}_t^{**}(s_t^c)$, $t = 1, 2, \ldots$, where $\mathbf{P}_t(s_t^*)$ has the balanced temptation property relative to s_t^c, the supergame has a balanced temptation equilibrium if*

$$\sum_{\tau=t}^{\infty} \alpha_{i\tau} P_{i\tau}(s_\tau^*) > \alpha_{it}\phi_{it}(s_{it}^*) + \sum_{\tau=t+1}^{\infty} \alpha_{i\tau} P_{i\tau}(s_t^c),$$

$$i = 1, \ldots, n, \qquad t = 1, 2, \ldots. \tag{8.21}$$

□ First note that by theorem 8.4, s_t^* exist which have the balanced temptation property relative to s_t^c. The condition in eq. (8.21) is that in each time period, for all players, it is more profitable to use the strategy outlined in eq. (8.6) than to do anything else. □

As long as $\mathscr{S}_t^c$ has more than one element for at least some t or if some of the constituent games have more than one point possessing the balanced temptation property relative to s_t^c, then the balanced temptation equilibrium is not a solution if the players are denied the opportunity to communicate. This is discussed in ch. 7 §4.

4. Applications to oligopoly models

There is one particular lesson of very great importance which comes from theorem 8.2. That is the possibility of a noncooperative equilibrium at which the firms receive Pareto optimal profits. Such an equilibrium requires no collusion in order to be maintained and the fact of its existence ought to dispel the erroneous notion that if firms are making profits above the Cournot equilibrium levels, then they are necessarily colluding. I do not wish to maintain that collusion cannot or does not exist or that profits above Cournot levels must necessarily stem from non-cooperative behavior of the sort studied in §2 and §3. The point is that such profits are not evidence, by themselves, on the presence or absence of collusion. It is an empirical question of great importance to determine whether firms do collude, and, if they do, in what fashion.

There is another unanswered question concerning these supergame equilibria which is both theoretical and empirical. That is how the players in a game or firms in a market might manage to jointly choose equilibrium strategies for the same equilibrium when equilibrium is not unique. One obvious way is for the firms to agree in advance to a particular equilibrium. It is pointed out above (in ch. 7 §4) that such an agreement is self enforcing precisely because it is an agreement on a noncooperative equilibrium. For countries like the United States in which collusion among firms is not legal, the necessary communication is believable even though collusion requiring outside enforcement is not believable. The latter is not believable because there is no outside mechanism or agency which may be brought to bear on a violator. The former type of agreement only requires verbal assent from everyone due to being self enforcing; so there need be no related documents or records. Again, it is an empirical question if such agreements are made.

Related to agreements on noncooperative equilibria is the question whether they constitute collusion, or noncollusive behavior. One might wish to call them collusive because they result from explicit (though nonbinding) agreement; or one might prefer to call them noncollusive because the equilibrium is noncooperative and noncooperative equilibria should not be called collusive. The answer does not seem to me clear cut. Perhaps in some circumstances one makes more sense and in others, the other.

Another related question is whether they can be attained other than by accident or agreement. Here no solution presents itself. That is, I know of no theoretical work which can be used to tell how a group of firms could behave noncooperatively, without communication, and choose or find their way to a Pareto optimal supergame noncooperative equilibrium. It is a mistake to rule out such a possibility merely because we do not know at present how to account for it. Again, I am not convinced one way or the other. It seems wisest to keep an open mind.

5. Concluding comments

The supergames approach taken in this chapter is an alternative to the reaction function models of ch. 5. In a sense it is more successful in that noncooperative equilibria are found for the supergames which are additional to those which are possible in the single-period models. It is clear that these equilibria, other than the balanced temptation equilibrium, may be found to exist in oligopoly models as readily as in the game models of the present chapter. Showing existence of the balanced temptation equilibrium in the oligopoly models is not an easy matter and has not been done. The problem lies in satisfying G9 and G11 or some suitable variant of them. Meanwhile, not enough is known of the characteristics of the profit possibility frontier in oligopoly models.

Of very great interest in the application of the supergames results to oligopoly models is the existence (for suitable values of the discount parameters) of a noncooperative equilibrium in which the payoffs are Pareto optimal even though the noncooperative equilibria of the single period model are not Pareto optimal. It should be further noted that the supergame model deals relatively easily with supergames in which the payoff functions change arbitrarily from period to period; whereas, the reaction function models require stationary payoffs. It is, of course, no problem to apply a suitable version of theorem 8.2 to oligopoly models with arbitrarily changing payoff functions.

Although the reaction function approach is quite old and holds considerable fascination for people working the oligopoly vein, it appears that the application of the supergames approach strikes richer ore.

CHAPTER 9

NONCOOPERATIVE SUPERGAMES HAVING A TIME DEPENDENT STRUCTURE

1. *Time dependent supergames*

1.1. *An overview*

The sense in which time dependence is lacking in the models of ch. 8 is that the payoff to each constituent game depends directly on actions taken in the time period associated with that game and on nothing else. It is important to distinguish between time dependent structure, which is now being considered, and time dependent behavior. The supergame strategies given by eq. (8.6) provide an example of time dependent behavior in games which are not structurally time dependent. Reaction functions in the models of ch. 5 are another example. To see what is meant by a time dependent structure, imagine a supergame in which the players take an action in each period. That action is the constituent game strategy. Say that there is a payoff in each period; and that it is a function of the actions of both the current and the immediately past period. In such a supergame the time dependence is structural and is therefore present no matter how the players choose to behave. The oligopoly model of ch. 6 §6 has a time dependent structure.

In ch. 8 §2.1 it is seen that supergame strategies can be quite complicated; and that interesting results follow from considering some of these time dependent strategies. Indeed, the converse is also true; namely that if attention were restricted to strategies which lacked time dependence, obtainable results would be only a trivial extension of the results for one period models. When a supergame strategy is of the form $\boldsymbol{\sigma}_i = (s_{i1}, s_{i2}, \ldots)$, where there is no time dependence in behavior, so in period t, s_{it} is chosen no matter what had been done by anyone before, the strategy is called *simple*. Restricting to simple strategies in time independent supergames, it is obvious that $\boldsymbol{\sigma} = (s_1, s_2, \ldots)$ corresponds to a noncooperative equilibrium if and only if s_t corresponds to a noncooperative equilibrium for the tth constituent game ($t = 1, 2, \ldots$).

Results do not fall out so easily for time dependent supergames, even with strategies restricted to be simple. It may not be obvious what is to be meant by a noncooperative equilibrium for the constituent game; however, should that be settled satisfactorily, the existence of constituent game noncooperative equilibria need not establish existence of noncooperative equilibria for the supergame itself. These questions are addressed in §2 for stationary supergames. Another natural question which arises for stationary supergames is whether there are *steady state* equilibria. These are noncooperative equilibria for which the players choose the same constituent game strategies in each time period. These are also considered in §2, where conditions are given under which the steady state equilibrium exists and is unique. A further result is that, in such models, any noncooperative equilibrium strategy converges to the steady state constituent game strategy. That is, if $\boldsymbol{\sigma}' = (s'_1, s'_2, s'_3, \ldots)$ is a noncooperative equilibrium then s'_t converges to s^c as $t \to \infty$ and s^c is the steady state constituent game strategy.

In §3 nonstationary supergames are investigated still restricting attention to simple strategies. For one class of games which are the direct generalization of the stationary games of §2, existence of noncooperative equilibrium is proved. Then a different class of games is investigated for which a *weak noncooperative equilibrium* is found. It is not until §4 that other strategies than simple strategies are considered. It is attempted to generalize the balanced temptation equilibrium to time dependent supergames. §§1–4 are based on Friedman (1974). §5 is devoted to the applications of the results of earlier sections to oligopoly models; and §6 contains summary comments. The remainder of the present section presents the model which is used throughout much of the chapter.

1.2. The game model

As in ch. 8, the supergame is defined by building it up from constituent games which are played in sequence. There is, for each player, a strategy set $\mathcal{S}_{it}$ containing the available strategies for the tth time period, and a payoff function, $P_{it}(s_t, s_{t-1})$, giving the period t payoff of the ith player. Note that the payoff in period t also depends on the strategy vector chosen in period $t-1$. The supergame payoff of the ith player is

$$\sum_{t=1}^{\infty} \alpha_{it} P_{it}(s_t, s_{t-1}) = F_i(\boldsymbol{\sigma}), \qquad i = 1, \ldots, n. \tag{9.1}$$

Assumptions which are used in §2 include:

S1. *There are a finite number of players, n.*

S2. *$\mathcal{S}_{it}$, the strategy set of the ith player in the tth time period, is a compact, convex subset of $\mathcal{R}^m$, $i = 1, \ldots, n$, $t = 1, 2, \ldots$. The strategy sets are bounded independently of t. Thus $\|\mathcal{S}_{it}\| \leq \bar{m} < \infty$, $i = 1, \ldots, n$, $t = 1, 2, \ldots$.*[1]

S3. *$P_{it}(s_t, s_{t-1})$ is continuous on $\mathcal{S}_t \times \mathcal{S}_{t-1}$. For any $(s_t, s_{t-1}) \in \mathcal{S}_t \times \mathcal{S}_{t-1}$, $|P_{it}(s_t, s_{t-1})| \leq \bar{P} < \infty$, $i = 1, \ldots, n$, $t = 1, 2, \ldots$.*

S4. *The payoff functions $P_{it}(s_{it}, \bar{s}_{it}, s_{i,t-1}, \bar{s}_{i,t-1})$ are concave with respect to $(s_{it}, s_{i,t-1}) \in \mathcal{S}_{it} \times \mathcal{S}_{i,t-1}$, $i = 1, \ldots, n$, $t = 1, 2, \ldots$.*

S5. *The discount parameters of the player are declining over time and bounded in sum. $1 = \alpha_{i1} > \alpha_{it} > \alpha_{i,t+1}$, $t = 2, 3, \ldots$. $\sum_{t=1}^{\infty} \alpha_{it} < \infty$, $i = 1, \ldots, n$.*

These assumptions repeat assumptions made in ch. 8, with necessary changes to adapt them to time dependent supergames. S1 is identical to G1 and S5 to G13. Apart from the assumption of concavity of the payoff functions in the player's own strategies in S4, rather than quasi-concavity, the remaining assumptions differ from assumptions used in ch. 8 only by the need to account for the time dependence.

In time dependent supergames, as in the reaction function oligopoly models, there is an initial condition which appears in the first-period payoff function. If first-period payoff is of the form $P_{i1}(s_1, s_0)$ and the players are regarded as choosing s_{i1}, s_{i2}, etc., then s_0 is an initial condition for the model. A supergame is described by the set of players, $\mathcal{N}$, the strategy sets, $\mathcal{S}_t$, $t = 1, 2, \ldots$, the initial condition s_0 and the payoff functions, $\boldsymbol{P}_t$, $t = 1, 2, \ldots$. A simple supergame strategy, $\boldsymbol{\sigma}_i$, is of the form $\boldsymbol{\sigma}_i = (s_{i0}, s_{i1}, s_{i2}, \ldots)$ where s_{i0} is part of the initial conditions; hence

[1]Recall the definition of the norm $\|\mathcal{S}_{it}\|$. $\|\mathcal{S}_{it}\| = \sup_{s', s'' \in \mathcal{S}_{it}} \|s' - s''\|$, where $\|s' - s''\| = \max_j |s'_j - s''_j|$.

not an object of choice by the ith player. The set of simple supergame strategies available to the ith player is $\mathscr{S}_i = s_{i0} \times \mathscr{S}_{i1} \times \mathscr{S}_{i2} \times \cdots$, and the joint strategy set of the n players is $\mathscr{S} = \mathscr{S}_1 \times \mathscr{S}_2 \times \cdots \times \mathscr{S}_n$.[2] Thus a supergame in which players are restricted to simple strategies is characterized by the set of players $\mathscr{N}$, the initial condition s_0, the strategy sets for the constituent games $\mathscr{S}_t$, the payoff functions for the constituent games $\boldsymbol{P}_t$ and the discount parameters $\boldsymbol{\alpha}_t$. This may be written concisely: $\{(\mathscr{N}, s_0, \mathscr{S}_t, \boldsymbol{P}_t, \boldsymbol{\alpha}_t),\ t = 1, \ldots, \infty\}$.

2. *Noncooperative equilibria in stationary supergames*

2.1. Simple strategy equilibria and other characteristics of the games

A supergame satisfying S1–S5 has a simple strategy noncooperative equilibrium; however, that theorem is not formally stated and proved until §3. In this section, a similar theorem is stated for stationary supergames. Proof is omitted because the theorem is a special case of theorem 9.6 in §3. There is an alternative way to write the supergame payoff function which is introduced later in this section and a useful result is gotten from it which is helpful subsequently.

Theorem 9.1. *A stationary supergame satisfying S1–S5 has a noncooperative equilibrium in the set of simple strategies.*

In connection with theorem 9.1, it is worth noting that existence of an equilibrium in simple strategies does not require that the players be formally restricted to the set of simple strategies. Upon a little reflection, it is clear that for any given strategies for the other players, if the ith player has a best reply, then he has a best reply in the set of simple strategies. This is not to say that his best replies consist only of simple strategies; but merely that his best replies to a given configuration of strategy choices of his rivals always include at least one simple strategy. Therefore if a game has an equilibrium when players are restricted to simple strategies, such an equilibrium is also an equilibrium when the game is changed by enlarging the set of available strategies.

[2]The notation $\mathscr{S}_5$ could mean $s_{5,0} \times \mathscr{S}_{5,1} \times \cdots$, the strategy set of the 5th player or $\mathscr{S}_{1,5} \times \mathscr{S}_{2,5} \times \cdots \times \mathscr{S}_{n,5}$, the strategies of the 5th time period. Context should make clear which is meant.

As this section deals only with stationary supergames, the discount parameters are of the form $\alpha_{it} = \alpha_i^{t-1}$. The payoff structure of the game may be written[3]

$$\begin{aligned}\sum_{t=1}^{\infty} \alpha_i^{t-1} P_i(s_t, s_{t-1}) = F_i(\boldsymbol{\sigma}) \\ &= [P_i(s_1, s_0) + \alpha_i P_i(s_2, s_1)] \\ &\quad + \alpha_i^2[P_i(s_3, s_2) + \alpha_i P_i(s_4, s_3)] + \cdots \\ &= \sum_{t=1}^{\infty} \alpha_i^{2t-2}[P_i(s_{2t-1}, s_{2t-2}) + \alpha_i P_i(s_{2t}, s_{2t-1})].\end{aligned} \tag{9.2}$$

Letting

$$P_i^*(s_{t+1}, s_t, s_{t-1}) = P_i(s_t, s_{t-1}) + \alpha_i P_i(s_{t+1}, s_t), \tag{9.3}$$

eq. (9.2) may be rewritten as

$$\begin{aligned}F_i(\boldsymbol{\sigma}) &= \sum_{t=1}^{\infty} \alpha_i^{2t-2} P_i^*(s_{2t}, s_{2t-1}, s_{2t-2}) \\ &= P_i(s_1, s_0) + \sum_{t=1}^{\infty} \alpha_i^{2t-1} P_i^*(s_{2t+1}, s_{2t}, s_{2t-1}),\end{aligned} \tag{9.4}$$

or

$$F_i(\boldsymbol{\sigma}) = \frac{1}{2}\left[P_i(s_1, s_0) + \sum_{t=1}^{\infty} \alpha_i^{t-1} P_i^*(s_{t+1}, s_t, s_{t-1})\right]. \tag{9.5}$$

The function $P_i^*(s_{t+1}, s_t, s_{t-1})$ is that part of the supergame payoff which involves s_t. If the P_i^* are regarded as payoff functions for a game in which the s_t are strategies, with s_{t+1} and s_{t-1} taken as fixed, such a constituent game has a noncooperative equilibrium.

Theorem 9.2. *Given S1–S4, a game having $\mathscr{S}_t$ as its strategy sets and P^*, defined by eq. (9.3), as its payoff functions, has a noncooperative equilibrium.*

□ Proof is easy and may be shown by seeing that the game satisfies G1–G3 and G6. There are n players (G1), the strategy sets are compact convex subsets of a Euclidean space (G2) and the payoff functions are continuous and bounded (G3). These follow directly from S1–S3 respectively. Because the sum of two concave functions is concave, the concavity

[3]Time subscripts may be omitted from the payoff functions because the game is stationary.

of $P_i^*(s_{t+1}, s_t, s_{t-1})$ in s_{it} follows from the concavity of $P_i(s_t, s_{t-1})$ in s_{it} and $s_{i,t-1}$.[4] Thus, the game has a noncooperative equilibrium. □

The equilibria established in theorem 9.2 can be represented by a correspondence: $\phi(s_{t+1}, s_{t-1}) = \{s_t | s_t$ is a noncooperative equilibrium, given s_{t+1} and $s_{t-1}\}$. For the moment, a very strong assumption on ϕ is made – that it is a continuous function which satisfies a Lipschitz condition.

S6. *ϕ is a continuous function which satisfies the following Lipschitz condition:*

$$\|\phi(s_{t+1}, s_{t-1}) - \phi(s'_{t+1}, s'_{t-1})\| \leq k_1\|s_{t+1} - s'_{t+1}\| + k_2\|s_{t-1}, s'_{t-1}\|, \tag{9.6}$$

where $k_1 + k_2 \leq k < 1$.

ϕ is intimately related to simple strategy noncooperative equilibria, as the following lemma shows.

Lemma 9.3. *Let $\sigma^* = (s_0^*, s_1^*, s_2^*, \ldots)$ be a simple strategy noncooperative equilibrium for a stationary supergame satisfying S1–S5. Then $s_t^* \in \phi(s_{t+1}^*, s_{t-1}^*)$, $t = 1, 2, \ldots$. Conversely, if σ^* satisfies the conditions*

$$s_t^* \in \phi(s_{t+1}^*, s_{t-1}^*), \qquad t = 1, 2, \ldots, \tag{9.7}$$

then σ^* is a noncooperative equilibrium strategy vector.

□ The first part of the lemma, which asserts that $s_t^* \in \phi(s_{t+1}^*, s_{t-1}^*)$, $t = 1, 2, \ldots$, if σ^* is an equilibrium strategy vector, is proved first. Assume the lemma false; and that σ^* is an equilibrium strategy vector such that $s_t^* \in \phi(s_{t+1}^*, s_{t-1}^*)$ does not hold for some t. Then there would be at least one period t' and one player i such that s_{it}^* would not be the maximizer of $P_i^*(s_{t'+1}^*, s_{it'}, \bar{s}_{it'}^*, s_{t'-1}^*)$. Were that so, another choice of $s_{it'}$ could be found which would increase $P_i^*(s_{t'+1}^*, s_{t'}^*, s_{t'-1}^*)$; however, from eq. (9.3) it is clear that $P_i^*(s_{t+1}^*, s_t^*, s_{t-1}^*)$ would be unaffected for $t > t' + 1$ and $t < t' - 1$. Looking at eq. (9.4) it is apparent that the supergame payoff of the ith player must rise; therefore, σ_i^* cannot have been a best reply for him against the strategies of the others, which contradicts the initial assumption that σ^* is an equilibrium strategy vector.

Now assume that although $s_t^* \in \phi(s_{t+1}^*, s_{t-1}^*)$, nonetheless, σ^* is not a supergame strategy vector associated with a noncooperative equilibrium.

[4]On the concavity of the sum of concave functions, see Berge (1963, theorem 4, p. 191).

Then for some player i there is a strategy $\boldsymbol{\sigma}'_i$ which yields a greater payoff to the ith player, given $\bar{\boldsymbol{\sigma}}_i^*$ than does $\boldsymbol{\sigma}'_i$. Let $F_i(\boldsymbol{\sigma}'_i, \bar{\boldsymbol{\sigma}}_i^*) - F_i(\boldsymbol{\sigma}^*) = \epsilon$. Using the boundedness of P_i, T may be found such that

$$\sum_{t=T}^{\infty} \alpha_i^{t-1} |P_i(s_t, s_{t-1})| < \tfrac{1}{4}\epsilon, \tag{9.8}$$

for any $s_t \in \mathscr{S}$, $t = T, T+1, \ldots$. Look now at payoffs over the first $T+1$ periods with s_0^*, s_{T+1}^* and all the $\bar{s}_{it}^*$, $(t = 1, \ldots, T)$, fixed,

$$\begin{aligned} &P_i(s_{i1}, \bar{s}_{i1}^*, s_0^*) + \alpha_i P_i(s_{i2}, \bar{s}_{i2}^*, s_{i1}, \bar{s}_{i1}^*) + \cdots \\ &+ \alpha_i^{T-1} P_i(s_{iT}, \bar{s}_{iT}^*, s_{i,T-1}, \bar{s}_{i,T-1}^*) + \alpha_i^T P_i(s_{T+1}^*, s_{iT}, \bar{s}_{iT}^*) \\ &= F_i^{**}(s_{i1}, \ldots, s_{iT}). \end{aligned} \tag{9.9}$$

From the definition of $\boldsymbol{\phi}$ it is clear that $(s_{i1}^*, \ldots, s_{iT}^*)$ is a local maximum for F_i^{**}; and from theorem 12 of Fenchel (1953, p. 63) it is known that a concave function defined on a convex subset of a finite dimensional vector space has at most one local maximum, and the local maximum is also the global maximum. Now return to the superior supergame strategy $\boldsymbol{\sigma}'_i$. By the way T is chosen, it follows that

$$F_i^{**}(s'_{i1}, \ldots, s'_{iT}) - F_i^{**}(s_{i1}^*, \ldots, s_{iT}^*) > \tfrac{1}{2}\epsilon, \tag{9.10}$$

which contradicts the assertion that $(s_{i1}^*, \ldots, s_{iT}^*)$ maximizes F_i^{**}. Therefore $s_t^* \in \boldsymbol{\phi}(s_{t+1}^*, s_{t-1}^*)$, $t = 1, \ldots,$ implies that $\boldsymbol{\sigma}^*$ is an equilibrium strategy vector. □

2.2. *Steady state equilibria for stationary supergames*

A steady state equilibrium is an equilibrium in which the players continually repeat the same moves (constituent game strategies) period after period. It is shown in this section that under S1–S6 a unique steady state noncooperative equilibrium, using simple strategies, exists. Such an equilibrium does not exist relative to any arbitrary initial condition s_0. In particular, if the steady state constituent game strategy is s^c, then for a stationary supergame satisfying S1–S6 to have a steady state equilibrium, the initial condition must be s^c.

Theorem 9.4. *If a stationary supergame satisfies S1–S6, then there is a unique constituent game strategy s^c such that if $s_0 = s^c$, then $s_t = s^c$, $t = 1, 2, \ldots,$ is a steady state noncooperative equilibrium for the supergame.*

□ The s^c of the theorem must satisfy $s^c = \boldsymbol{\phi}(s^c, s^c)$. Because $\boldsymbol{\phi}$ is a function from $\mathcal{S} \times \mathcal{S}$ into $\mathcal{S}$, it becomes a function from $\mathcal{S}$ into $\mathcal{S}$ when the restriction $s_{t+s} = s_{t-1}$ is placed on it.[5] Under this condition it has a fixed point, and indeed, a unique one as $\boldsymbol{\phi}$ is a contraction. Therefore, the steady state equilibrium exists and is unique. □

The unique steady state s^c of theorem 9.4 has another significance for games satisfying S1–S6. If $\boldsymbol{\sigma}' = (s'_0, s'_1, \ldots)$ is any simple strategy noncooperative equilibrium vector for the supergame, then, as t goes to infinity, s'_t converges to s^c.

Theorem 9.5. *If $\boldsymbol{\sigma}' = (s'_0, s'_1, s'_2, \ldots)$ is a simple strategy noncooperative equilibrium strategy vector for a supergame which satisfies S1–S6, then* $\lim_{t\to\infty} s'_t = s^c$.

□ Because $\boldsymbol{\sigma}'$ is a noncooperative equilibrium strategy, $s'_t = \boldsymbol{\phi}(s'_{t+1}, s'_{t-1})$, $t = 1, 2, \ldots$. Now a sequence of functions $g_t(s_{t+1}, s_0)$ is defined which contain the same information as $\boldsymbol{\phi}$, though in a different form. For an equilibrium strategy $\boldsymbol{\sigma}'$, $s'_t = g_t(s'_{t+1}, s'_0)$, $t = 1, 2, \ldots$. This sequence of functions is used to prove the theorem. The first member of the sequence is

$$s_1 = \boldsymbol{\phi}(s_2, s_0) = g_1(s_2, s_0). \tag{9.11}$$

To get $g_2(s_3, s_0)$, consider substituting from eq. (9.11) into $\boldsymbol{\phi}(s_3, s_1)$. This gives

$$s'_2 = \boldsymbol{\phi}(s_3, g_1(s_2, s_0)). \tag{9.12}$$

For fixed s_3 and s_0, eq. (9.12) is a function taking s_2 to s'_2. Looked at in this way, $\boldsymbol{\phi}$ has a fixed point which is unique if g_1 is a contraction. Assume uniqueness temporarily and denote the fixed point by

$$s_2 = g_2(s_3, s_0). \tag{9.13}$$

Continue to define functions g_t inductively. Say $g_{t-1}(s_t, s_0)$ has been defined and is a contraction. Now let

$$s'_t = \boldsymbol{\phi}(s_{t+1}, s_{t=1}) = \boldsymbol{\phi}(s_{t+1}, g_{t-1}(s_t, s_0)). \tag{9.14}$$

Regarding eq. (9.14) as a function taking s_t to s'_t for fixed s_{t+1} and s_0, the

[5]The correspondence $\boldsymbol{\phi}$ of §2.1 becomes the function $\boldsymbol{\phi}$ in the present section.

fixed point of the function may be denoted

$$s_t = g_t(s_{t+1}, s_0). \tag{9.15}$$

$g_1 = \boldsymbol{\phi}$ and is clearly a contraction; hence, g_2 exists. It is now shown inductively that all the g_t exist and are contractions by showing that if g_{t-1} exists and is a contraction, then the same holds for g_t. Let λ_{t-1} and μ_{t-1} be the Lipschitz conditions on g_{t-1} with respect to s_t and s_0. From $s'_t = \boldsymbol{\phi}(s_{t+1}, g_{t-1}(s_t, s_0))$ it is seen that for an arbitrary $s'^1 = s^1_{t+1},\ g_{t-1}(s^1_t, s^1_0)$ and $s'^2_t = \boldsymbol{\phi}(s^2_{t+1},\ g_{t-1}(s^2_t, s^2_0))$,

$$\begin{aligned}\|s'^1_t - s'^2_t\| &= \|\boldsymbol{\phi}(s^1_{t+1}, g_{t-1}(s^1_t, s^1_0)) - \boldsymbol{\phi}(s^2_{t+1}, g_{t-1}(s^2_t, s^2_0))\| \\ &\le k_1\|s^1_{t+1} - s^2_{t+1}\| + k_2\|g_{t-1}(s^1_t, s^1_0) - g_{t-1}(s^2_t, s^2_0)\| \\ &\le k_1\|s^1_{t+1} - s^2_{t+1}\| + k_2[\lambda_{t-1}\|s^1_t - s^2_t\| + \mu_{t-1}\|s^1_0 - s^2_0\|].\end{aligned} \tag{9.16}$$

When $s'^1_t = s^1_t$ and $s'^2_t = s^2_t$, eq. (9.16) may be written as

$$\|s^1_t - s^2_t\| \le \|s^1_{t+1} - s^2_{t+1}\| k_1/(1 - k_2\lambda_{t-1}) + \|s^1_0 - s^2_0\| k_2\mu_{t-1}/(1 - k_2\lambda_{t-1}). \tag{9.17}$$

If $\lambda_{t-1} + \mu_{t-1} < 1$, then $k_1/(1 - k_2\lambda_{t-1}) + k_2\mu_{t-1}/(1 - k_2\lambda_{t-1}) \le k_1 + k_2 < 1$, implying that g_t is a contraction whenever g_{t-1} is a contraction.

The newly constructed functions g_t are now used to show that s'_t converges to s^c as t goes to infinity. Recall that the g_t satisfy a Lipschitz condition of the form

$$\|g_t(s^1_{t+1}, s^1_0) - g_t(s^2_{t+1}, s^2_0)\| \le \lambda_t\|s^1_{t+1} - s^2_{t+1}\| + \mu_t\|s^1_0 - s^2_0\|, \quad t = 1, 2, \ldots, \tag{9.18}$$

with $\lambda_t + \mu_t \le k$. In relation to an equilibrium strategy $\boldsymbol{\sigma}'$ an associated sequence showing the distance of each s'_t from s^c can be constructed,

$$\delta_t = \|s'_t - s^c\|, \qquad t = 1, 2, \ldots. \tag{9.19}$$

Using eq. (9.18) with $s^1_t = s'_t$ and $s^2_t = s^c$, and defining $\lambda_0 = 1$ an upper bound may be found for δ_1,

$$\begin{aligned}\delta_1 &\le \lambda_1\delta_2 + \mu_1\delta_0 \\ &\le \lambda_1(\lambda_2\delta_3 + \mu_2\delta_0) + \mu_1\delta_0 \\ &\le \prod_{t=1}^{T}\lambda_1\delta_{T+1} + \delta_0\sum_{t=1}^{T}\mu_t\prod_{\tau=0}^{t-1}\lambda_\tau.\end{aligned} \tag{9.20}$$

As $\lim_{T\to\infty} \Pi_{t=1}^{T} \lambda_t \delta_{T+1} = 0$, eq. (9.20) implies

$$\delta_1 \leq \delta_0 \sum_{t=1}^{\infty} \mu_t \prod_{\tau=0}^{t-1} \lambda_\tau$$
$$= \lim_{T\to\infty} \delta_0[\mu_1 + \lambda_1(\mu_2 + \lambda_2(\mu_3 + \lambda_3(\ldots(\mu_T)\ldots)))] \leq k\delta_0. \quad (9.21)$$

By induction, $\delta_t \leq k^t\delta_0$; therefore, $\lim_{t\to\infty} s'_t = s^c$. □

So far in this chapter we have dealt with a fairly general class of time dependent supergames. The two principal limitations placed on the models, in comparison with the game models of chs. 7 and 8, are the invariance of the payoff structure over time, the replacement of quasi-concavity with concavity, and the condition that $\boldsymbol{\phi}$ is a function obeying a Lipschitz condition. All these, taken together, allow proof of existence of a noncooperative equilibrium and the very appealing result that, no matter what the initial condition, for any simple strategy noncooperative equilibrium, as time proceeds, single-period choices converge to a particular steady state single-period strategy vector, s^c.

In supergames with time dependence, it is not possible to merely string together single period noncooperative equilibria to obtain a noncooperative equilibrium for the supergame. In fact, it is not really obvious what should be taken as the counterpart of a single-period game in a time dependent supergame. On the face of it, the payoff functions $P_i(s_t, s_{t-1})$, with s_{t-1} given seem a natural choice as the payoff functions associated with the single-period game of period t. But then a noncooperative equilibrium for the supergame has, in general, no observable connection with independent noncooperative equilibria for the individual constituent games.

If, on the other hand, the payoff functions $P_i^*(s_{t+1}, s_t, s_{t-1})$ are taken to be the payoff functions associated with the single period game of period t, the supergame payoff function may be written as they are shown in eq. (9.5). It is then true that a noncooperative equilibrium for the supergame is a sequence of noncooperative equilibria for the constituent games; although, not just any sequence. It must be a sequence satisfying the relation $s'_t = \boldsymbol{\phi}(s'_{t+1}, s'_{t-1})$.

The concavity assumption is needed because a sum of concave functions is concave; whereas, a sum of quasi-concave functions need not be quasi-concave; and the existence theorems need quasi-concavity. The restriction on $\boldsymbol{\phi}$ is somewhat relaxed in the next section; however, not in an entirely satisfactory fashion.

3. Equilibrium in nonstationary supergames

The section is divided into two subsections. The first contains a generalization of theorem 9.1 to nonstationary supergames; and the second, using different assumptions than S1–S5, introduces a variant of the noncooperative equilibrium, called a *weak noncooperative equilibrium*. Conditions are given under which this new equilibrium exists.

3.1. Noncooperative equilibrium in nonstationary supergames

Giving up stationarity means the supergame payoff functions are given by eq. (9.1),

$$\sum_{t=1}^{\infty} \alpha_{it} P_{it}(s_t, s_{t-1}) = F_i(\boldsymbol{\sigma}), \qquad i = 1, \ldots, n. \tag{9.1}$$

Theorem 9.1 is a consequence of theorem 9.6:

Theorem 9.6. *A supergame satisfying S1–S5 has a noncooperative equilibrium in the set of simple strategies.*

□ Proof is accomplished by showing that S1–S5 imply that G1, G3, G6–G8 are satisfied, which means the theorem holds as a consequence of theorem 7.7. S1 and G1 are the same (saying that n is finite). G7 and G8, taken together, are equivalent to S2, which states that $\mathcal{S}_i$ (the ith player's set of simple supergame strategies) is closed, bounded and convex.

Turning to G3, F_i is clearly bounded because the P_{it} are bounded by $\bar{P}$ independently of t and the sum of the discount parameters is bounded; hence,

$$\sum_{t=1}^{\infty} \alpha_{it} P_{it}(s_t) \leq \bar{P} \sum_{t=1}^{\infty} \alpha_{it} < \infty, \tag{9.22}$$

for any choice of s_t, $t = 1, 2, \ldots$. Continuity of the F_i may be proved by showing that for any $\epsilon > 0$ there is a $\delta > 0$ such that, if $\|\boldsymbol{\sigma}' - \boldsymbol{\sigma}''\| < \delta$ then $|F_i(\boldsymbol{\sigma}') - F_i(\boldsymbol{\sigma}'')| < \epsilon$. It suffices to show the continuity of each F_i separately. First choose $\epsilon > 0$. There is a finite t^* such that

$$\bar{P} \sum_{\tau=t+1}^{\infty} \alpha_{i\tau} < \epsilon/2, \qquad t > t^*. \tag{9.23}$$

Because P_{it} is continuous and $\mathcal{S}_{it} \times \mathcal{S}_{i,t-1}$ is compact, for any $\epsilon_t > 0$ there is $\delta_t > 0$ such that if $\|s'_t - s''_t\| < \delta_t$ and $\|s'_{t-1} - s''_{t-1}\| < \delta_t$ then $|P_{it}(s'_t, s'_{t-1}) - P_{it}(s''_t, s''_{t-1})| < \epsilon_t$. Now choose

$$\epsilon_t = \epsilon/(2t^*\alpha_{it}) > 0, \qquad t = 1, \ldots, t^*, \tag{9.24}$$

and let $\delta = \min_{t \leq t^*} \alpha_{it}\delta_t > 0$. If $\|\boldsymbol{\sigma}' - \boldsymbol{\sigma}''\| < \delta$ then

$$\begin{aligned} |F_i(\boldsymbol{\sigma}') - F_i(\boldsymbol{\sigma}'')| &\leq \sum_{t=1}^{\infty} \alpha_{it} |P_{it}(s'_t, s'_{t-1}) - P_{it}(s''_t, s''_{t-1})| \\ &< \sum_{t=1}^{t^*} \alpha_{it} |P_{it}(s'_t, s'_{t-1}) - P_{it}(s''_t, s''_{t-1})| + \epsilon/2 \\ &\leq \sum_{t=1}^{t^*} \alpha_{it}\epsilon/(2t^*\alpha_{it}) + \epsilon/2 = \epsilon. \end{aligned} \tag{9.25}$$

Thus F_i is continuous.

G6 follows from S4 and from F_i being a sum of concave functions. F_i is, of course, concave rather than merely quasi-concave. Thus G1, G3, G6–G8 are implied by S1–S5; so the theorem is proved as a consequence of theorem 7.7. □

In the manner of eq. (9.3) the supergame payoff function may be written in another form with the help of

$$P^*_{it}(s_{t+1}, s_t, s_{t-1}) = P_{it}(s_t, s_{t-1}) + (\alpha_{i,t+1}/\alpha_{it})P_{i,t+1}(s_{t+1}, s_t). \tag{9.26}$$

Then, following eq. (9.5),

$$F_i(\sigma) = \frac{1}{2}\left[P_{i1}(s_1, s_0) + \sum_{t=1}^{\infty} \alpha_{it} P^*_{it}(s_{t+1}, s_t, s_{t-1})\right]. \tag{9.27}$$

From theorem 9.2, it is known that a game having $\mathscr{S}_t$ for strategy sets and the P^*_{it} as payoff functions has a noncooperative equilibrium. Let $\boldsymbol{\phi}_t(s_{t+1}, s_{t-1})$ denote the set of noncooperative equilibria for such a game. Lemma 9.3 generalizes to the nonstationary supergame.

Lemma 9.7. *Let $\boldsymbol{\sigma}^* = (s^*_0, s^*_1, s^*_2, \ldots)$ be a simple strategy noncooperative equilibrium for a supergame satisfying S1–S5. Then $s^*_t \in \boldsymbol{\phi}_t(s^*_{t+1}, s^*_{t-1})$, $t = 1, 2, \ldots$. Conversely, if $\boldsymbol{\sigma}^*$ satisfies the conditions $s^*_t \in \boldsymbol{\phi}_t(s^*_{t+1}, s^*_{t-1})$, then $\boldsymbol{\sigma}^*$ is a noncooperative equilibrium strategy vector.*

□ The proof of lemma 9.3 may be repeated with t subscripts added to $\boldsymbol{\phi}$ and the P^*_{it} substituted for the P^*_i. □

Lemma 9.8. *$\boldsymbol{\phi}_t$ is upper semicontinuous.*

□ Let $\{(s_{t+1}^l, s_{t-1}^l)\}$ be a convergent sequence of points in $\mathcal{S}_{t+1} \times \mathcal{S}_{t-1}$, whose limit is (s_{t+1}^0, s_{t-1}^0). Let $\{s_t^l\}$ be any associated sequence of points where $s_t^l \in \phi_t(s_{t+1}^l, s_{t-1}^l)$. ϕ_t is upper-semicontinuous if, when s_t^l has a limit, s_t^0, then $s_t^0 \in \phi_t(s_{t+1}^0, s_{t-1}^0)$. Assume $s_t^0 \notin \phi_t(s_{t+1}^0, s_{t-1}^0)$. Then, for at least one player, i,

$$\max_{s_{it} \in \mathcal{S}_{it}} P_{it}^*(s_{t+1}^0, \bar{s}_{it}^0, s_{it}, s_{t-1}^0) - P_{it}^*(s_{t+1}^0, s_t^0, s_{t-1}^0) = \epsilon_1 > 0. \tag{9.28}$$

By continuity, for any $\epsilon_2 > 0$ there is $\delta > 0$ such that if

$$\|(s'_{t+1}, s'_t, s'_{t-1}) - (s''_{t+1}, s''_t, s''_{t-1})\| < \delta, \tag{9.29}$$

then

$$|P_{it}^*(s'_{t+1}, s'_t, s'_{t-1}) - P_{it}^*(s''_{t+1}, s''_t, s''_{t-1})| < \epsilon_2. \tag{9.30}$$

Because $(s_{t+1}^l, s_t^l, s_{t-1}^l)$ is a sequence whose limit is $(s_{t+1}^0, s_t^0, s_{t-1}^0)$, there is a finite l^* such that for $l > l^*$,

$$\|(s_{t+1}^l, s_t^l, s_{t-1}^l) - (s_{t+1}, s_t^0, s_{t-1}^0)\| < \delta, \tag{9.31}$$

which implies

$$|P_{it}^*(s_{t+1}^l, s_t^l, s_{t-1}^l) - P_{it}^*(s_{t+1}^0, s_t^0, s_{t-1}^0)| < \epsilon_2, \qquad l > l^*. \tag{9.32}$$

Choose $\epsilon_2 < \epsilon_1$. This contradicts the assertion that $s_t^0 \notin \phi_t(s_{t+1}^0, s_{t-1}^0)$; therefore ϕ_t is upper-semicontinuous. □

3.2. *Weak noncooperative equilibria*

Now a somewhat different class of games is taken up. For these games, existence of a *weak noncooperative equilibrium* is examined. What characterizes the weak noncooperative equilibrium is that no player can change any one of his constituent game strategies and increase his supergame payoff. It is possible that changing two or more of his constituent game strategies results in an increased supergame payoff. S4 is weakened.

S7. *$P_{it}^*(s_{t+1}, s_t, s_{t-1})$ is a quasi-concave function of s_{it}.*

S7 represents a considerable weakening of S4. Also, S5 is not used in this section, which means it is possible for F_i to be unbounded.

Under S1–S3 and S7, theorem 9.2 still holds, and may be stated without further proof.

Theorem 9.9. *Given S1–S3 and S7, a game having $\mathscr{S}_t$ as its strategy sets and $\mathbf{P}_t^*$, defined by eq. (9.26), as its payoff functions, has a noncooperative equilibrium.*

As previously, the set of noncooperative equilibria for this game is denoted $\boldsymbol{\phi}_t(s_{t+1}, s_{t-1})$. It also remains true that if σ^* is a noncooperative equilibrium strategy vector for the supergame, then $s_t^* \in \boldsymbol{\phi}_t(s_{t+1}^*, s_{t+1}^*)$, $t = 1, 2, \ldots$. The converse, that if $s_t^* \in \boldsymbol{\phi}_t(s_{t+1}^*, s_{t-1}^*)$ for all t, then $\boldsymbol{\sigma}^*$ is a noncooperative equilibrium, need not hold. This is because the sum of quasi-concave functions need not be quasi-concave. Upper semicontinuity of the $\boldsymbol{\phi}_t$ still holds. An assumption on the $\boldsymbol{\phi}_t$ is needed which parallels S6.

S8. *Each correspondence $\boldsymbol{\phi}_t$ has a selection $\boldsymbol{f}_t$ which obeys the following Lipschitz conditions:*

$$\|f_t(s'_{t+1}, s'_{t-1}) - f_t(s'_{t+1}, s''_{t-1})\| \leq k_{t-}\|s'_{t-1} - s''_{t-1}\|, \tag{9.33}$$

$$\|f_t(s'_{t+1}, s'_{t-1}) - f_t(s''_{t+1}, s'_{t-1})\| \leq k_{t+}\|s'_{t+1} - s''_{t+1}\|, \tag{9.34}$$

with $k_{t+} + k_{t-} < 1$ for any s'_{t+1}, $s''_{t+1} \in \mathscr{S}_{t+1}$, any s'_{t-1}, $s''_{t-1} \in \mathscr{S}_{t-1}$ and $t = 2, 3, \ldots$. For $t = 1$,

$$\|f_1(s'_2, s_0) - f_1(s''_2, s_0)\| < \|s'_2 - s''_2\|. \tag{9.35}$$

$\boldsymbol{\sigma}^*$ is a simple supergame strategy vector associated with a weak noncooperative equilibrium if $s_t^* \in \mathscr{S}_t$ and $s_t^* \in \boldsymbol{\phi}_t(s_{t+1}^*, s_{t-1}^*)$, $t = 1, 2, \ldots$.

Theorem 9.10. *If a supergame satisfies S1–S3, S7 and S8 then there exists a simple strategy vector $\boldsymbol{\sigma}^*$ which is associated with a weak noncooperative equilibrium – that is, $s_t^* \in \boldsymbol{\phi}_t(s_{t+1}^*, s_{t-1}^*)$, $t = 1, 2, \ldots$.*

□ Using S8 there is a sequence of functions, $f_t(s_{t+1}, s_{t-1})$, defined on every point of $\mathscr{S}_{t+1} \times \mathscr{S}_{t-1}$, which are continuous and for which $f_t(s_{t+1}, s_{t-1}) \in \boldsymbol{\phi}_t(s_{t+1}, s_{t-1})$. The $\boldsymbol{f}_t$ need not be unique. It is clear that a strategy vector $\boldsymbol{\sigma}^*$ which satisfies $s_t^* = f_t(s_{t+1}^*, s_{t-1}^*)$, $t = 1, 2, \ldots$, is a weak noncooperative equilibrium. Such a strategy vector must be shown to exist. The method of proof is to use the $\boldsymbol{f}_t$ to define a new sequence of functions, $g_t(s_{t+1})$, which have the property that if $s_t^* = g_t(s_{t+1}^*)$ for $t = 1, 2, \ldots$, then $s_t^* = f_t(s_{t+1}^*, s_{t-1}^*)$. It is shown that there exists a $\boldsymbol{\sigma}^*$ such that $s_t^* = g_t(s_{t+1}^*)$ for all $t \geq 1$; which, in turn, means there is a weak noncooperative equilibrium.

The first member of the g_t sequence is

$$g_1(s_2) = f_1(s_2, s_0). \tag{9.36}$$

This representation of f_1 is clearly possible, as s_0 is a fixed initial condition. It is now shown that if $g_{t-1}(s_t)$ is defined and obeys a Lipschitz condition with ratio $k < 1$, then there is a function $g_t(s_{t+1})$ which obeys a Lipschitz condition with ratio $k < 1$ and, for any $s^0_{t+1} \in \mathcal{S}_{t+1}$ there is a $s^0_{t-1} \in \mathcal{S}_{t-1}$ such that

$$f_t(s^0_{t+1}, s^0_{t-1}) = f_t(s^0_{t+1}, g_{t-1}(s^0_t)) = s^0_t, \tag{9.37}$$

for some $s^0_t \in \mathcal{S}_t$. $g_t(s^0_{t+1}) = s^0_t$.

To derive g_t, first consider $f_t(s_{t+1}, g_{t-1}(s_t))$. Clearly, for *any* fixed $s_{t+1} \in \mathcal{S}_{t+1}$, f_t is a contraction mapping from $\mathcal{S}_t$ into $\mathcal{S}_t$; hence, for fixed s_{t+1}, f_t has a unique fixed point (see Bartle (1964, theorem 16.16)). Denote by $g_t(s_{t+1})$ the fixed point of f_t for s_{t+1}. It remains to show that g_t obeys a Lipschitz condition with ratio $k(<1)$; choose $s'_{t+1}, s''_{t+1} \in \mathcal{S}_{t+1}$ and

$$s'_t = g_t(s'_{t+1}),\ s''_t = g_t(s''_{t+1}),\ s'_{t-1} = g_{t-1}(s'_t), \tag{9.38}$$

and $s''_{t-1} = g_{t-1}(s''_t)$.

Then

$$\begin{aligned} \|s'_t - s''_t\| &= \|f_t(s'_{t+1}, s'_{t-1}) - f_t(s''_{t+1},\ s''_{t-1})\| \\ &\leq k_{t+}\|s'_{t+1} - s''_{t+1}\| + k_{t-}\|s'_{t-1} - s''_{t-1}\| \\ &\leq k_{t+}\|s'_{t+1} - s''_{t+1}\| + k_{t-}\|s'_t - s''_t\|. \end{aligned} \tag{9.39}$$

Therefore

$$\begin{aligned} \|s'_t - s''_t\| &\leq \|s'_{t+1} - s''_{t+1}\| k_{t+}/(1 - k_{t-}) \\ &< \|s'_{t+1} - s''_{t+1}\|. \end{aligned} \tag{9.40}$$

If there is some strategy $\boldsymbol{\sigma}^*$ for which $s^*_t = g_t(s^*_{t+1})$, $t = 1, 2, \ldots$, then the theorem is proved, as $s^*_t = g_t(s^*_{t+1})$, $t = 1, 2, \ldots$, implies $s^*_t = f_t(s^*_{t+1}, s^*_{t-1})$, $t = 1, 2, \ldots$, which implies $s^*_t \in \boldsymbol{\phi}_t(s^*_{t+1}, s^*_{t-1})$, $t = 1, 2, \ldots$.

To show such a strategy exists, let $\boldsymbol{\sigma}^0 = (s_0, s^0_1, s^0_2, \ldots)$ be an arbitrary strategy and define $\boldsymbol{\sigma}^k = (s_0, s^k_1, s^k_2, \ldots)$ as follows:

$$s^k_t = s^0_t, \qquad t > k, \tag{9.41}$$

$$s^k_t = g_t(s^k_{t+1}), \qquad t = k, k-1, \ldots, 1. \tag{9.42}$$

The elements of the sequence $\boldsymbol{\sigma}^k$, $k = 0, 1, 2, \ldots$, being members of a compact space, have a cluster point, $\boldsymbol{\sigma}^*$. $\boldsymbol{\sigma}^*$ satisfies the conditions of the theorem. In fact $\boldsymbol{\sigma}^* = \lim_{k\to\infty} \boldsymbol{\sigma}^k$ as the g_t are all contractions. □

4. *Balanced temptation equilibria for stationary supergames*

There may be several ways in which the balanced temptation property might be specified and the balanced temptation equilibrium defined for time dependent supergames. One of these is worked out below. Although the supergame model to be used in this section is stationary, it does not appear possible to show existence of a balanced temptation equilibrium for which payoffs are Pareto optimal. The balanced temptation equilibrium which is developed has a *local* Pareto optimality under which it is impossible to increase the payoffs of a given period by changing the constituent game strategy vector of that period.

Let $\mathcal{S}(s_0)$ denote the set of simple supergame strategies which are Pareto optimal, given s_0. Formally

$$\mathcal{S}(s_0) = \{\boldsymbol{\sigma} \in \mathcal{S} | \boldsymbol{F}(\boldsymbol{\sigma}') \not\geq \boldsymbol{F}(\boldsymbol{\sigma}) \text{ for all } \boldsymbol{\sigma}' \in \mathcal{S}\}. \tag{9.43}$$

Let $\mathcal{T}(s_0)$ denote all elements of $\mathcal{S}_1$ which are $t = 1$ components of at least one element of $\mathcal{S}(s_0)$,

$$\mathcal{T}(s_0) = \{s_1' \in \mathcal{S}_1 | s_1' = s_1 \text{ for at least one } (s_0, s_1, s_2, \ldots) \in \mathcal{S}(s_0)\}. \tag{9.44}$$

Finally, let

$$\mathcal{H}(s_0) = \{\boldsymbol{P}(s_1, s_0) | s_1 \in \mathcal{T}(s_0), \text{ and for all } s_1' \in \mathcal{T}(s_0), \boldsymbol{P}(s_1', s_0) \not> \boldsymbol{P}(s_1, s_0)\}. \tag{9.45}$$

$\mathcal{H}(s_0)$ is the set of period 1 payoffs which are Pareto optimal, given that period one strategy choice is restricted to $\mathcal{T}(s_0)$. $\mathcal{S}(s_{t-1})$, $\mathcal{T}(s_{t-1})$ and $\mathcal{H}(s_{t-1})$ may be defined analogously. For example, $\mathcal{S}(s_{t-1})$ is the set of simple supergame strategies which are Pareto optimal for the supergame which proceeds from period t onward and has s_{t-1} as its initial condition. Now let

$$\boldsymbol{\sigma}^c(s_{t-1}) = (s_{t-1}, s_t^c(s_{t-1}), s_{t+1}^c(s_{t-1}), \ldots) \tag{9.46}$$

denote a simple strategy which is a noncooperative equilibrium for the supergame which proceeds from period t onward and has s_{t-1} as its initial condition.[6]

An assumption is needed which is a counterpart to G9:

S9. *For any* $(\rho_1, \ldots, \rho_n) > 0$ *with* $\Sigma_{i=1}^n \rho_i = 1$, *and any* $s_{t-1} \in \mathcal{S}_t$, *there is*

[6]To simplify the notation, $s_t^c = s_t^c(s_{t-1})$, where no confusion should result.

$s_t \in \mathcal{T}(s_{t-1})$ such that

$$[P_i(s_t, s_{t-1}) - P_i(s_t^c, s_{t-1})]/\sum_{j=1}^{n} [P_j(s_t, s_{t-1}) - P_j(s_t^c, s_{t-1})] = \rho_i, \quad i = 1, \ldots, n. \tag{9.47}$$

and $\mathbf{P}(s_t, s_{t-1}) \in \mathcal{H}(s_{t-1})$.

A supergame strategy $\boldsymbol{\sigma} = (s_0, s_1, \ldots)$ is said to possess *local Pareto optimality* if $\boldsymbol{\sigma} \in \mathcal{S}$ and $s_t \in \mathcal{T}(s_{t-1})$, $t = 1, 2, \ldots$. $\boldsymbol{\sigma}$ is said to be *globally Pareto optimal* if there is no $\boldsymbol{\sigma}' \in \mathcal{S}$ such that $\boldsymbol{F}(\boldsymbol{\sigma}') \gg \boldsymbol{F}(\boldsymbol{\sigma})$. The meaning of global Pareto optimality is quite clear and usual; however, local Pareto optimality requires a closer look. The concept is *local* in the sense that the Pareto optimality is within each single time period only; however, within the single time period the locally Pareto optimal choice, s_t, is further restricted to be Pareto optimal only with respect to strategy choices in $\mathcal{T}(s_{t-1})$. It is this latter condition which guarantees that any globally Pareto optimal strategy is also locally Pareto optimal. The converse is not true in general.

The *balanced* temptation property is now defined in a manner analogous to its definition for games lacking time dependence. $\boldsymbol{\sigma}^0$ has the balanced temptation property if:

$$\begin{aligned}
&[P_i(s'_{it}, \bar{s}^0_{it}, s^0_{t-1}) - P_i(s^0_t, s^0_{t-1})]/[P_i(s^0_t, s^0_{t-1}) - P_i(s^c_t, s^0_{t-1})] \\
&= [P_j(s'_{jt}, \bar{s}^0_{jt}, s^0_{t-1}) - P_j(s^0_t, s^0_{t-1})]/[P_j(s^0_t, s^0_{t-1}) - P_j(s^c_t, s^0_{t-1})], \\
&i, j = 1, \ldots n, \quad t = 1, 2, \ldots,
\end{aligned} \tag{9.48}$$

and s'_{it} is defined by the condition

$$P_i(s'_{it}, \bar{s}^0_{it}, s^0_{t-1}) = \max_{s_{it} \in \mathcal{S}_i} P_i(s_{it}, \bar{s}^0_{it}, s^0_{t-1}). \tag{9.49}$$

The balanced temptation property is defined here in a local way. The ratio which must be equal for all players is the extra gain to be made in a given period, divided by the amount by which current period payoff under $\boldsymbol{\sigma}^0$ exceeds what the payoff would be if the current period marked the first period of the use of a simple strategy noncooperative equilibrium strategy vector. This latter amount does not measure the per period foregone payoffs which result from using s'_{it} because a reversion to simple strategy noncooperative play would begin in period $t + 1$ with $(s'_{it}, \bar{s}^0_{it})$ as the initial condition. Also, even after such a reversion, there is no reason to suppose that per period payoffs are constant. These considerations do raise questions about whether the balanced temptation equilibrium, defined for time dependent games, is sufficiently interesting.

Theorem 9.11. *For a stationary supergame satisfying S1–S5 and S9 there exists a simple supergame strategy* $\boldsymbol{\sigma}^*$ *which has the balanced temptation property and which is locally Pareto optimal.* $\boldsymbol{\sigma}^*$ *depends on the initial condition* s_0.

□ Proof may be shown by construction and appeal to theorem 8.4. From the axioms it is clear that there is a noncooperative simple strategy $\boldsymbol{\sigma}^c(s_0)$ and a set $\mathcal{H}(s_0)$ of points, some of which dominate $(P_1(s_1^c, s_0), \ldots, P_n(s_1^c, s_0))$. From theorem 8.4 it is known there is $s_1^0 \in \mathcal{S}(s_0)$ which has the balanced temptation property. Proceeding inductively, having found s_τ^0, $\tau = 1, 2, \ldots, t-1$, $\boldsymbol{\sigma}^c(s_{t-1}^0)$ is known, as well as $\mathcal{S}(s_{t-1}^0)$. There is then $s_t^0 \in \mathcal{S}(s_{t-1}^0)$ which has the balanced temptation property; hence, there is $\boldsymbol{\sigma}^0$ which has the balanced temptation property. □

Now supergame strategies can be formulated which may form a noncooperative equilibrium exhibiting the balanced temptation property. Let $\boldsymbol{\sigma}^0$ satisfy the conditions of theorem 9.11. A supergame strategy $\boldsymbol{\sigma}^*$, not among the simple strategies, may be defined as follows: for the ith player,

$$s_{i1} = s_{i1}^0, \tag{9.50}$$

$$s_{it} = s_{it}^0 \quad \text{if } s_\tau = s_\tau^0, \quad \tau = 1, \ldots, t-1. \tag{9.51}$$

If

$$s_\tau = s_\tau^0, \quad \tau = 1, \ldots, t-2 \tag{9.52}$$

and

$$s_{t-1} \neq s_{t-1}^0, \tag{9.53}$$

then

$$s_{it} = s_{it}^c \tag{9.54}$$

and

$$s_{i\tau} = s_{i\tau}^c(s_{t-1}), \quad \tau = t+1, t+2, \ldots . \tag{9.55}$$

The following theorem is proved:

Theorem 9.12. *The strategies* $\boldsymbol{\sigma}^*$, *defined by eqs. (9.50)–(9.55) correspond to a noncooperative equilibrium having the balanced temptation* *property if*

$$\sum_{\tau=1}^{\infty} \alpha_i^{\tau-t} P_i(s_\tau^0, s_{\tau-1}^0) > P_i(s'_{it}, \bar{s}_{it}^0, s_{t-1}^0) + \alpha_i F_i(\boldsymbol{\sigma}^c(s'_{it}, s_{it}^0)),$$

$$i = 1, \ldots, n, \quad t = 1, 2, \ldots, \quad s_0^0 = s_0. \tag{9.56}$$

There are two points to be noted. First, $\boldsymbol{\sigma}^0$ need not be unique; and, second, there is no assurance that the condition in eq. (9.56) is met for a $\boldsymbol{\sigma}^0$ having the balanced temptation property.

5. Applications to oligopoly models

The results of §2 and §3 on simple strategy equilibria can be applied to oligopoly models. As a first step, the differentiated products price model must be reformulated to incorporate time dependence. Two possibilities present themselves: (a) demand in period t depends on current prices and on the sales of period $t-1$ and (b) demand in period t depends on current prices and on the prices of period $t-1$. (a) is probably the easier to justify, as it would apply to instances in which the demand of a given period depends on either inventories of the good in the hands of consumers or on recent consumption levels. Consider each separately.

If a good has a long life, as does any consumer durable, people who purchase in one time period tend to be out of the market for several periods to follow. Or at least, the probability that a given consumer purchases a new unit should rise as the newest one in his possession ages. Thus the larger are sales in period $t-1$, the lower are sales in period t, prices being given.

Recent consumption of a nondurable can have either a positive or negative effect. One possibility is that the more a good is consumed, the more people wish to consume further. This would apply to goods which are habit forming or for which taste tends to be acquired and which are, as a result, more appreciated the more they are used. The reverse occurs if people "get their fill" of a good quite easily.

In terms of modeling a quantity effect, if

$$q_{it} = F_i(\boldsymbol{p}_t, \boldsymbol{q}_{t-1}), \tag{9.57}$$

then, substituting $F_i(\boldsymbol{p}_{t-1}, \boldsymbol{q}_{t-2})$ for $\boldsymbol{q}_{t-1}$ in eq. (9.57), it is apparent that all prices going back to the initial time period of the model are determinants of current demand. This may be analytically troublesome. It is certainly outside the range of the models considered earlier in this chapter.[7]

Turning now to the second alternative under which current demand depends on both current and previous period prices, demand is given by

$$q_{it} = F_i(\mathbf{p}_t, \mathbf{p}_{t-1}), \qquad i = 1, \ldots, n. \tag{9.58}$$

Period t profit is

$$\pi_i(\mathbf{p}_t, \mathbf{p}_{t-1}) = p_{it}F_i(\mathbf{p}_t, \mathbf{p}_{t-1}) - C_i(F_i(\mathbf{p}_t, \mathbf{p}_{t-1})), \qquad i = 1, \ldots, n. \tag{9.59}$$

Although much remains to be verified, this model looks more likely to fit the conditions of §2 and §3. It is less satisfactory regarding the interpretation of the effect of $\mathbf{p}_{t-1}$ on q_{it}. Whenever I think of possible interpretations, nearly all of them are the ones already given for the effect of previous period sales; however, clearly, $\mathbf{p}_{t-1}$ can be only a proxy variable in these cases.

Another justification for a previous price effect is that knowledge of both present and past prices cause price expectations for the future which affect current period demand. For example, if today's prices exceed yesterday's, consumers might buy more than they would if yesterday's prices were as high as today's. The reason for this behavior is that they predict higher prices for tomorrow in the former situation; hence, they buy extra today to anticipate some of their future demand. These considerations are so much beyond the explicit models under discussion that I prefer to think of the effect of $\mathbf{p}_{t-1}$ as a proxy for the effect of $\mathbf{q}_{t-1}$.

Now consider whether a model with demand functions given by eq. (9.58) and profit functions given by eq. (9.59) can be made to satisfy assumptions such as S1–S5. First, to make the game analogy explicit, p_{it} plays the role of s_{it} and $\pi_i(\mathbf{p}_t, \mathbf{p}_{t-1})$ plays the role of $P_i(\mathbf{s}_t, \mathbf{s}_{t-1})$. Supergame payoff is defined in the obvious way. It is trivially obvious that all assumptions except S4 are satisfied easily. S4 would be satisfied if the following conditions were met,[8]

$$\begin{aligned} &\pi_i^{ii}(\mathbf{p}_t, \mathbf{p}_{t-1}) < 0, \\ &\pi_i^{i+n,i+n}(\mathbf{p}_t, \mathbf{p}_{t-1}) < 0, \\ &\pi_i^{ii}\pi_i^{i+n,i+n} - [\pi_i^{i,i+n}]^2 > 0, \end{aligned} \tag{9.60}$$

[7]For linear demand, Selten (1965, 1968) deals with a model of this character.

[8]Note that the $i+n$th variable is $p_{i,t-1}$ so the partial derivatives are with respect to combinations of p_{it} and $p_{i,t-1}$.

for $\boldsymbol{p}_t$ and $\boldsymbol{p}_{t-1}$ in some appropriate set. Concavity has previously been assumed only on that part of the price space in which the current price of the firm is at least as great as its marginal cost.

Setting out the concavity conditions in detail:[9]

$$\pi_i^{ii} = (2 - C_i''F_i^i)F_i^i + (p_{it} - C_i')F_i^{ii} < 0, \qquad i = 1, \ldots, n, \tag{9.61}$$

$$\pi_i^{i+n,i+n} = -C_i''F_i^{i+n}F_i^{i+n} + (p_{it} - C_i')F_i^{i+n,i+n} < 0, \qquad i = 1, \ldots, n, \tag{9.62}$$

$$\begin{aligned}\pi_i^{ii}\pi_i^{i+n,i+n} - [\pi_i^{i,i+n}]^2 = {} & [p_{it} - C_i']^2[F_i^{ii}F_i^{i+n,i+n} - (F_i^{i,i+n})^2] \\ & - C_i''(p_{it} - C_i')(F_i^{i+n}F_i^i)\begin{bmatrix} F_i^{ii} & -F_i^{i,i+n} \\ -F_i^{i,i+n} & F_i^{i+n,i+n}\end{bmatrix}\begin{bmatrix} F_i^{i+n} \\ F_i^i \end{bmatrix} \\ & + 2(p_{it} - C_i')F_i^iF_i^{i+n,i+n} - [2(p_{it} - C_i')F_i^{i,i+n} + F_i^{i+n}]F_i^{i+n} \\ & > 0, \qquad i = 1, \ldots, n.\end{aligned} \tag{9.63}$$

If price is at least as large as marginal cost ($p_{it} \geq C_i'$), marginal cost is nondecreasing ($C_i'' \geq 0$) and F_i is concave then the conditions in eqs. (9.61) and (9.62) are met. As to the final condition, the first three of the four terms are nonnegative; however, they go to zero as $p_{it} - C_i'$ goes to zero. The sign of the last term is unclear; but as price nears marginal cost, it must become negative unless F_i^{i+n} should happen to go to zero also (and sufficiently rapidly).

It is clear that the concavity condition is not yet assured. One way to obtain it is to find a way to restrict further the set of "interesting" prices. Consider forcing marginal revenue ($p_{it} + F_i/F_i^i$) to be at least as large as marginal cost, assuming a positive fixed cost, and requiring nonnegative profits for each period. It only remains to be sure that, for any $\boldsymbol{p}_{t-1}$ and $\bar{\boldsymbol{p}}_{it}$, there are current prices for which current profit is positive. With these additional conditions, the firm has, for each period, a nonempty set of prices for which its current profits are nonnegative; and, on that set of prices, current profit is a concave function. If the firms are forced to keep prices within the range where marginal cost is no more than marginal revenue, then a noncooperative equilibrium exists for the model. On the other hand, if firms are allowed to choose prices anywhere in the interval from zero to p_i^+, it is unclear whether it is always in their interest to choose within the smaller interval. Probably more conditions would be needed on the model to assure an equilibrium with the wider range of prices available.

[9]The arguments of the functions are suppressed to make the notation more concise.

There is a result which is easy to see if S6 is added to the assumptions. If a stationary oligopoly is to have a steady state equilibrium, it surely must be at prices which are such that marginal revenue exceeds marginal cost and profits are positive. Imagine prices for which such conditions do not hold for at least one firm. Then the firm could find another steady state price which would be superior.

Existence of weak noncooperative equilibrium seems to pose the same problems for time dependent oligopoly as does noncooperative equilibrium; though perhaps it is easier to assure quasi-concavity of $\pi_i(\boldsymbol{p}_t, \boldsymbol{p}_{t-1}) + \alpha_i\pi_i(\boldsymbol{p}_{t+1}, \boldsymbol{p}_t)$ than concavity of π_i.

With respect to the balanced temptation equilibrium, the same problem crops up here as in the time independent models. That is the satisfaction of S9, the condition which assures needed characteristics for the profit possibility frontier.

While the details are not fully worked out in this section, it is clear that the time dependent supergame results can find fruitful application in oligopoly models; but not without some effort.

6. Concluding comments

On generalizing the supergame model of ch. 8 to allow time dependence, existence of a noncooperative equilibrium based upon single-period results is lost. There is a vestige of remaining connection through the correspondences $\boldsymbol{\phi}_t$. It would not be wholly unreasonable to regard the payoff function P_i^* as the analog of the single-period payoff function. Then a sequence of single-period noncooperative equilibria for the supergame, which then satisfy $\boldsymbol{s}_t^* \in \boldsymbol{\phi}_t(\boldsymbol{s}_{t+1}^*, \boldsymbol{s}_{t-1}^*)$ for all t, form a weak noncooperative equilibrium for the supergame. Furthermore, any simple strategy noncooperative equilibrium for the supergame is also a weak noncooperative equilibrium. Thus it is also a sequence of single game noncooperative equilibria. The converse, of course, is not true in general; however, it is true under S1–S5.

Interpreting $P_i^*(s_{t+1}, s_t, s_{t-1})$ as the analog of the single-period payoff function is plausible, but not the only possibility. In favor of it is that it is the whole segment of the supergame payoff which involves s_t. Against it is that it incorporates payoffs to be actually received in both periods t and $t+1$. If the model were generalized so that the payoff in period t were to depend on several past periods' strategies rather than just one, then $\boldsymbol{P}^*$

would have to be defined thus,

$$P_i^*(s_{t+k}, \ldots, s_{t-k}) = \sum_{l=0}^{k} \alpha_i^l P_i(s_{t+l}, \ldots, s_{t+l-k}). \tag{9.64}$$

The reader must decide for himself whether he finds this satisfactory.

It is pleasing that the time dependent supergame can be formulated and results obtained for nonstationary as well as stationary games. In addition, for stationary games there are results on steady state equilibria. Most of the results appear to be adaptable to oligopoly models.

CHAPTER 10

NONCOOPERATIVE STOCHASTIC GAMES

1. The stochastic game model

1.1. Stochastic games and supergames

The model used in this chapter is a supergame model without time dependence, having a countable number of distinct constituent games. Instead of encountering the games in a specified order, the occurrence of a game is randomly determined with probabilities which depend both on the current constituent game and on the constituent game strategies which have been chosen. Randomness only enters the model in determining the *transition* from one constituent game to the next. In general, it is possible for a given constituent game to be realized (i.e. played) more than once, or never. Unlike chs. 8 and 9, attention is restricted entirely to simple strategies and the equilibrium whose existence is proved is stationary. Of course, admitting randomness into the model in the integral, interesting way in which it appears broadens the scope of supergame models in a very desirable fashion.

In the remainder of the present section, the axioms for the constituent games are given, then the rest of the model is described. This includes the rules of transition from one state (constituent game) to another, a discussion of strategies and description of the payoff function. The final part of this section contains a result which limits the set of strategies which must be considered. §2 takes up the question of existence of equilibrium in a model in which the number of states is countable. It is shown that a stationary simple strategy noncooperative equilibrium exists. The equilibrium strategies are stationary in the sense that the action chosen by a player in a given time period does not depend on what period it is. The choice depends on which constituent game is being played; however, any time a particular game should recur, the same actions would be chosen. In §3 comparisons are made between the model and results of the chapter and chs. 8 and 9. In §4 applications to oligopoly are considered.

1.2. Axioms for the constituent games

As in earlier chapters, it is assumed there are n players who are unchanging over time. They play a sequence of constituent games which are described below. The constituent games are indexed by $k = 1, 2, \ldots$. k is called the *state* and $\mathbf{\Omega}$ is the set of states. When $\mathbf{\Omega}$ is finite and there are K states, $\mathbf{\Omega} = \{1, 2, \ldots, K\}$; and when $\mathbf{\Omega}$ is countable, but not finite, $\mathbf{\Omega}$ is the set of all positive integers $\{1, 2, \ldots\}$. $\mathbf{S}_{ik}$ is the strategy set of the ith player in the kth state.[1] As before, $\mathcal{S}_k = \mathcal{S}_{1k} \times \cdots \times \mathcal{S}_{nk}$, the joint strategy set for the kth state, and the elements of $\mathcal{S}_{ik}$ are s_{ik}. $s_k = (s_{1k}, \ldots, s_{nk})$. The payoff function for the ith player in the kth state is $P_{ik}(s_k)$. The axioms for constituent games are, in addition to S1:

S10. *$\mathcal{S}_{ik} \subset \mathcal{R}^m$ is a compact, convex set for $i = 1, \ldots, n$ and $k \in \mathbf{\Omega}$.*

S11. *The payoff functions, $P_{ik}(s_k)$ are continuous on $\mathcal{S}_k$. They are bounded by $|P_{ik}(s_k)| \leq \bar{P} < \infty$, $i = 1, \ldots, n$, $k \in \mathbf{\Omega}$.*

S12. *$P_{ik}(s_{ik}, \bar{s}_{ik})$ is a concave function of s_{ik}, $i = 1, \ldots, n$, $k \in \mathbf{\Omega}$.*

These assumptions are so much like assumptions made in earlier chapters that comment on them is superfluous; however, it is worthwhile to place these assumptions in relation to the work which they exposit. The topic of stochastic games begins with Shapley (1953a) which deals with finite two-person zero sum stochastic games. Rogers (1969) generalizes Shapley's results to two persons, variable sum. The equilibrium is the noncooperative equilibrium of Nash (1951) though the results do not generalize Nash because the latter uses an n-person model, while Roger's game is two person. Rogers does indicate how the n-person game might be tackled. Sobel (1971) has an existence theorem for an n-person noncooperative stochastic game, thus providing a generalization of Nash as well as of Rogers and Shapley.

All of these models share with one another the assumption that the

[1]The terms *state* and *constituent game* are used interchangeably.

number of states (constituent games) is finite. Unlike the model of this chapter, they all assume that the constituent games are finite. That is, in any constituent game, a player has a finite number of pure strategies from which to choose; and, of course, mixed strategies are available. For the exposition of this chapter, mixed strategies are not allowed. Because a player's strategy set is convex and his payoff concave in his own strategy, any mixed strategy is (weakly) dominated by a pure strategy. Sobel (1973) generalizes further his earlier results by allowing payoff functions defined on compact, convex sets, essentially like those specified in S10–S11; and allowing the number of states to be uncountable. The treatment below does not follow Sobel in going beyond a countable number of states because the exposition becomes very much more complex as rather simple, easily understood probabilities are replaced with measurable function. Staying with a countable number of states, it is possible to give a fully rigorous presentation of a very useful and interesting model. Meanwhile, the interested reader is definitely encouraged to read Sobel (1973).

1.3. The transition law, supergame strategies and supergame payoffs

1.3.1. The transition law

The transition law consists of a listing of the probability that a given state will be visited next, given the present state and the presently chosen constituent game payoff. Apart from obvious conditions about probabilities being nonnegative and summing to one, the probabilities are required to be continuous functions of the constituent game strategy for any current state. The transition law is written $q(k'|k, s_k)$ and is the probability that k' is the next state, given that k is the current state and s_k the current constituent game strategy.

S13. *$q(k'|k, s_k) \geq 0$, $\Sigma_{k' \in \Omega}\, q(k'|k, s_k) = 1$ for $k \in \boldsymbol{\Omega}$ and $s_k \in \mathcal{S}_k$. For fixed k, $\lambda \in [0, 1]$ and s_k, $s'_k \in \mathcal{S}_k$, $q(k'|k, \lambda s_k + (1-\lambda)s'_k) = \lambda q(k'|k, s_k) + (1-\lambda)q(k'|k, s'_k)$.*

Note the linearity of q in S13 is for fixed k, meaning that if there were some s_k available in more than one state, the transition probability associated with s_k could change thoroughly arbitrarily from one of the states to another. Assuming $q(k'|k, s_k)$ to be linear in s_k may seem somewhat restrictive for the present model; however, in a finite model, such as Sobel (1971), linearity with respect to mixed strategies holds when arbitrary probabilities are assigned to each (joint) pure strategy choice.

The transition law is interesting because the players' actions affect the probabilities. Were that not so, the game would not differ in any significant way from the nontime dependent supergames previously studied. The players in the game are taken to be maximizers of expected profit, and, because of the nature of the transition law, the action of a player today affects not only the payoff received today, but also the probability of being in any given game tomorrow. As some games may be more advantageous for a player than others, the effect of today's action on these probabilities must be taken into account.

1.3.2. Supergame strategies

A supergame strategy is, of course, a complete prescription on how to meet every possible contingency, based, in principle, on all available information. Imagine the position of a player in period t, before he has made his move (chosen his action or single-period strategy). He knows the *history* of the game to this point. Letting $\boldsymbol{h}_t$ denote the history at time t, it may be written

$$\boldsymbol{h}_t = (k_1, s_{k_1}, k_2, s_{k_2}, \ldots, k_{t-1}, s_{k_{t-1}}, k_t), \tag{10.1}$$

where $s_{k_\tau} \in \mathcal{S}_{k_\tau}$, $\tau = 1, \ldots, t-1$. k_τ is the actual state which obtains in periods $\tau = 1, \ldots, t$, and the s_{k_τ} are the observed actions. As the player knows all past states and actions as well as the current state, he can then choose any s_{ik_t} to associate with any possible history, $\boldsymbol{h}_t$. In general, a strategy for the supergame is of the form $\boldsymbol{\sigma} = (\boldsymbol{\theta}_{i1}(\boldsymbol{h}_1), \boldsymbol{\theta}_{i2}(\boldsymbol{h}_2), \ldots)$. The function $\boldsymbol{\theta}_{it}(\boldsymbol{h}_t)$, called a *policy*, can be any function which associates an element of $\mathcal{S}_{ik_t}$ with each value of $\boldsymbol{h}_t$. This parallels the discussion of supergame strategies in ch. 8 §2.1.

The simple strategies are of special interest. They are of the form $\boldsymbol{\theta}_{it}(\boldsymbol{h}_t) = \boldsymbol{\theta}_{it}(k_t)$, $t = 1, 2, \ldots$, where the policy followed in any time period is a function of the current state alone. Other details of the history – past states and actions – play no role. Indeed it turns out that attention can be restricted to an even smaller class of strategies consisting of the stationary simple strategies. These may be written $\boldsymbol{\theta}_i(k_t)$, $t = 1, 2, \ldots$, and have the characteristic that the policy followed by the ith player is the same in every time period. That is, if $k_t = k_{t'}$ then $\boldsymbol{\theta}_i(k_t) = \boldsymbol{\theta}_i(k_{t'})$. The action chosen at time t depends only on the state and not on the time period itself. $\boldsymbol{\theta}_i$ is sometimes called the policy and sometimes the strategy of the player.

The reason why stationary simple strategies are paramount is that if all players except the ith use such strategies, then among the optimal strategies available to the ith player is a stationary simple strategy. In the exposition to follow, general strategies are denoted $\boldsymbol{\sigma}$ and both stationary simple strategies and the policies associated with them by $\boldsymbol{\theta} = (\boldsymbol{\theta}_1, \ldots, \boldsymbol{\theta}_n)$. The set of all stationary policies for the ith player, consisting of all functions whose domain is the set of states and which associate with a state, k, any element of $\mathcal{S}_{ik}$, is denoted $\boldsymbol{\Theta}_i$. Then $\boldsymbol{\Theta} = \boldsymbol{\Theta}_1 \times \boldsymbol{\Theta}_2 \times \cdots \times \boldsymbol{\Theta}_n$.

1.3.3. Supergame payoffs

If a supergame strategy is specified for each player, the expected payoff to the ith player over the whole game depends on the particular state in which the supergame begins. Because of this it is convenient to form an expected payoff function. Let $\boldsymbol{q}(k, s_k)$ denote the probability distribution which governs the transition from state k when the action taken in state k is s_k. Thus

$$\boldsymbol{q}(k, s_k) = (q(1|k, s_k), q(2|k, s_k), \ldots). \tag{10.2}$$

Under a policy $\boldsymbol{\theta}$ which associates the action s_k with the state $k (k \in \boldsymbol{\Omega})$, $\boldsymbol{P}_{i\theta}$ is the vector of single period payoffs associated with the various states under $\boldsymbol{\theta}$,

$$\boldsymbol{P}_{i\theta} = (P_{i1}(s_1), P_{i2}(s_2), P_{i3}(s_3), \ldots). \tag{10.3}$$

For a given policy $\boldsymbol{\theta}$ a transition matrix may be defined using eq. (10.2):

$$\boldsymbol{q}_{\theta} = \begin{bmatrix} q(1, s_1) \\ q(2, s_2) \\ \cdot \\ \cdot \\ \cdot \end{bmatrix} = \begin{bmatrix} q(1|1, s_1) & q(2|1, s_1) \ldots \\ q(1|2, s_2) & q(2|2, s_2) \ldots \\ \cdot & \cdot \\ \cdot & \cdot \\ \cdot & \cdot \end{bmatrix}. \tag{10.4}$$

$\boldsymbol{q}(k, s_k)\boldsymbol{P}_{i\theta}$ is the undiscounted expected payoff in the next period when the policy to be used in the next period is $\boldsymbol{\theta}$, the current state is k and the current action is s_k. The vector of such payoffs is $\boldsymbol{q}_{\theta}\boldsymbol{P}_{i\theta}$. If the policy were to be used in many successive periods, the associated, undiscounted, expected payoff vectors associated with them would be $\boldsymbol{q}_{\theta}\boldsymbol{q}_{\theta}\boldsymbol{P}_{i\theta}$,

$\boldsymbol{q}_{\theta}\boldsymbol{q}_{\theta}\boldsymbol{q}_{\theta}\boldsymbol{P}_{i\theta}$, $\boldsymbol{q}_{\theta}\boldsymbol{q}_{\theta}\boldsymbol{q}_{\theta}\boldsymbol{q}_{\theta}\boldsymbol{P}_{i\theta}$, etc. These may be written more concisely as $\boldsymbol{q}_{\theta}^{\tau}\boldsymbol{P}_{i\theta}$, which is a vector whose kth component is the undiscounted expected payoff which the ith player will receive τ periods later, given that the policy $\boldsymbol{\theta}$ is currently being followed and will continue to be followed. Over the course of the supergame, if $\boldsymbol{\theta}$ is used throughout, the discounted supergame payoff vector of the ith player is

$$\boldsymbol{F}_{i\theta} = \sum_{t=1}^{\infty} \alpha_i^{t-1}\boldsymbol{q}_{\theta}^{t-1}\boldsymbol{P}_{i\theta}, \qquad i = 1, \ldots, n. \tag{10.5}$$

The vector of expected supergame payoffs of the ith player associated with an arbitrary strategy vector $\boldsymbol{\sigma}$, is denoted $\boldsymbol{F}_{i\sigma} = (F_{i\sigma 1}, F_{i\sigma 2}, \ldots, F_{i\sigma K})$.

1.4. Any strategy is weakly dominated by a stationary simple strategy

The central result of the chapter, which is proved in §2.2, is that a stochastic supergame satisfying S1 and S10–S13 has a stationary simple strategy noncooperative equilibrium. Due to a result of Blackwell, attention may be restricted for the remainder of the chapter to the set of stationary simple strategies $\boldsymbol{\Theta}$.

An optimal strategy is one for which the associated expected payoff is at least as high as under any other strategy for any of the states in $\boldsymbol{\Omega}$. Let $\bar{\boldsymbol{\theta}}_i$ be the stationary policies being followed by the other players and let $\boldsymbol{\sigma}_i^*$ be a best reply to $\bar{\boldsymbol{\theta}}_i$. To be a best reply, or optimal strategy, it must be that for any other strategy, $\boldsymbol{\sigma}_i$, simple or otherwise,

$$\boldsymbol{F}_{i,(\sigma_i^*,\bar{\theta}_i)} \geq \boldsymbol{F}_{i,(\sigma_i,\bar{\theta}_i)}. \tag{10.6}$$

Proof is not given for the following lemma, which can be found in Blackwell (1965).[2]

Lemma 10.1. *If a best reply to $\bar{\boldsymbol{\theta}}_i$ exists for the ith player in a stochastic supergame satisfying S1 and S10–S13 and having a countable set of states $\boldsymbol{\Omega}$, then an optimal stationary simple strategy exists – that is, among the best replies is an element of $\boldsymbol{\Theta}_i$.*

[2]This lemma is stated in terms of the model of this chapter. Blackwell is dealing with a straight dynamic programming problem which is a one-person decision problem; however, with the other players' behavior given by stationary simple strategies, the remaining model for the ith player is a subset of the models with which he deals. Our lemma 10.1 is his theorem 6c.

This lemma does not assert that a best reply does exist; but only that if there is a best reply, then there is one which is stationary and simple. The stationary character of the process makes it intuitively plausible that there should be an optimal stationary policy if an optimal policy exists. Essentially, this is because when the second period arrives and the second state is known, the decision-maker's situation is identical to what it would have been in the first period with that state as the initial state. The particular history prior to the current state is irrelevant.

2. Noncooperative equilibrium in the stochastic game

It is correct, though an oversimplification, to say that proving existence of a noncooperative equilibrium comes down to showing that the best reply mapping has a fixed point; and that to prove the existence of the fixed point, it is sufficient to show that the best reply mapping is upper semicontinuous and has convex image sets. These properties are shown in §2.2; however, there remains a fundamental question of what is to be meant by a *best reply* for a player. Assume that players other than the *i*th are using $\bar{\boldsymbol{\theta}}_i$. One unambiguous notion of best reply is that contained in eq. (10.6); namely, that $\boldsymbol{\theta}_i^*$ is a best reply to $\bar{\boldsymbol{\theta}}_i$ if the expected payoff to player i is at least as large as under any alternative strategy it might choose, for each and every state. Intuitively it is plausible that no best reply in this sense exist. Perhaps one strategy maximizes the expected payoff in state 1; but not in other states; and, in order to maximize the expected payoff in some other state, unconditional optimality for state 1 must be lost? Fortunately, it is possible to prove that a best reply in the sense of eq. (10.6) does exist. This is done in §2.1.

2.1. Decision from the viewpoint of one player

Throughout this section it is assumed that all players other than the *i*th are using known, unchanging stationary simple strategies $\bar{\boldsymbol{\theta}}_i$. Thus the problem being addressed is that of finding a best reply for the *i*th player to the strategies of the others. This is a dynamic programming problem and some of the dynamic programming literature provides needed results. To simplify the notation throughout §2.1, $\bar{\boldsymbol{\theta}}_i$ is usually suppressed, as is the subscript *i*. The only player under consideration is player *i*; hence, where, say, $\boldsymbol{\theta}$ is written, it refers to a policy of the *i*th player. In lemma 10.1 (§1.4) it is noted that among the best replies is a stationary simple strategy; then, in the present section, it is shown that a best reply, or optimal policy, in the sense of eq. (10.6) exists. The exposition follows Denardo (1971). In §2.1.1 it is proved that an optimal payoff exists, and, in §2.1.2 the optimal payoff is shown to be attainable by a policy available to the player.

2.1.1. Existence of an optimal payoff function

$\boldsymbol{F}_{\boldsymbol{\theta}}$ may be regarded as a return function which associates an expected payoff with each state k (for fixed $\boldsymbol{\theta}$). As such, it is a function from $\boldsymbol{\Omega}$ to the real line, $\mathscr{R}$. An optimal payoff function $\boldsymbol{F}^*$ is a return function which gives, for each possible initial state, the least upper bound on attainable payoffs. $\boldsymbol{F}^*$ may also be thought of as a function from $\boldsymbol{\Omega}$ to $\mathscr{R}$. It is defined as follows:[3]

$$F_k^* = \sup_{\theta = \Theta} F_{\theta k}, \qquad k \in \boldsymbol{\Omega}. \tag{10.7}$$

As a first step, it is now shown that associated with any stationary policy is a return function. Let $\mathscr{V}$ be the set of all bounded functions from $\boldsymbol{\Omega}$ to $\mathscr{R}$, the real line. For $\boldsymbol{F} \in \mathscr{V}$, let a norm be given by $\|\boldsymbol{F}\| = \sup_{k \in \boldsymbol{\Omega}} |F_k|$, and let $d(\boldsymbol{F}^0, \boldsymbol{F}^1) = \sup_{k \in \boldsymbol{\Omega}} |F_k^0 - F_k^1|$ be a distance for $\mathscr{V}$.

Lemma 10.2. *If $\boldsymbol{\theta}$ is a stationary policy which has an associated return function $\boldsymbol{F}_{\boldsymbol{\theta}}$, then $\boldsymbol{F}_{\boldsymbol{\theta}} \in \mathscr{V}$.*

□ To prove this lemma it suffices to show that $\|\boldsymbol{F}_{\boldsymbol{\theta}}\|$ must be bounded. By S11, the constituent game payoffs are bounded in absolute value by $\bar{P}$; hence supergame payoff cannot exceed $\bar{P}/(1-\alpha)$ in absolute value. Thus, if an optimal payoff function exists, it is a member of $\mathscr{V}$, the set of bounded functions from $\boldsymbol{\Omega}$ to $\mathscr{R}$. □

Lemma 10.3. *For any stationary policy $\boldsymbol{\theta}$, there exists a unique payoff function $\boldsymbol{F}_{\boldsymbol{\theta}} \in \mathscr{V}$.*

□ The method of proof is to show that the payoff function associated with $\boldsymbol{\theta}$ is the fixed point of a contraction mapping which carries elements

[3]At this point, the return function associated with $\boldsymbol{\theta}$, $\boldsymbol{F}_{\boldsymbol{\theta}}$, has not been shown to exist; therefore the definition in eq. (10.7) is really contingent on showing that existence, which is done in lemma 10.3.

of $\mathcal{V}$ into $\mathcal{V}$. As such, it is a unique fixed point. For an arbitrary element $\boldsymbol{F}$ of $\mathcal{V}$, define a mapping by

$$\boldsymbol{F}' = \boldsymbol{P}_{\theta} + \alpha \boldsymbol{q}_{\theta}\boldsymbol{F} = \boldsymbol{Y}_{\theta}(\boldsymbol{F}). \tag{10.8}$$

F'_k may be interpreted as the return in state k when the policy $\boldsymbol{\theta}$ is used in period 1 and the discounted payoff from period 2 onward is given by $\boldsymbol{F}$. If $\boldsymbol{F}' \neq \boldsymbol{F}$, then $\boldsymbol{F}$ is not the discounted payoff function associated with $\boldsymbol{\theta}$; but if $\boldsymbol{F}' = \boldsymbol{F}$, then it is.

The mapping $\boldsymbol{Y}_{\theta}$ is a contraction, carrying points of $\mathcal{V}$ into $\mathcal{V}$; hence, it has a unique fixed point. This fixed point is the payoff function associated with $\boldsymbol{\theta}$. To see that $\boldsymbol{Y}_{\theta}$ is a contraction, look at $d(\boldsymbol{Y}_{\theta}(\boldsymbol{F}^0), \boldsymbol{Y}_{\theta}(\boldsymbol{F}^1))$ for arbitrary $\boldsymbol{F}^0, \boldsymbol{F}^1 \in \mathcal{V}$:

$$\begin{aligned} d(\boldsymbol{Y}_{\theta}(\boldsymbol{F}^0), \boldsymbol{Y}_{\theta}(\boldsymbol{F}^1)) &= d((\boldsymbol{P}_{\theta} + \alpha \boldsymbol{q}_{\theta}\boldsymbol{F}^0), (\boldsymbol{P}_{\theta} + \alpha \boldsymbol{q}_{\theta}\boldsymbol{F}^1)) \\ &= \alpha d(\boldsymbol{q}_{\theta}\boldsymbol{F}^0, \boldsymbol{q}_{\theta}\boldsymbol{F}^1) \\ &\leq \alpha d(\boldsymbol{F}^0, \boldsymbol{F}^1). \end{aligned} \tag{10.9}$$

It is also clear from lemma 10.2 that $\boldsymbol{Y}_{\theta}$ carries elements of $\mathcal{V}$ into $\mathcal{V}$; and, in particular, if the subset of $\mathcal{V}$ is considered which consists of functions bounded by $\bar{P}/(1-\alpha)$, then members of this subset are carried into the same subset by $\boldsymbol{Y}_{\theta}$. Let $\boldsymbol{F}_{\theta}$ denote the unique fixed point of $\boldsymbol{Y}_{\theta}$. It is also the unique return function associated with $\boldsymbol{\theta}$. □

Using the return functions from lemma 10.3, it is now possible to assert the existence of the optimum return function which is defined in eq. (10.7). The optimum return function is also the upper envelope of the family of return functions. Two points about it should be emphasized. The first is that, so far, it has not been shown to be attainable. Secondly, though, for a given state, it is possible to find a policy which yields a return arbitrarily close to the optimum for that state, it has not been shown that there is a single policy which can come within a prescribed distance of the optimum return, regardless of state. By the way $\boldsymbol{F}^*$ is defined, it is clear that payoffs exceeding it cannot be found, even with respect to a single state.

2.1.2. Attainability of the optimum payoff

To say that the optimum return is attainable means there is some policy, $\boldsymbol{\theta}^*$ for which $\boldsymbol{F}_{\theta^*} = \boldsymbol{F}^*$. Showing attainability is done in several steps, the first of which establishes a characteristic of the mappings $\boldsymbol{Y}_{\boldsymbol{\theta}}$. This is done in lemma 10.4, after which a maximization operator is introduced which is used to find a return function which is shown to be the optimum return function. The maximization operator is proved in lemma 10.5 to be a contraction; and its unique fixed point is found, in several additional steps, to be the optimal return function. Thus it is shown in theorem 10.9 that the optimum return can be attained by a simple stationary policy.

Lemma 10.4. *For any $\boldsymbol{\theta} \in \boldsymbol{\Theta}$ and $\boldsymbol{F} \in \mathcal{V}$, $d(\boldsymbol{F}_{\boldsymbol{\theta}}, \boldsymbol{F}) \leq d(\boldsymbol{Y}_{\boldsymbol{\theta}}(\boldsymbol{F}), \boldsymbol{F})/(1-\alpha)$.*

□ Letting $\boldsymbol{Y}^2_{\boldsymbol{\theta}}(\boldsymbol{F}) = \boldsymbol{Y}_{\boldsymbol{\theta}}(\boldsymbol{Y}_{\boldsymbol{\theta}}(\boldsymbol{F}))$ be the composition of $\boldsymbol{Y}_{\boldsymbol{\theta}}$ with itself and $\boldsymbol{Y}^m_{\boldsymbol{\theta}}$ the m-fold composition, then

$$\begin{aligned} d(\boldsymbol{Y}^m_{\boldsymbol{\theta}}(\boldsymbol{F}), \boldsymbol{F}) &\leq \sum_{j=1}^{m} d(\boldsymbol{Y}^j_{\boldsymbol{\theta}}(\boldsymbol{F}), \boldsymbol{Y}^{j-1}_{\boldsymbol{\theta}}(\boldsymbol{F})) \\ &\leq \sum_{j=1}^{m} \alpha^{j-1} d(\boldsymbol{Y}_{\boldsymbol{\theta}}(\boldsymbol{F}), \boldsymbol{F}) \\ &\leq d(\boldsymbol{Y}_{\boldsymbol{\theta}}(\boldsymbol{F}), \boldsymbol{F})/(1-\alpha). \end{aligned} \tag{10.10}$$

Because $d(\boldsymbol{Y}^m(\boldsymbol{F}_{\boldsymbol{\theta}}), \boldsymbol{F}_{\boldsymbol{\theta}}) \to 0$ as $m \to \infty$, then $d(\boldsymbol{Y}^m(\boldsymbol{F}), \boldsymbol{F}) \to d(\boldsymbol{F}_{\boldsymbol{\theta}}, \boldsymbol{F})$, and eq. (10.10) reduces to

$$d(\boldsymbol{F}_{\boldsymbol{\theta}}, \boldsymbol{F}) \leq d(\boldsymbol{Y}_{\boldsymbol{\theta}}(\boldsymbol{F}), \boldsymbol{F})/(1-\alpha), \tag{10.11}$$

which is what was to be proved. □

Now let $\boldsymbol{M}$ be a maximization operator which is a function whose range and domain are in $\mathcal{V}$.

$$\boldsymbol{M}(\boldsymbol{F}_k) = \sup_{s \in \mathcal{S}_k} (\boldsymbol{P}_{sk} + \alpha \boldsymbol{q}(k, s)\boldsymbol{F}), \qquad k \in \boldsymbol{\Omega}. \tag{10.12}$$

For any return function $\boldsymbol{F}$, $\boldsymbol{M}(\boldsymbol{F})$ is the return function associated under $\boldsymbol{M}$. $\boldsymbol{M}(\boldsymbol{F})$ may be interpreted in this way: $\boldsymbol{M}(\boldsymbol{F}_k)$ is the least upper bound on the player's expected return if $\boldsymbol{F}$ is the return function which prevails as of period 2 and the action chosen in period 1 is selected to yield a period 1 payoff arbitrarily close to the upper bound on period 1 payoffs. Loosely speaking, $\boldsymbol{M}(\boldsymbol{F}_k)$ is the highest payoff which can be associated with current state k if the current action may be freely chosen, but the policy yielding $\boldsymbol{F}$ must be followed from the second period on.

Lemma 10.5. *$\boldsymbol{M}$ maps $\mathcal{V}$ into $\mathcal{V}$ and is a contraction.*

□ To see that $\boldsymbol{M}$ maps functions in $\mathscr{V}$ to functions in $\mathscr{V}$, note that if $\boldsymbol{F} \in \mathscr{V}$ is bounded in absolute value by $\bar{P}/(1-\alpha)$, then $\boldsymbol{M}(\boldsymbol{F})$ obeys the same bound:

$$\sup_{s \in \mathscr{S}_k} |P_{sk} + \alpha \boldsymbol{q}(k, s)\boldsymbol{F}| \leq \sup_{s \in \mathscr{S}_k} |P_{sk}| + \alpha \sup_{s \in \mathscr{S}_k} |\boldsymbol{q}(k, s)\boldsymbol{F}| = \bar{P}/(1+\alpha). \tag{10.13}$$

That $\boldsymbol{M}$ is a contraction can be seen from

$$\begin{aligned} d(\boldsymbol{M}(\boldsymbol{F}^0), \boldsymbol{M}(\boldsymbol{F}^1)) &= \sup_{k \in \boldsymbol{\Omega}} d(\sup_{s \in \mathscr{S}_k} (P_{sk} + \alpha \boldsymbol{q}(k, s)\boldsymbol{F}^0), \\ &\quad \sup_{s \in \mathscr{S}_k} (P_{sk} + \alpha \boldsymbol{q}(k, s)\boldsymbol{F}^1)) \\ &\leq \sup_{k \in \boldsymbol{\Omega}} d(\sup_{s \in \mathscr{S}_k} P_{sk}, \sup_{s \in \mathscr{S}_k} P_{sk}) \\ &\quad + \alpha \sup_{k \in \boldsymbol{\Omega}} d(\sup_{s \in \mathscr{S}_k} \boldsymbol{q}(k, s)\boldsymbol{F}^0, \sup_{s \in \mathscr{S}_k} \boldsymbol{q}(k, s)\boldsymbol{F}^1) \\ &= \alpha d(\boldsymbol{q}\boldsymbol{F}^0, \boldsymbol{q}\boldsymbol{F}^1) \leq \alpha d(\boldsymbol{F}^0, \boldsymbol{F}^1). \end{aligned} \tag{10.14}$$

Two mappings have been shown to be contractions, and in both instances, the discount parameter has played a crucial role in the proof. The mapping $\boldsymbol{M}$ is a *myopic improvement function.* That is, it gives a least upper bound on the highest attainable payoff function when the player can do what he likes in the present period but is stuck with what he gets in the future – or when he takes the view that he will not change his long-run plan from that which gives him $\boldsymbol{F}$, but will choose his current action so as to maximize supergame payoffs given that his future actions will be according to the old plan. In the theorem which follows, it is proved that there is exactly one return function which cannot be improved upon by the myopic improvement function $\boldsymbol{M}$.

Theorem 10.6. *There is a unique* $\boldsymbol{F}^{**} \in \mathscr{V}$ *for which* $\boldsymbol{F}^{**} = \boldsymbol{M}(\boldsymbol{F}^{**})$.

□ The proof follows directly from lemma 10.5. $\boldsymbol{M}$ is a contraction which maps a closed, bounded convex set into itself; hence, it has a unique fixed point which may be denoted $\boldsymbol{F}^{**}$. □

From the way that $\boldsymbol{F}^{**}$ is defined it is clear that it must be *close* to being attainable. At least there must be a policy which is capable of coming arbitrarily close to $\boldsymbol{F}^{**}$ in many (i.e. any finite number of) states. Additionally, $\boldsymbol{F}^{**}$ must be bounded above by $\boldsymbol{F}^{*}$, the optimum return function defined in eq. (10.7). What is to be shown in the remainder of this section is that $\boldsymbol{F}^{**}$ is actually attainable by some policy, $\boldsymbol{\theta}^*$; and, $\boldsymbol{F}^{**} = \boldsymbol{F}^*$. Thus the optimum return function is attainable.

Theorem 10.7. *For $\epsilon > 0$, there exists a policy $\boldsymbol{\theta}$ such that $d(\boldsymbol{Y}_{\boldsymbol{\theta}}(\boldsymbol{F}^{**}), \boldsymbol{F}^{**}) \leq \epsilon(1-\alpha)$; and any such policy satisfies $d(\boldsymbol{F}_{\boldsymbol{\theta}}, \boldsymbol{F}^{**}) \leq \epsilon$. If $d(\boldsymbol{Y}_{\boldsymbol{\theta}}(\boldsymbol{F}^{**}), \boldsymbol{F}^{**}) = 0$ then $\boldsymbol{F}_{\boldsymbol{\theta}} = \boldsymbol{F}^{**}$.*

□ A policy $\boldsymbol{\theta}$ which satisfies the conditions of the theorem is constructed. For each k, choose $\boldsymbol{\theta}(k)$ so that

$$P_{\boldsymbol{\theta}(k)k} + \alpha \boldsymbol{q}(k, \boldsymbol{\theta}(k))\boldsymbol{F}^{**} > F_k^{**} - \epsilon(1-\alpha). \tag{10.15}$$

From the definition of $\boldsymbol{F}^{**}$ it is clear this can be done. The left side of eq. (10.15) is $\boldsymbol{Y}_{\boldsymbol{\theta}}(\boldsymbol{F}^{**})$; so, recalling that $\boldsymbol{Y}_{\boldsymbol{\theta}}(\boldsymbol{F}^{**}) \leq \boldsymbol{M}(\boldsymbol{F}^{**}) = \boldsymbol{F}^{**}$, the first statement of the theorem is proved.

To see that $d(\boldsymbol{F}_{\boldsymbol{\theta}}, \boldsymbol{F}^{**}) \leq \epsilon$, recall that for any $\boldsymbol{F}$ and any $\boldsymbol{\theta}$,

$$d(\boldsymbol{F}_{\boldsymbol{\theta}}, \boldsymbol{F}) \leq d(\boldsymbol{Y}_{\boldsymbol{\theta}}(\boldsymbol{F}), \boldsymbol{F})/(1-\alpha). \tag{10.16}$$

Substituting $\boldsymbol{F}^{**}$ for $\boldsymbol{F}$ in eq. (10.16), and using $d(\boldsymbol{Y}_{\boldsymbol{\theta}}(\boldsymbol{F}^{**}), \boldsymbol{F}^{**}) \leq \epsilon(1-\alpha)$, yields

$$d(\boldsymbol{F}_{\boldsymbol{\theta}}, \boldsymbol{F}^{**}) \leq d(\boldsymbol{Y}_{\boldsymbol{\theta}}(\boldsymbol{F}^{**}), \boldsymbol{F}^{**}), \boldsymbol{F}^{**})/(1-\alpha) \leq \epsilon, \tag{10.17}$$

which proves the second statement of the theorem. That $d(\boldsymbol{Y}_{\boldsymbol{\theta}}(\boldsymbol{F}^{**}), \boldsymbol{F}^{**}) = 0$ implies $\boldsymbol{F}_{\boldsymbol{\theta}} = \boldsymbol{F}^{**}$ is obvious from the definition of the distance measure. The distance can be zero if and only if $F_{\theta k} = F_k^{**}$ for all $k \in \boldsymbol{\Omega}$. □

Corollary 10.8. *There is a policy $\boldsymbol{\theta}$ such that $\boldsymbol{F}_{\boldsymbol{\theta}} = \boldsymbol{F}^{**}$.*

□ Construct a sequence of policies $\boldsymbol{\theta}^m$ having the property that, for an arbitrary $\epsilon > 0$,

$$d(\boldsymbol{F}_{\boldsymbol{\theta}^m}, \boldsymbol{F}^{**}) \leq \epsilon/2^m, \qquad m = 1, 2, \ldots. \tag{10.18}$$

From theorem 10.7, this is clearly possible. The $\boldsymbol{\theta}^m$ are elements of a compact set; hence the sequence has a cluster point. Call such a cluster point $\boldsymbol{\theta}^*$. Then $\boldsymbol{F}_{\boldsymbol{\theta}^*} = \boldsymbol{F}^{**}$. □

It is now possible to show that the optimum return function $\boldsymbol{F}^*$ is attainable. This is done by proving that $\boldsymbol{F}^* = \boldsymbol{F}^{**}$.

Theorem 10.9. *There is a policy $\boldsymbol{\theta}$ such that $\boldsymbol{F}_{\boldsymbol{\theta}} = \boldsymbol{F}^*$.*

□ It is already known that $\boldsymbol{F}^{**} \leq \boldsymbol{F}^*$. The method of proof is to show that $\boldsymbol{F}^* \leq \boldsymbol{F}^{**}$, thus implying $\boldsymbol{F}^* = \boldsymbol{F}^{**}$ and that $\boldsymbol{\theta}^*$ is the policy which attains $\boldsymbol{F}^*$. Recall that for any $\boldsymbol{\theta} \in \boldsymbol{\Theta}$ and any $\boldsymbol{F}^0, \boldsymbol{F}^1 \in \mathcal{V}$, if $\boldsymbol{F}^0 \geq \boldsymbol{F}^1$ then $\boldsymbol{Y}_{\boldsymbol{\theta}}(\boldsymbol{F}^0) \geq \boldsymbol{Y}_{\boldsymbol{\theta}}(\boldsymbol{F}^1)$. This may be seen with the help of eq. (10.8), which defines $\boldsymbol{Y}_{\boldsymbol{\theta}}$,

$$P_{\boldsymbol{\theta}(k),k} + \alpha \boldsymbol{q}(k, \boldsymbol{\theta}(k))\boldsymbol{F}^0 - [P_{\boldsymbol{\theta}(k),k} + \alpha \boldsymbol{q}(k, \boldsymbol{\theta}(k))\boldsymbol{F}^1] = \alpha \boldsymbol{q}(k, \boldsymbol{\theta}(k))(\boldsymbol{F}^0 - \boldsymbol{F}^1). \qquad (10.19)$$

Equation (10.19) is clearly nonnegative when $\boldsymbol{F}^0 \geq \boldsymbol{F}^1$. Furthermore, comparing the definitions of $\boldsymbol{Y}_{\boldsymbol{\theta}}$ and $\boldsymbol{F}^{**}$,

$$\boldsymbol{Y}_{\boldsymbol{\theta}}(\boldsymbol{F}^{**}) \leq \boldsymbol{F}^{**} \qquad (10.20)$$

for any $\boldsymbol{\theta}$; hence,

$$\boldsymbol{Y}_{\boldsymbol{\theta}}^m(\boldsymbol{F}^{**}) \leq \boldsymbol{F}^{**}, \qquad m = 1, 2, \ldots. \qquad (10.21)$$

Also,

$$d(\boldsymbol{Y}_{\boldsymbol{\theta}}^m(\boldsymbol{F}^{**}), \boldsymbol{F}_{\boldsymbol{\theta}} \to 0 \text{ as } m \to \infty, \qquad \boldsymbol{\theta} \in \boldsymbol{\Omega}. \qquad (10.22)$$

Eq. (10.22) implies that $\boldsymbol{F}_{\boldsymbol{\theta}} \leq \boldsymbol{F}^{**}$ for any $\boldsymbol{\theta} \in \boldsymbol{\Theta}$, which implies that

$$\sup_{\theta \in \boldsymbol{\Omega}} F_{\boldsymbol{\theta} k} \leq F_k^{**}, \qquad k \in \boldsymbol{\Omega}. \qquad (10.23)$$

But the left-hand side of eq. (10.23) is F_k^*, hence $\boldsymbol{F}^* \leq \boldsymbol{F}^{**}$. The latter, with $\boldsymbol{F}^{**} \geq \boldsymbol{F}^*$, implies $\boldsymbol{F}^* = \boldsymbol{F}^{**}$, which means that $\boldsymbol{F}_{\boldsymbol{\theta}^*} = \boldsymbol{F}^*$. □

2.2. Noncooperative equilibrium for the n-person stochastic game

The results of §2.1 provide a natural definition of the best reply mapping. Letting $\mathscr{r}_i(\bar{\boldsymbol{\theta}}_i)$ denote the set of best replies to $\bar{\boldsymbol{\theta}}_i$ for player i, it is now known that $\mathscr{r}_i(\bar{\boldsymbol{\theta}}_i)$ is not empty. Furthermore, for any $\bar{\boldsymbol{\theta}}_i \in \bar{\boldsymbol{\Theta}}_i$, some of the best replies are in $\boldsymbol{\Theta}_i$. In the remainder of this chapter, best replies outside of $\boldsymbol{\Theta}_i$ continue to be ignored, and the range of $\mathscr{r}_i$ is taken to be in $\boldsymbol{\Theta}_i$. Proving existence of a noncooperative equilibrium is accomplished if it is shown that the mapping $\mathscr{r}(\boldsymbol{\theta}) = (\mathscr{r}_1(\bar{\boldsymbol{\theta}}_1), \ldots, \mathscr{r}_n(\bar{\boldsymbol{\theta}}_n))$ is upper semicontinuous, has compact, convex image sets, and maps a compact, convex set into itself. The upper semicontinuity of $\mathscr{r}$ follows directly from continuity with respect to $\boldsymbol{\theta}$ of the expected payoff functions $\boldsymbol{F}_{i\boldsymbol{\theta}}$; however, the results of §2.1 do not require continuity of $\boldsymbol{F}_{\boldsymbol{\theta}}$ with respect to $\boldsymbol{\theta}$. §2.2.1 looks into continuity of the $\boldsymbol{F}_{i\boldsymbol{\theta}}$. In §2.2.2 the convexity of the image sets of $\mathscr{r}$ is proved and in §2.2.3 the existence of equilibrium is proved.

2.2.1. Continuity of the expected payoff function

Under policies $\boldsymbol{\theta}$, the payoff function of the ith player is given by

$$\boldsymbol{F}_{i\theta} = \sum_{t=1}^{\infty} \alpha_i^{t-1} \boldsymbol{q}_\theta^{t-1} \boldsymbol{P}_{i\theta}. \tag{10.5}$$

Continuity of $\boldsymbol{F}_{i\theta}$ with respect to $\boldsymbol{\theta}$ is almost immediate when $\boldsymbol{\Omega}$ is finite.

Theorem 10.10. *Under S1 and S10–S13, when there are K ($< \infty$) states, the expected payoff functions $\boldsymbol{F}_{i\theta}$ are continuous with respect to $\boldsymbol{\theta}$, ($i = 1, \ldots, n$).*

□ It is sufficient to prove the result for one i. First, from S11, $\boldsymbol{P}_{i\theta}$ is continuous in $\boldsymbol{\theta}$; and from S13, $\boldsymbol{q}_\theta$ is continuous in $\boldsymbol{\theta}$. Furthermore, both $\boldsymbol{P}_{i\theta}$ and $\boldsymbol{q}_\theta$ are bounded. Thus eq. (10.5) may be rewritten

$$\boldsymbol{F}_{i\theta} = [\boldsymbol{I} - \alpha_i \boldsymbol{q}_\theta]^{-1} \boldsymbol{P}_{i\theta}, \tag{10.24}$$

and, with $0 \le \alpha_i < 1$, the continuity of $\boldsymbol{q}_\theta$ with respect to $\boldsymbol{\theta}$ implies, along with the dominant diagonal (see McKenzie (1960)) condition on $[\boldsymbol{I} - \alpha_i \boldsymbol{q}_\theta]$, that $[\boldsymbol{I} - \alpha_i \boldsymbol{q}_\theta]^{-1}$ is continuous with respect to $\boldsymbol{\theta}$; therefore, $\boldsymbol{F}_{i\theta}$ is a continuous function of $\boldsymbol{\theta}^*$. □

Continuity of $\boldsymbol{F}_{i\theta}$ with respect to $\boldsymbol{\theta}$ when the number of states is not finite may be expressed in either of two ways. The first is to state that each $F_{i\theta k}$ is a continuous function of $\boldsymbol{\theta}$. The second is to specify that for arbitrary $\boldsymbol{\lambda} = (\lambda_1, \lambda_2, \ldots)$, where $\lambda_k \ge 0$, $\Sigma_{k=1}^{\infty} \lambda_k = 1$, the function $\Sigma_{k=1}^{\infty} \lambda_k F_{i\theta k}$ is continuous with respect to $\boldsymbol{\theta}$. This includes the first statement of continuity as a special case. $\boldsymbol{\lambda}$ may be thought of as a probability distribution over initial states. The result which is proved using $\boldsymbol{\lambda}$ is that in a game in which the initial state is first chosen according to $\boldsymbol{\lambda}$, the expected payoff, $\Sigma_{k=1}^{\infty} \lambda_k F_{i\theta k}$, is a continuous function of $\boldsymbol{\theta}$.[5]

Theorem 10.11. *Under S1 and S10–S13, with countable $\boldsymbol{\Omega}$, and probabilities $\boldsymbol{\lambda}$ over the initial choice of states, the expected payoff function $\Sigma_{k\in\boldsymbol{\Omega}} \lambda_k F_{i\theta k}$ is a continuous function of $\boldsymbol{\theta}$.*

□ Let $\boldsymbol{\theta}^l$, $l = 1, 2, \ldots$, be a sequence of policies which converge to $\boldsymbol{\theta}^0$,

[5] A side matter worth keeping in mind is that the optimal $\boldsymbol{\theta}_i$ (given $\bar{\boldsymbol{\theta}}_i$) is independent of the initial probabilities $\boldsymbol{\lambda}$. This is the main result of §2.1.

and let $\boldsymbol{q}_l$ be the transition matrix associated with $\boldsymbol{\theta}^l$ for $l = 0, 1, 2, \ldots$. The constituent game strategies associated with $\boldsymbol{\theta}^l$ are $\{s_k^l\}$. It is shown below that as $\boldsymbol{\theta}^l$ converges to $\boldsymbol{\theta}^0$, $\Sigma_{k\in\Omega} \lambda_k F_{i\theta^l k}$ converges to $\Sigma_{k\in\Omega} \lambda_k F_{i\theta^0 k}$. Choose ξ_l, γ_l, δ_l and k_l, $l = 1, 2, \ldots$, to satisfy the following conditions: (a) $\xi_l, \delta_l > 0$, $l = 1, 2, \ldots$; (b) $\lim_{l\to\infty} \xi_l = 0$; (c) $\frac{1}{4} \geq \delta_1 > \delta_2 > \delta_3 > \cdots$; (d) $\lim_{l\to\infty} \delta_l = 0$; and (e) $1 - 4\delta_l < \gamma_l < 1 - 3\delta_l$, $l = 1, 2, \ldots$. For each ξ_l and δ_l there is a finite $l(\delta_l)$ and finite $k_l \geq l$ such that, for $m \geq l(\delta_l)$,

$$|P_{i\theta^m k} - P_{i\theta^0 k}| \leq \xi_l, \tag{10.25}$$

$$\sum_{k'\in\Omega} |q(k'|k, s_k^m) - q(k'|k, s_k^0)| \leq \delta_l, \tag{10.26}$$

for $k = 1, \ldots, k_l$, and

$$\sum_{k'=1}^{k_l} q(k'|k, s_k^m) \geq \gamma_l, \tag{10.27}$$

for $k = 1, \ldots, l$. δ_l is an upper bound on the summed deviations of the elements of one of the first k_l rows of $\boldsymbol{q}_m$ from the corresponding elements of $\boldsymbol{q}_0$ for $m \geq l(\delta_l)$. γ_l is a lower bound on the sum of the first k_l elements of one of the first k_l rows of $\boldsymbol{q}_m$. That the various δ_l, γ_l and k_l can be found follows from the convergence of the $\boldsymbol{q}_l$ to $\boldsymbol{q}_0$.

δ_l, γ_l and k_l place an upper bound on the size of the sum of the absolute values of the deviations of the elements of each of the first l rows of $\boldsymbol{q}_m^t$ from the corresponding elements of $\boldsymbol{q}_0^t$, and a lower bound on the sum of the first k_l elements of each of the first l rows of $\boldsymbol{q}_m^t$ ($m \geq l(\delta_l)$). These bounds are developed below after it is shown where they are needed. To that end, let

$$\boldsymbol{q}_m^t = \begin{bmatrix} q_{11}^{mt} & q_{12}^{mt} & \cdots \\ q_{21}^{mt} & q_{22}^{mt} & \cdots \\ \cdot & \cdot & \\ \cdot & \cdot & \\ \cdot & \cdot & \end{bmatrix}, \qquad m = 0, 1, 2, \ldots, \quad t = 1, 2, \ldots, \tag{10.28}$$

and let

$$\epsilon_{jmt} = \sum_{k\in\Omega} |q_{jk}^{mt} - q_{jk}^{0t}|, \qquad j, t = 1, 2, \ldots, \quad m = 0, 1, 2, \ldots, \tag{10.29}$$

Note that $\epsilon_{jm1} \leq \delta_l$ for $j \leq k_l$ and $m \geq l(\delta_l)$. Now, for $m \geq l(\delta_l)$, consider the difference in expected payoff under $\boldsymbol{\theta}^m$ and $\boldsymbol{\theta}^0$. It is

$$\left|\sum_{t=1}^{\infty} \alpha_i^{t-1}\lambda \boldsymbol{q}_m^{t-1}\boldsymbol{P}_{i\theta^m} - \sum_{t=1}^{\infty} \alpha_i^{t-1}\lambda \boldsymbol{q}_0^{t-1}\boldsymbol{P}_{i\theta^0}\right|, \tag{10.30}$$

which is not greater than

$$\begin{aligned} &\xi_l + 2\bar{P}\sum_{t=2}^{\infty} \alpha_i^{t-1}\epsilon_{j,m,t-1}\sum_{k'=1}^{l}\lambda_{k'} + 2\bar{P}\sum_{t=2}^{\infty}\alpha_i^{t-1}\sum_{k'=l+1}^{\infty}\lambda_{k'} \\ &\le \xi_l + 2\bar{P}\sum_{t=2}^{\infty}\alpha_i^{t-1}\epsilon_{j,m,t-1} + (2\bar{P}\alpha_i/(1-\alpha_i))\sum_{k'=l+1}^{\infty}\lambda_{k'}. \end{aligned} \tag{10.31}$$

As $l \to \infty$ the first and third terms in eq. (10.31) go to zero. To see that the second term goes to zero, upper bounds must be found for the ϵ_{jmt} for $j \le l$, $m \ge l(\delta_l)$ and $t \ge 2$. Such a bound is

$$\epsilon_{j,m,t} \le 3^{t-1}\delta_l + \sum_{k'=1}^{t-2} 3^{t-k'-1}(1-\gamma_l^{k'}). \tag{10.32}$$

That the bounds given in eq. (10.32) hold may be seen by noting, first, that

$$\sum_{k=1}^{k_l} q_{jk}^{mt} \ge \gamma_l^t, \qquad j = 1, \ldots, l. \tag{10.33}$$

Second, if the $\epsilon_{j,m,t-1}$ obey a common bound of $\epsilon_{m,t-1}$ for $j = 1, \ldots, l$, then the $\epsilon_{j,m,t}$ obey a common bound of

$$\begin{aligned} \epsilon_{mt} &\le 2\epsilon_{m,t-1} + \epsilon_{m,t-1}^2 + (1-\gamma_l^{t-1}) \\ &< 3\epsilon_{m,t-1} + (1-\gamma_l^{t-1}) \\ &\le 3^{t-1}\delta_l + \sum_{k'=1}^{t-2} 3^{t-k'-1}(1-\gamma_l^{k'}), \qquad t \ge 2. \end{aligned} \tag{10.34}$$

Thus the middle term in eq. (10.31) is bounded by

$$\begin{aligned} &2\bar{P}\left[\delta_l\sum_{t=2}^{\infty}\alpha_i^{t-1}3^{t-2} + \sum_{t=3}^{\infty}\alpha_i^{t-1}\sum_{k'=1}^{t-3}3^{t-k'-2}(1-\gamma_l^{k'})\right] \\ &= 2\bar{P}\alpha_i[\delta_l/(1-3\alpha_i\delta_l) + \alpha_i(1/(1-\alpha_i) - \gamma_l/(1-\alpha_i\gamma_l))/(1-3\alpha_i)]. \end{aligned} \tag{10.35}$$

As $l \to \infty$, $\delta_l \to 0$ and $\gamma_l \to 1$; hence, the expression in square brackets in eq. (10.35) goes to zero. Therefore, the payoff function $\Sigma_{k\in\Omega}\lambda_k F_{i\theta k}$ is continuous in $\boldsymbol{\theta}$. □

2.2.2. Convexity of the image sets of the best reply mapping

The continuity proved in theorem 10.11 assures that the best reply mapping, $\imath$, is upper-semicontinuous. Proof is not given in the present chapter because it would be essentially the same as the proof of this property in theorem 7.4. It is shown in the present section that the set of best replies to a given strategy is convex. That is, if $\boldsymbol{\theta}^0, \boldsymbol{\theta}^1 \in \imath(\boldsymbol{\theta}^2)$, then for $\lambda \in (0, 1)$, $\lambda\boldsymbol{\theta}^0 + (1-\lambda)\boldsymbol{\theta}^1 \in \imath(\boldsymbol{\theta}^2)$ as well.

Lemma 10.12. *Under S1 and S10–S13, with countable* $\boldsymbol{\Omega}$, *the sets* $\imath(\boldsymbol{\theta})$ *are convex for any* $\boldsymbol{\theta} \in \boldsymbol{\Theta}$.

□ $\imath(\boldsymbol{\theta})$ is convex if each of the $\imath_i(\bar{\boldsymbol{\theta}}_i)$ is convex for any $\bar{\boldsymbol{\theta}}_i \in \bar{\boldsymbol{\theta}}_i$; therefore, the proof may, without loss of generality, be stated only for player 1. Let $(\boldsymbol{\theta}_1^0, \boldsymbol{\theta}_2^0, \ldots, \boldsymbol{\theta}_n^0) = (\boldsymbol{\theta}_1^0, \bar{\boldsymbol{\theta}}_1^0)$ and $\boldsymbol{\theta}^1 = (\boldsymbol{\theta}_1^1, \boldsymbol{\theta}_2^0, \ldots, \boldsymbol{\theta}_n^0) = (\boldsymbol{\theta}_1^1, \bar{\boldsymbol{\theta}}_1^0)$ be two arbitrary policies which differ only with respect to the choice of player 1; and let $\boldsymbol{\psi} = (\boldsymbol{\psi}_1, \bar{\boldsymbol{\theta}}_1^0)$, where $\boldsymbol{\psi}_1 = \lambda\boldsymbol{\theta}_1^0 + (1-\lambda)\boldsymbol{\theta}_1^1$ for $\lambda \in (0, 1)$. The lemma is proved if $\boldsymbol{\theta}_1^0, \boldsymbol{\theta}_1^1 \in \imath_1(\bar{\boldsymbol{\theta}}_1^0)$ implies $\boldsymbol{\psi}_1 \in \imath_1(\bar{\boldsymbol{\theta}}_1^0)$. Let the single-period payoffs associated with $\boldsymbol{\theta}^0$, $\boldsymbol{\theta}^1$ and $\boldsymbol{\psi}$ be, respectively, $\boldsymbol{P}_{10}$, $\boldsymbol{P}_{11}$ and $\boldsymbol{P}_{1\psi}$ and let the three transition matrices be $\boldsymbol{q}_0$, $\boldsymbol{q}_1$ and $\boldsymbol{q}_\psi$. Note that $\sum_{t=1}^{\infty} \alpha_1^{t-1}\boldsymbol{q}^{t-1}$ has a limit which may be denoted $[\boldsymbol{I} - \alpha_1\boldsymbol{q}]^{-1}$ for any transition matrix $\boldsymbol{q}$. That $\boldsymbol{\theta}_1^0$ and $\boldsymbol{\theta}_1^1$ are both optimal means $\boldsymbol{F}_{1\theta^0} = \boldsymbol{F}_{1\theta^1}$ or

$$\begin{aligned}\sum_{t=1}^{\infty} \alpha_1^{t-1}\boldsymbol{q}_0^{t-1}\boldsymbol{P}_{10} &= [\boldsymbol{I} - \alpha_1\boldsymbol{q}_0]^{-1}\boldsymbol{P}_{10}\\ &= \sum_{t=1}^{\infty} \alpha_1^{t-1}\boldsymbol{q}_1^{t-1}\boldsymbol{P}_{11}\\ &= [\boldsymbol{I} - \alpha_1\boldsymbol{q}_1]^{-1}\boldsymbol{P}_{11}.\end{aligned} \tag{10.36}$$

It must be shown that

$$[\boldsymbol{I} - \alpha_1\boldsymbol{q}_\psi]^{-1}\boldsymbol{P}_{1\psi} = [\boldsymbol{I} - \alpha_1\boldsymbol{q}_0]^{-1}\boldsymbol{P}_{10}. \tag{10.37}$$

Using the concavity of $\boldsymbol{P}_{1\theta}$,

$$\begin{aligned}\boldsymbol{F}_{1\psi} &= [\boldsymbol{I} - \alpha_1\boldsymbol{q}_\psi]^{-1}\boldsymbol{P}_{1\psi}\\ &= \lambda[\boldsymbol{I} - \alpha_1\boldsymbol{q}_\psi]^{-1}\boldsymbol{P}_{10} + (1-\lambda)[\boldsymbol{I} - \alpha_1\boldsymbol{q}_\psi]^{-1}\boldsymbol{P}_{11}\\ &\quad + [\boldsymbol{I} - \alpha_1\boldsymbol{q}_\psi]^{-1}\boldsymbol{P}_\epsilon,\end{aligned} \tag{10.38}$$

where $\boldsymbol{P}_\epsilon \geq 0$. From eq. (10.37),

$$\boldsymbol{P}_{11} = [\boldsymbol{I} - \alpha_1\boldsymbol{q}_1][\boldsymbol{I} - \alpha_1\boldsymbol{q}_0]^{-1}\boldsymbol{P}_{10}. \tag{10.39}$$

Substituting into eq. (10.38) yields

$$\begin{aligned}\boldsymbol{F}_{1\psi} &= \lambda[\boldsymbol{I}-\alpha_1\boldsymbol{q}_\psi]^{-1}\boldsymbol{P}_{10}+(1-\lambda)[\boldsymbol{I}-\alpha_1\boldsymbol{q}_\psi]^{-1}[\boldsymbol{I}-\alpha_1\boldsymbol{q}_1]\\ &\quad\times[\boldsymbol{I}-\alpha_1\boldsymbol{q}_0]^{-1}\boldsymbol{P}_{10}+[\boldsymbol{I}-\alpha_1\boldsymbol{q}_\psi]^{-1}\boldsymbol{P}_\epsilon\\ &= [\boldsymbol{I}-\alpha_1\boldsymbol{q}_\psi]^{-1}\{\lambda[\boldsymbol{I}-\alpha_1\boldsymbol{q}_0]+(1-\lambda)[\boldsymbol{I}-\alpha_1\boldsymbol{q}_1]\}[\boldsymbol{I}-\alpha_1\boldsymbol{q}_0]^{-1}\boldsymbol{P}_{10}\\ &\quad+[\boldsymbol{I}-\alpha_1\boldsymbol{q}_\psi]^{-1}\boldsymbol{P}_\epsilon. \end{aligned} \tag{10.40}$$

Using S13, the expression in curly brackets in eq. (10.40) becomes $\boldsymbol{I}-\alpha_1\boldsymbol{q}_\psi$, and eq. (10.40) becomes

$$\begin{aligned}\boldsymbol{F}_{1\psi} &= [\boldsymbol{I}-\alpha_1\boldsymbol{q}_0]^{-1}\boldsymbol{P}_{10}+[\boldsymbol{I}-\alpha_1\boldsymbol{q}_\psi]^{-1}\boldsymbol{P}_\epsilon\\ &= \boldsymbol{F}_{1\theta^0}+[\boldsymbol{I}-\alpha_1\boldsymbol{q}_\psi]^{-1}\boldsymbol{P}_\epsilon. \end{aligned} \tag{10.41}$$

Noting that all elements of $[\boldsymbol{I}-\alpha_1\boldsymbol{q}_\psi]^{-1}$ and $\boldsymbol{P}_\epsilon$ are nonnegative, $\boldsymbol{F}_{1\psi}\geq \boldsymbol{F}_{1\theta^0}$; but strict inequality for any coordinate contradicts the optimality of $\boldsymbol{\theta}_1^0$ and $\boldsymbol{\theta}_1^1$. Therefore, equality holds, and the image sets of the best reply mapping are seen to be convex. □

It may be noted from the proof of lemma 10.12 that if strong concavity of $\boldsymbol{P}_{1\theta}$ were to hold, then $\boldsymbol{P}_\epsilon\neq 0$ (unless $\boldsymbol{\theta}^0=\boldsymbol{\theta}^1$) and $\boldsymbol{\imath}$ would be a (single valued) function, i.e. for any $\bar{\boldsymbol{\theta}}_1$, the optimal policy would be unique.

2.2.3. Existence of equilibrium

The noncooperative equilibrium is defined in a familiar way. The strategies given by the simple policies $\boldsymbol{\theta}^*=(\boldsymbol{\theta}_1^*,\ldots,\boldsymbol{\theta}_n^*)$ are strategies associated with a noncooperative equilibrium if (a)

$$\boldsymbol{\theta}_i^*\in\boldsymbol{\Theta}_i,\qquad i=1,\ldots,n, \tag{10.42}$$

and (b)

$$F_{i\theta^*k}=\sup_{\theta_i\in\Theta}F_{i,(\theta_i,\bar{\theta}_i^*),k},\qquad k\in\boldsymbol{\Omega},\quad i=1,\ldots,n. \tag{10.43}$$

Theorem 10.13. *A stochastic supergame satisfying S1, S10–S13 and having a countable set of states Ω has a noncooperative equilibrium with simple stationary strategies.*

□ It has already been shown that the best reply mapping is upper-semicontinuous and has convex image sets. By its definition, it is clear that it also maps the set $\boldsymbol{\Theta}$ into itself. The only remaining question is whether the set $\boldsymbol{\Theta}$ is compact. Because the individual constituent game strategy sets are compact, the set $\boldsymbol{\Theta}$ is compact in the product topology, using the Tychonoff product theorem (see Dunford and Schwartz (1957, p. 32)). For the super-semicontinuity of $\mathscr{r}$ to hold, the stochastic game payoffs must be continuous in the policies relative to the product topology; however, the proof of theorem 10.11 is carefully constructed so that continuity does hold relative to the product topology. Thus, the fixed point theorem of Bohnenblust and Karlin (1950) may be used to assert that $\mathscr{r}$ has a fixed point; hence, the stochastic game has a noncooperative equilibrium.[6] □

3. A comparison of the stochastic supergame model to other supergame models

There is a way to view the stochastic supergame model as a generalization of the model of ch. 8 and of some of the models of ch. 9; although most of the interesting results of chs. 8 and 9 are not special cases of theorem 10.13. Take first the model of ch. 8 in which the supergame lacks time dependence. Most obviously, if the transition law is completely degenerate, so that $q(k'|k, s_k) = 1$ when $k' = 1$ and equals zero otherwise, the game is essentially a stationary supergame without time dependence. The only way the game might differ is that no specification has been made here concerning the first period. However, after the first period, the state is always state 1 no matter what the choices of the players or the initial state.

The model can be made to cover the nonstationary games of ch. 8 by specifying a transition law such as

$$q(k+1|k, s_k) = 1, \qquad k \in \boldsymbol{\Omega}, \tag{10.44}$$

$$q(k'|k, s_k) = 0, \qquad k, k' \in \boldsymbol{\Omega}, \qquad k' \neq k+1. \tag{10.45}$$

Abstracting from first-period start up complications, eqs. (10.44) and (10.45) specify a transition law in which chance plays no role, and in which a finite sequence of games is played in a preset order. The equilibrium of theorem 10.13 is a simple strategy equilibrium of the form $\boldsymbol{\sigma} = (s_1^c, s_2^c, \ldots)$.

[6]See ch. 7 §3.4.

The models of ch. 8 also have equilibria for which the associated strategies are not simple strategies. It is not clear whether, and how, these may be adapted to the stochastic supergame model. Needless to say, the same comment applies to the balanced temptation equilibrium.

For the time dependent supergames of ch. 9, it is again possible to specialize the transition law in a way which eliminates the stochastic element to get either a stationary or a nonstationary time dependent supergame. To see the simpler game first, assume that for all k, the strategy sets $\mathcal{S}_{ik}$ are identical. Actually, for the supergame model of this chapter, only an approximation to the time dependent supergame can be given. First, partition the strategy set of the constituent game into a large number of very small regions $\mathcal{S}_1^*, \mathcal{S}_2^*, \ldots$, which are pairwise disjoint and whose union is the whole strategy set. In each of these regions, numbered $k = 1, 2, \ldots$, choose a point s_k. The transition law is

$$q(k'|k, s_k) = 1 \text{ if } s_k \in \mathcal{S}_{k'}^*, \tag{10.46}$$

$$q(k'|k, s_k) = 0 \text{ if } s_k \notin \mathcal{S}_k^*. \tag{10.47}$$

The payoff functions are

$$P_{ik}(s_k) = P_i(s_k, s_k^*). \tag{10.48}$$

By making the sets $\mathcal{S}_k^*$ smaller and more numerous, a closer and closer approximation is obtained to the stationary time dependent supergame. The key element is, of course, that the payoff function in a given time period depends on the current and previous period actions. This model could be modified so that the payoff functions are different, known in advance, from one period to another.

With the model as outlined above, it would probably be awkward, and perhaps impossible, to get the results on steady state equilibria which are obtained in ch. 9 §2.2. Very likely the extensions of the balanced temptation equilibrium, which involves strategies which are not simple, in ch. 8 §4 would also pose problems; however, the generality of the stochastic game model is very great and there is the possibility of getting all the results on noncooperative games which have been presented in chs. 7–10 to be special cases of it.

4. Applications to oligopoly

Adaptation of oligopoly models to conform to the stochastic game axioms presents no special problems beyond those encountered in adapting the simple noncooperative games of ch. 7 or the supergames of chs. 8 and 9 if, for the latter, the axioms which are special to deriving balanced temptation equilibria are left out of account.

With respect to the broadening of the models to allow stochastic elements, the transition law which makes transition probabilities depend on the current state and current action, is quite general.[7] It would be possible to have stochastic elements in the firms' demand or cost functions or both. The stochastic element for a firm could depend on the actions of any of the other firms as well as on its own. One required element is that when the firms make their decisions in any period t, the payoff functions have already been determined. Uncertainty only enters with respect to what the future payoff functions will be.

[7]An application of stochastic games to oligopoly can be found in Kirman and Sobel (1974).

CHAPTER 11

COOPERATIVE GAMES – THE VALUE APPROACH

1. An overview of cooperative games

The distinguishing feature of cooperative games is that the players are able to change the rules of the game as part of their play. Or, what comes to the same thing, they can restrict the sets of strategies to which they have access.[1] The means of accomplishing these changes is the binding agreement. In general, it is the members of a coalition who have the power to make binding agreements. These agreements bind only those in the coalition.[2] Thus, players 1, 3, 5 and 6 may agree that they will use, respectively, s'_1, s'_3, s'_5 and s'_6. They cannot force, say, player 2 to refrain from using any particular strategy in his strategy set; though their actions may make certain choices unattractive to him.

There are two good reasons which players might have for being willing to see their sets of available strategies reduced. The first is that an agreement among players which jointly limits their actions could force payoffs which are greater for each player than he would get in a noncooperative equilibrium. It is easy to give an example. Imagine a game in which the noncooperative equilibrium s^c is unique and interior to the set of attainable payoffs. Say the strategy vector s^* gives payoffs on the payoff frontier with $P_i(s^*) > P_i(s^c)$, $i = 1, \ldots, n$. For all the players to agree on s^* means that the ith player agrees to reduce his strategy set from $\mathcal{S}_i$ to $\{s_i^*\}$. In this restricted game, the noncooperative equilibrium is (trivially) at s^*. Compared with playing the original game noncooperatively it is clear that the restriction of available actions is in the interest of each player. What agreement can be expected to emerge, or ought to emerge are questions with which chs. 11 and 12 are concerned.

[1]In some models, for example those in which mixed strategies play a role, the set of strategies available to a coalition may be larger than the combinations of strategies of the individual members. See §4.2.1.

[2]Recall that binding agreements include commitments made by only one player.

The second reason for restricting one's strategies is to make a threat be credible. One player may threaten the others, saying that if a certain point is not agreed upon, then he will follow an action which will be costly to the others. It may easily be that the threatened action would not be in the player's interest to carry out if the desired agreement were not made. On the other hand, if the remaining players believe he would actually carry out the threat, they might be more inclined to go along with the agreement he wants. Assuming the others cannot be duped, the threatening player's threat is not credible unless he can put himself in a position which forces him to carry it out if his desired agreement is not made. If he can irrevocably commit himself to use the threat under prescribed circumstances, then he may be able to better his position over what it would otherwise be.

With the exception of some work of Aumann's (1959, 1960, 1961), all the models in chs. 11 and 12 are of *one-shot* or single-period games. Two approaches are taken. The first is the *value* approach which is studied in the remainder of the present chapter. The second is the *core* and is studied in ch. 12. Underlying the value approach is the aim of arriving at a vector of *values*, of which the ith component is the value of the game to the ith player. The vector of values should be both attainable and on the payoff frontier, as well as forming an outcome which is *desirable* or *just*. The models and results are those of Nash (1950, 1953), Shapley (1953b) and Harsanyi (1963). Nash restricts himself to two-person games in which utility is not transferable between players. That is, one player cannot give one unit of utility to the other at a cost of one unit to himself. Such transfers could be made if utility for each player were measured by money with one unit of utility equal to one unit of money for each. Shapley looks at n-person games with transferable utility and has developed a notion of value which is associated with his name (the *Shapley value*). Harsanyi has developed results for n-person games in which utility is not transferable. If Harsanyi's model is restricted to having transferable utility, which is less general than nontransferable, his results coincide with those of Shapley. On the other hand, letting n equal 2, he gets the same results as Nash.

The core is a generalization of Edgeworth's (1881) contract curve. The principal result of ch. 12 is to show a class of games in normal form for which the core is not empty. For this purpose, the exposition follows Scarf (1967, 1971). In quite different papers, Farrell (1970) and Shitovitz (1973) examine the core of a market economy in which oligopoly is represented by means of having a few traders be far more important than the other traders. Their importance is built into the model by giving them much larger endowments than those held by the *competitive* traders.

§§ 2–4 report on the results of Nash, Shapley and Harsanyi, respectively. §5 contains concluding comments.

2. The Nash models for two-person cooperative games

Both Nash (1950) and (1953) present models in which the solution is partly determined by the threats which the players are able to make. The first model, found in Nash (1950), called the *bargaining model*, is one in which each player has only one threat action. In Nash (1953) each player may have many threats from which to choose. The fixed threat model is integrated into the variable threat model. Once threats are chosen, the solution is found by treating the model as if it were the fixed threat version. Of course, each player chooses his threat with an eye to making the solution point as advantageous as possible. In §2.1 the bargaining model is presented. §2.2 contains an exposition of the general two person model along the lines of Nash (1953), and §2.3 presents a different version of the latter model which is potentially more useful for applications to oligopoly.

2.1. The Nash bargaining model

2.1.1. The assumptions on the model

The bargaining model is best understood in this way. There are two people, each of whom has his own endowment of goods. Each can consume his own endowment, or the two can trade with one another and then each consumes the bundle of goods he possesses after trade. Each is also assumed to accept the von Neumann–Morgenstern axioms on utility; therefore, the set of conceivable trades is something like that depicted in fig. 11.1.[3] The axes measure the payoffs (i.e. the von Neumann–Morgenstern utility) of players 1 (P_1) and 2 (P_2). The shaded region gives all the attainable utility levels which the two can conceivably reach by trading, including levels making both worse off. The point $\boldsymbol{P}^T$ is the *status quo* or *no trade* point at which each consumes his own endowment. By

[3] For an exposition of von Neumann–Morgenstern utility see Luce and Raiffa (1957, ch. 2).

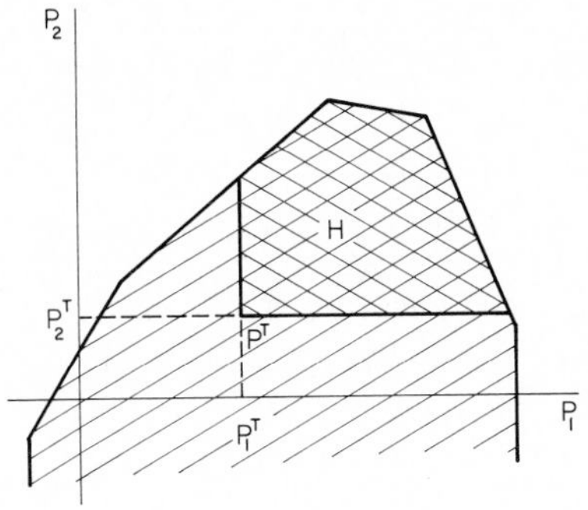

Fig. 11.1

refusing to trade, either player can force both to the point *T*. The cross hatched region, consisting of points which give at least as much utility to each player as the no trade point, is compact and convex. This set is called $\mathscr{H}$. The compactness of $\mathscr{H}$ must be assumed; however, convexity follows from von Neumann–Morgenstern utility. *Individual rationality*, the assumption that a player does not agree to an outcome which gives him a lower payoff than he can guarantee himself, is also taken for granted. In the present context, individual rationality requires that each player get at least as great a payoff as he could obtain by consuming his own endowment. With the impossibility of forcing the *i*th player to a payoff below P_i^T, his threat point payoff, it is irrelevant what are the characteristics of the attainable payoff set outside $\mathscr{H}$.

The assumptions on the structure of the model, stated in the preceding paragraph may be put formally:

C1. *The ith player has a strategy set $\mathscr{S}_i$ which is a compact and convex subset of $\mathscr{R}^m$, (i = 1, 2).*

C2. *There is a distinguished strategy vector s^T, called the threat strategy, such that $P_j(s_i^T, \bar{s}_i) = P_j(s^T)$ for all $\bar{s}_i \in \bar{\mathscr{S}}_i$ and for i, j = 1, 2.*

$$\mathscr{H} = \{\boldsymbol{P}(s) | s \in \mathscr{S}, \boldsymbol{P}(s) \geq \boldsymbol{P}(s^T)\}. \tag{11.1}$$

C3. *$\mathscr{H}$ is compact and convex.*

C1 postulates a *usual* compact and convex strategy set for each player. Note that continuity of the payoff functions need not even be assumed for this model. To provide a concrete interpretation of strategy, let the two strategy sets be identical, with a point in one of them being an allocation of the total endowments of the two players. Thus, if the ith player is thought to have an initial endowment of $\boldsymbol{w}_i = (w_{i1}, w_{i2}, \ldots, w_{ik})$, $i = 1, 2$, a strategy is an allocation to the two players of no more than $\boldsymbol{w}_1 + \boldsymbol{w}_2$ and is of the form $\boldsymbol{s}_1 = (s_{11}, s_{12}, \ldots, s_{1k}, s_{1,k+1}, \ldots, s_{1,2k})$. It would make sense that $s_{1j} \geq 0$ and feasibility requires that $s_{1j} + s_{1,j+k} \leq w_{1j} + w_{2j}$ – that is, negative amounts cannot be allocated and the totals allocated to the players cannot exceed the amounts they possess between them.

Now the payoff functions can be described. Let $u_i(\boldsymbol{x})$ be the utility to player i of consuming the commodity bundle $\boldsymbol{x}$, $\boldsymbol{x} \in \mathcal{R}^k$. Say that whenever $\boldsymbol{s}_1 = \boldsymbol{s}_2$, that is, when both players propose the same allocation, they actually make the trade which carries out that allocation. If $\boldsymbol{s}_1 \neq \boldsymbol{s}_2$, then assume no trade takes place, leaving each to consume his initial endowment. Then $P_i(\boldsymbol{s}) = u_i(s_{i1}, \ldots, s_{ik})$ when $\boldsymbol{s}_1 = \boldsymbol{s}_2$, and $P_i(\boldsymbol{s}) = u_i(\boldsymbol{w}_i) = P_i(\boldsymbol{s}^T)$ otherwise. $\boldsymbol{s}^T$ is naturally defined here as $\boldsymbol{s}^T = (\boldsymbol{w}_1, \boldsymbol{w}_2)$. A player's threat strategy is to propose that no trade take place. Being a cooperative game, in which players can talk to each other, there is no obstacle to their coming to a verbal agreement. To make a binding agreement means that they arrive at a pair $(\boldsymbol{s}_1, \boldsymbol{s}_2)$ such that $\boldsymbol{s}_1 = \boldsymbol{s}_2$, and that they go together to an *umpire* and announce their strategies simultaneously. The umpire has the power to enforce their agreement and the obligation to do so.

The preceding interpretation of the game is meant only as an example. What is central to the model is that there is a compact and convex set $\mathcal{H}$ of payoff pairs which are attainable for the two players. That one element of the set, called the threat point, $\boldsymbol{P}(\boldsymbol{s}^T)$ or $\boldsymbol{P}^T$, is minimal. That is, for any $\boldsymbol{P} \in \mathcal{H}$, $\boldsymbol{P} \geq \boldsymbol{P}^T$. And that to attain any point in $\mathcal{H}$ other than $\boldsymbol{P}^T$ requires agreement of the two players. In the absence of agreement on some point, the payoffs are necessarily $\boldsymbol{P}^T$. This is precisely like a trading situation in which no trade takes place unless the two traders agree on a particular, mutually satisfactory trade. This latter condition is embodied in C2.

2.1.2. The axioms for the Nash solution

The conditions known as the *Nash axioms*, which are given below, constitute a definition of the cooperative game equilibrium proposed by Nash. Being a unique equilibrium, it is in order to call it a solution. The Nash axioms are:

N1. *The solution should be unique.*

N2. *Let two games be given by $\mathcal{H}_1$ and $\mathcal{H}_2$. If $\mathcal{H}_1$ can be transformed into $\mathcal{H}_2$ by a positive linear transformation of P_1 and another of P_2, then the solutions to the two games are related by the same pair of transformations.*

N3. *The solution should be an undominated point $\mathbf{P}^*$ of $\mathcal{H}$, i.e. there should be no element $\mathbf{P}$ of $\mathcal{H}$ such that $\mathbf{P} > \mathbf{P}^*$.*

N4. *Let $\mathcal{H}_1$ and $\mathcal{H}_2$ be two games having the same threat point. If $\mathcal{H}_1 \subseteq \mathcal{H}_2$ and the solution of $\mathcal{H}_2$ is an element of $\mathcal{H}_1$, then both games have the same solution.*

N5. *If the set $\mathcal{H}$ is symmetric about a 45° line through the threat point, then the solution lies on that line.*

The axiom N2 states invariance of a solution with respect to positive linear transformations of utility. That is, if two traders, with their given endowments, find that the optimal trade is that player 1 give to player 2 three jackknives plus a frog in exchange for five oranges and a snake, then, that trade remains the solution even if the utility function of one or the other (or both) is subjected to positive linear transformation. The third axiom, N3, is the natural condition that the two players should not make as a final trade anything which could be improved upon for at least one without worsening the payoff of the other. The next axiom, N4, is called *independence of irrelevant alternatives.* Nash points out that its effect is to make the solution depend on local conditions in the neighborhood of the solution and not on aspects of the model outside its vicinity. Another way to look at the axiom is that if a game is *enlarged* by the addition of possible new trades, but the no trade point is unchanged, then the solution to the new, larger game is either one of the new points or it is the former solution. It is not some other point which was previously available.[4] The last axiom is one of symmetry. It could be said that N5 requires that the labelling of the axes (i.e. who happens to be player 1 and who is player 2) cannot affect the outcome. Taking N2 into account, the threat point may be placed at the origin. Then if the symmetry condition is

[4]The balanced temptation equilibrium (see ch. 8 §2) is an example of an equilibrium which does not satisfy N4. The best payoff a player i can get by choosing his best reply to $\bar{s}_i^*$ helps determine the equilibrium payoff, and it is not a characteristic which is local to $\mathbf{P}(s^*)$.

met $(P_1, P_2) \in \mathscr{H}$ whenever $(P_2, P_1) \in \mathscr{H}$. Under these conditions, it is difficult to see why an equilibrium should not lie on a 45° line through the origin. The players and their prospects appear identical.

2.1.3. *Existence of the solution*

The main result of this section is that a two-person trading game which satisfies C1–C3 has a Nash equilibrium satisfying N1–N5. The solution point is easily characterized and simple to compute. Denoting the solution by $\boldsymbol{P}^* = (P_1^*, P_2^*)$, it satisfies the condition

$$(P_1^* - P_1^T)(P_2^* - P_2^T) = \max_{p \in \mathscr{H}} (P_1 - P_1^T)(P_2 - P_2^T). \tag{11.2}$$

Equation (11.2) is often called the *Nash product*. The two terms which are multiplied are, respectively, the gain from trade of player 1 and the gain from trade of player 2; so the solution is that trade in $\mathscr{H}$ for which the product of gains from trade is maximized. Each player's gain is measured in his own utility units. The solution is illustrated in fig. 11.2. The Nash product has constant value along any rectangular hyperbola which is asymptotic to axes passing through the threat point; hence, the solution is found where such a rectangular hyperbola is tangent to the upper boundary of $\mathscr{H}$.

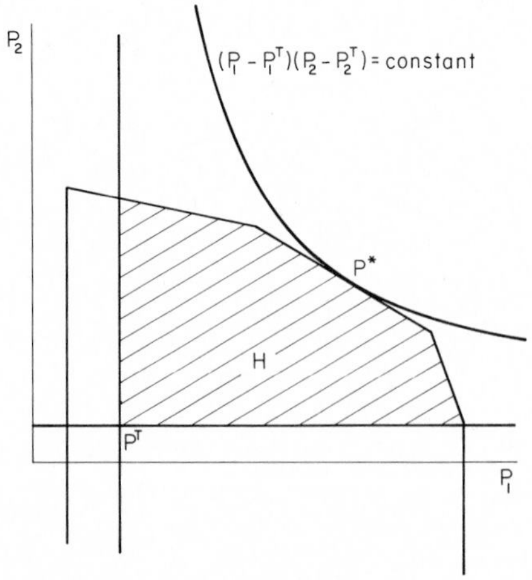

Fig. 11.2

Theorem 11.1. *A two-person trading game satisfying C1–C3 has a solution satisfying N1–N5. The solution also satisfies eq. (11.2).*

□ The proof proceeds by construction. First it is shown that the point $\boldsymbol{P}^*$ which satisfies eq. (11.2) also satisfies N1–N5. Then it is shown that no other point in $\mathcal{H}$ satisfies all five conditions. Because $\mathcal{H}$ is convex and the set of points lying on or above a rectangular hyperbola is strictly convex, $\boldsymbol{P}^*$ must be unique. If two points in $\mathcal{H}$ attain the same Nash product, then a convex combination of them is in $\mathcal{H}$ and must attain a higher value of the product.

That N2 is met may be seen by observing, first, that a change of origin for either of the payoffs has no effect on the Nash product because it is measured from $\boldsymbol{P}^T$ in any case; and, second, a multiplicative transformation changes all payoff products in proportion. That is, if the payoff of player 1 is transformed to

$$a + bP_1, \qquad b > 0, \tag{11.3}$$

the value of a has no effect on the Nash product. The product is multiplied by b; hence, it is maximized at the same point as previously.

N3 is clearly satisfied; for, if there were some $\boldsymbol{P}' \in \mathcal{H}$ such that $\boldsymbol{P}' > \boldsymbol{P}^*$ then $\boldsymbol{P}'$ would have a larger Nash product than $\boldsymbol{P}^*$. N4 is also satisfied; because if $\boldsymbol{P}^*$ maximizes the Nash product for a game $\mathcal{H}_2$, $\boldsymbol{P}^* \in \mathcal{H}_1$ where $\mathcal{H}_1 \subseteq \mathcal{H}_2$ and both share the same threat point, then $\boldsymbol{P}^*$ must maximize the Nash product on $\mathcal{H}_1$ as well. $\mathcal{H}_1$ differs from $\mathcal{H}_2$ by the removal of some alternatives. Finally, if $\mathcal{H}$ is symmetric about a 45° line through the threat point, then the maximum value of the Nash product must occur on the 45° line. To see this, assume the maximum value were to occur off the 45° line at (P_1', P_2'). Then by symmetry, the same product occurs at (P_2', P_1') which is also in $\mathcal{H}$; and a convex combination of them is even better, contradicting their assumed optimality. Thus $\boldsymbol{P}^*$, which maximizes the Nash product, satisfies N1–N5.

It remains to show that no other point in $\mathcal{H}$ can satisfy all the required conditions. For an arbitrary game, $\mathcal{H}$, transform the utility of player 1 so that $\boldsymbol{P}^*$, the point which maximizes the Nash product, lies on the 45° line through $\boldsymbol{P}^T$. Without loss of generality, $\boldsymbol{P}^T$ may be taken as the origin (0, 0). Now consider a game $\mathcal{H}'$ with its threat point at (0, 0) and with $\mathcal{H}'$ defined as

$$\mathcal{H}' = \{\boldsymbol{P} | \boldsymbol{P} \geq 0, P_1 + P_2 \leq P_1^* + P_2^*\}. \tag{11.4}$$

This game is illustrated in fig. 11.3, where the heavily shaded area is $\mathcal{H}$

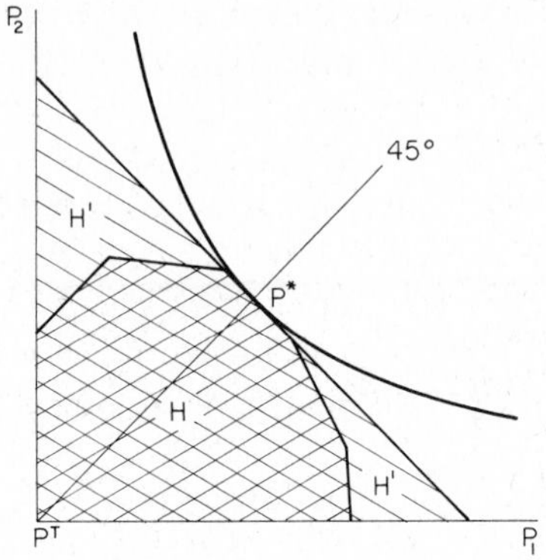

Fig. 11.3

(after being transformed to put $\boldsymbol{P}^*$ on the 45° line) and the heavily plus lightly shaded areas are $\mathcal{H}'$. By symmetry, N5, the solution to $\mathcal{H}'$ must be $\boldsymbol{P}^*$. In comparing $\mathcal{H}$ and $\mathcal{H}'$, they have the same threat point, $\mathcal{H} \subseteq \mathcal{H}'$ and the solution of $\mathcal{H}'$ is in $\mathcal{H}$; hence, by N4 $\boldsymbol{P}^*$ must also be the solution of $\mathcal{H}$. Finally, the solution of $\mathcal{H}$ is unchanged by linear transformations of a player's utility. Thus $\boldsymbol{P}^*$ is the solution of the original game, which proves the theorem. □

2.2. Nash's model for variable threat two-person cooperative games

In Nash (1953) two versions of the variable threat two-person game are given, called the axiomatic version and the negotiation version. There is a third, the arbitration. The exposition below uses the arbitration approach. This game is represented in normal form because explicit strategy choice is central to the play of the game, much as it is for noncooperative games. Each player is, following Nash (1953), assumed to have a finite number of pure strategies; and von Neumann–Morgenstern utility is assumed. Then, the strategy set of a player becomes the unit simplex in an $m_i - 1$ dimensional space where m_i is the number of pure strategies available to the ith player; and the ith player's payoff function is linear in s_1 and linear in s_2, making it bilinear in (s_1, s_2). The strategy sets and payoff functions are like those in Nash (1951) for $n = 2$ (see ch. 7 §3.3).[5]

[5]For the two players cooperating, they may jointly randomize over all pure strategy pairs; thus their joint strategy set is larger than $\mathcal{S}_1 \times \mathcal{S}_2$. See §4.2.1.

C4. *$\mathcal{S}_i$, the strategy set of the ith player, is the unit simplex in $\mathcal{R}^{m_i-1}$, (i = 1, 2).*

C5. *The payoff function of the ith player, $P_i(s_1, s_2)$, is bounded and linear in s_1 and s_2, (i = 1, 2).*

This model is sufficiently general that there need not be a *no trade point*. In the present game, it is possible that, for any fixed strategy choice of player 1, each distinct strategy choice of player 2 results in a different payoff pair from any other. A useful analogy here is a single period duopoly model in which each player chooses a price. For player one to threaten is likely to mean that he promises to choose a particularly low price if no agreement is reached; however, the profits associated with that low price are not known merely by knowing what player 1 will do. They also depend on the price choice of player 2. If both name threat prices, then a specific pair of threat payoffs is determined. So it is with the Nash model. If the players name as their threat strategies $\boldsymbol{s}^T = (s_1^T, s_2^T)$, then threat payoffs $\boldsymbol{P}(\boldsymbol{s}^T) = (P_1(\boldsymbol{s}^T), P_2(\boldsymbol{s}^T))$ are determined.

The workings of the arbitration model are this: the two players agree to the Nash axioms N1–N5 as yielding an acceptable solution for a bargaining game, or game with fixed threats; and they agree to each choose a threat strategy whose associated payoffs are to play the role of no trade point in order to determine a solution. This version of the model may be thought of as a three-stage process: (1) the players agree to an outcome satisfying N1–N5 when threat payoffs are fixed, (2) they agree to each choose a threat strategy which is used to give threat payoffs, and (3) final payoffs are found which satisfy N1–N5 relative to the previously determined threat payoffs.

There is a special relationship, illustrated in fig. 11.4, between the threat and solution points. It is that the straight line from the threat point to the solution point has the negative of the slope of the tangent to the payoff set at the threat point, i.e. *slope AB* = – *slope ab*. Where the solution point is at a corner of the Pareto optimal set, so a tangent is not defined, then a straight line through the solution point with the required slope does not touch the interior of $\mathcal{H}$.[6] Looking again at fig. 11.4, if any other point on the line $0A$, such as B, happened to be the threat point, the solution point would be unchanged because the same slope condition would be met.

[6]Let the Pareto optimal curve satisfy the differentiable function $P_2 = \phi(P_1)$ and assume the threat point is (0, 0). Then the solution point must be where $P_1\phi(P_1)$ is maximized. This occurs where $-\phi' = P_2/P_1$.

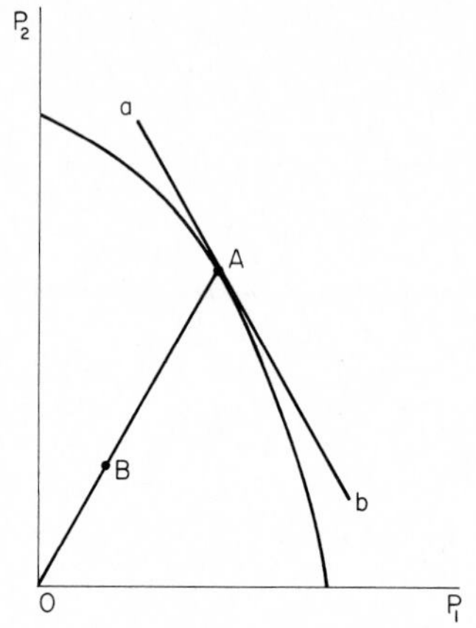

Fig. 11.4

In general, each point in a convex set $\mathscr{H}$, if taken to be the threat point, has a unique solution point associated with it. This is illustrated in fig. 11.5 for a polyhedral set $\mathscr{H}$ (which satisfies C4 and C5). Each point on a straight line (such as AF), taken as a threat point, has as its associated solution point the end point of the line which lies on the Pareto optimal

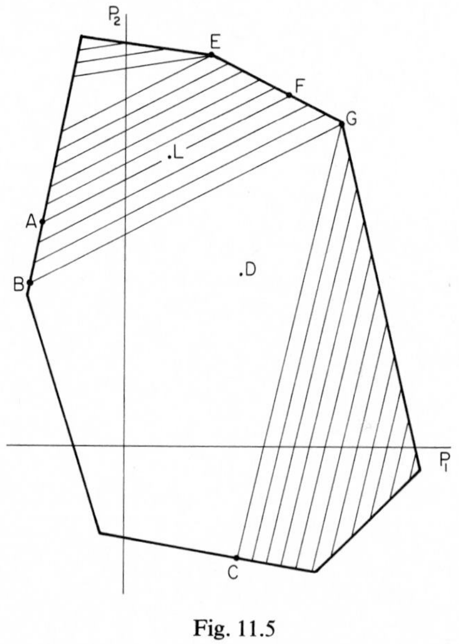

Fig. 11.5

curve (such as F). The slope of AF is the negative of the slope of EG. For a point like G, the associated points in $\mathscr{H}$ which, as threat points, would have G as their solution are all the points on the line GB, all the points on the line GC and all the points in $\mathscr{H}$ which lie between those two lines, such as D.

Thus it is clear that the more the threat point is high and to the left, the more player 2 is benefited. Conversely, player 1 is better off as the threat point moves down and to the right. Given the commitment to the Nash bargaining solution, the game really becomes noncooperative with each player choosing a threat strategy and the pair of threat strategies determining a final outcome on the Pareto frontier. This noncooperative game has a special character which puts it closer to a zero sum game than to a general noncooperative game. It is that the game is *strictly competitive*, which means that all outcomes are Pareto optimal. If (P_1', P_2') and (P_1'', P_2'') are two possible outcomes and $P_1'' < P_1'$, then $P_2' < P_2''$.

Now consider the nature of a pair of optimal, or equilibrium, threats (s_1^T, s_2^T). Say the point L in fig. 11.5 is attained by the strategies $\boldsymbol{s}'$. If $\boldsymbol{s}'$ were chosen as the threat strategies, then the resulting payoffs which the players receive would be at F. Is s_1' optimal for player 1 given that player 2 chooses s_2' as his threat strategy? The answer is *yes* if there is no $s_1 \in \mathscr{S}_1$ such that $\boldsymbol{P}(s_1, s_2')$ lies below the line AF. Otherwise, s_1' is not a best reply to s_2'. Letting ρ denote the slope of the line AF, $\boldsymbol{s}'$ is an equilibrium pair of threats if

$$P_2(\boldsymbol{s}') - \rho P_1(\boldsymbol{s}') = \min_{s_1 \in \mathscr{S}} [P_2(s_1, s_2') - \rho P(s_1, s_2')], \tag{11.5}$$

$$P_2(\boldsymbol{s}') - \rho P_1(\boldsymbol{s}') = \max_{s_2 \in \mathscr{S}} [P_2(s_1', s_2) - \rho P_1(s_1', s_2)]. \tag{11.6}$$

Not only does an optimal threat pair $\boldsymbol{s}^T$ exist, there is a unique solution for a game satisfying C4 and C5. While the optimal threat need not be unique, all optimal threats yield the same solution point.

Theorem 11.2. *Assume a two-person game satisfying C4 and C5, and in which players choose threat strategies which are used to determine a threat payoff from which a Nash bargaining solution is found as the outcome of the game. Such a game has an equilibrium pair of strategies.*

□ The solution point varies continuously as a function of the threat point $\boldsymbol{P}(\boldsymbol{s}^T)$, and the threat point varies continuously with $\boldsymbol{s}^T$; hence the solution point is a continuous function of $\boldsymbol{s}^T$. Therefore, the best reply mapping, carrying points of $\mathscr{S}$ into $\mathscr{S}$, is upper semicontinuous. Consider the set of best replies of player 2 to a given strategy of player 1. They must all yield threat payoffs which lead to the final outcome; hence, they must all maximize an expression like eq. (11.6) for an appropriate value of ρ. Because P_2 is concave in s_2 and P_1 is convex in s_2, $P_2 - P_1$ is concave in s_2, making the set of best replies convex.[7] A parallel argument may be made for player 1. The Kakutani fixed point theorem applies, proving there is an optimal, though not necessarily unique, pair of threat strategies s^T. □

Theorem 11.3. *The solution for a game satisfying the conditions of theorem 11.2 is unique.*

□ Let (P'_1, P'_2) and (P''_1, P''_2) be two solutions of the game and suppose $P'_1 \leq P''_1$. Suppose $\boldsymbol{P}'$ results from threat strategies s' and $\boldsymbol{P}''$ from s''. Let $\boldsymbol{P}^0$ be the solution which results from (s''_1, s'_2). $P^0_2 \leq P''_2$ because s''_2 is a best reply to s''_1. Due to the strictly competitive nature of the game, $P^0_1 \geq P''_1$. In addition, $P'_1 \geq P^0_1$ because s'_1 is a best reply against s'_2. Therefore, $P'_1 \geq P''_1$ and $P''_1 \leq P'_1$; hence, $P'_1 = P''_1$. Doing the same exercise with player 2 establishes the theorem.[8] □

For both the bargaining model and the general two-person game, the value to a player of the game is the payoff he receives at the solution point. Because a member of either class of game has a unique solution, the game has a value in an unambiguous sense. The value of the game is the vector of solution payoffs, and the value to a specific player is the payoff to him.

2.3. *A variant of the Nash model for two-person cooperative games*

Finite games are not convenient for oligopoly applications. Fortunately, there is no obstacle to formulating a variant of the Nash model which is more useful for such purposes. This is done in a way which preserves the existence and uniqueness results of §2.2. Some familiar assumptions are made:

[7]Of course, both P_1 and P_2 are linear in s_2.

[8]The uniqueness of this theorem applies more broadly. In particular, the payoffs associated with noncooperative equilibrium for any strictly competitive two-person game are unique, including, of course, the payoffs at the saddle point equilibrium of a two-person zero sum game.

C6. *The payoff function of the ith player, $P_i(s)$, is continuous and bounded, ($i = 1, 2$).*

The set of attainable payoffs may be defined in a familiar way,

$$\mathcal{H} = \{\mathbf{P}(s) | s \in \mathcal{S}\}. \tag{11.7}$$

The set of Pareto optimal payoffs is

$$\mathcal{H}^* = \{\mathbf{P} | \mathbf{P} \in \mathcal{H} \text{ and there is no } \mathbf{P}' \in \mathcal{H} \text{ such that } \mathbf{P}' > \mathbf{P}\}. \tag{11.8}$$

C7. *For any $\mathbf{P} \in \mathcal{H}$, $\mathbf{P}^* \in \mathcal{H}^*$, $\mathbf{P}^* > \mathbf{P}$ and $\lambda \in (0, 1)$, $\lambda\mathbf{P} + (1-\lambda)\mathbf{P}^* \in \mathcal{H}$.*

C1 is also assumed. These assumptions are similar to G2 and G3 of ch. 7 §2.1 and G8 of ch. 8 §2.1.[9] Now the ρ of eqs. (11.5) and (11.6) may be defined for all points $\mathbf{P} \in \mathcal{H}$ and $\mathbf{P} \notin \mathcal{H}^*$ (i.e., all points which are attainable, but not Pareto optimal). For such $\mathbf{P}$, let $\mathbf{P}^*$ be the associated point in $\mathcal{H}^*$ which would be the solution if $\mathbf{P}$ were a fixed threat; then

$$\rho(\mathbf{P}) = (P_2^* - P_2)/(P_1^* - P_1). \tag{11.9}$$

C8. *For all $\mathbf{P}$ for which ρ is defined.*

$$P_2(s) - \rho(\mathbf{P}(s))P_1(s) \tag{11.10}$$

is concave in s_2 and convex in s_1.[10]

The last condition assures that the image sets of the best reply function are convex. Proof of the counterpart to theorem 11.2 for games satisfying C1 and C6–C8 is so similar that it need not be repeated. Similarly, the proof of theorem 11.3 can be used without alteration for a similar theorem on the present model.

Theorem 11.4. *Assume a two-person game satisfying C1 and C6–C8, and in which the players choose threat strategies which are used to determine a threat payoff from which a Nash bargaining solution is found as the outcome of the game. Such a game has an equilibrium pair of strategies. The equilibrium strategies need not be unique, but all such strategy pairs are associated with the same solution point.*

[9] C7 is slightly weaker than G8.

[10] Where second partial derivatives exist, these conditions are that $P_2^2 - \rho P_1^2 \leq 0$ and $P_2^1 - \rho P_1^1 \geq 0$. This may be seen by taking the appropriate second derivatives of eq. (11.10) and doing some simplifying with the help of the first-order maximization conditions.

3. *The Shapley value for n-person cooperative games with transferable utility*

The exposition in this section follows Shapley (1953b). One point of divergence between the present exposition and the original is that the game to be defined below has a finite number of players. Shapley defines a game which may have an infinite number of players; however, his results are for games in which only a finite subset of the players *matter*. To say that a player does not matter means that (a) by himself he can get zero and (b) any coalition of which he is not a member gets exactly as much as it would get if he joined it.

3.1. Shapley's axioms for the value and the class of games to be considered

The meaning of transferable utility is best seen after Shapley's definition of the game is given. Letting $\mathcal{N}$ denote the set of all players and $\mathcal{K}$, $\mathcal{L}$, $\mathcal{M}$, coalitions (i.e. subsets of $\mathcal{N}$), a *characteristic function*, $v(\mathcal{K})$, is a *set function* associating with any subset of $\mathcal{N}$ a real number. $v(\mathcal{K})$ is the payoff which the coalition $\mathcal{K}$ can assure for itself. k, l, and m are the number of members of $\mathcal{K}$, $\mathcal{L}$, and $\mathcal{M}$, respectively. The notation $\mathcal{K}-\mathcal{L}$ denotes the coalition composed of players who are in $\mathcal{K}$, but not in $\mathcal{L}$. Thus, $\mathcal{K}-\mathcal{L}=\mathcal{K}\cap\tilde{\mathcal{L}}$ is the complement in $\mathcal{K}$ of $\mathcal{L}$. It is sometimes necessary to distinguish between a subset and a proper subset; so $\mathcal{K}\subseteq\mathcal{N}$ means that $\mathcal{K}$ is contained in $\mathcal{N}$, and $\mathcal{K}\subset\mathcal{N}$ means that $\mathcal{K}$ is contained in $\mathcal{N}$ and is not equal to $\mathcal{N}$.

C9. *The number of players, n, is finite.*

C10. $v(0)=0$.

C11. $v(\mathcal{K})\geq v(\mathcal{K}\cap\mathcal{L})+v(\mathcal{K}-\mathcal{L})$ *for all* $\mathcal{K}$, $\mathcal{L}\subseteq\mathcal{N}$.

C10 asserts that the empty coalition gets nothing. This assumption may be thought of as a convenient normalization of the payoffs. C11 is called *superadditivity*. It asserts that when two disjoint coalitions combine, they can get at least as much as the two could get separately. This is analogous to scale economies in the absence of costs of organization. That each coalition, $\mathcal{K}$, has a payoff it can guarantee itself, $v(\mathcal{K})$, which is a scalar and which can be divided among the members any way they agree on captures the essence of comparable and transferable utility. The units of $v(\mathcal{K})$ are, in effect, universal utility units.

For a game, v, Shapley seeks a *value* $\boldsymbol{\phi}(v)$ which associates a number, $\phi_i(v)$, with each player i in the game.[11] $\boldsymbol{\phi}(v)$ is, then, the solution of the game. The conditions which $\boldsymbol{\phi}$ is to meet are symmetry, Pareto optimality and additivity for independent games. Before these are given formally, a class of games based on v is defined. Consider the set of all permutations of the numbers 1, 2, . . . , n. Denote by π a function which permutes 1, 2, . . . , n into another member of the set of permutations. So $\pi(\mathcal{K})$ is the set of numbers into which $\mathcal{K}$ is mapped under the permutation π. If, for example, π reverses the ordering of the players, then $\pi(\{i\}) = n + 1 - i$. Then πv is a game which is defined as follows:

$$\pi v(\pi \mathcal{K}) = v(\mathcal{K}), \qquad \text{for all } \mathcal{K} \subseteq \mathcal{N}. \tag{11.11}$$

That is, when the players in coalition $\mathcal{K}$ are renumbered, they are still able to attain the same payoff as previously. Equation (11.11) defines a class of games which are symmetric in the sense that what one player or group of players can get is independent of what they are called. There is no essential difference between v and πv. Shapley's conditions are:

Sh1. *For each* π, $\phi_{\pi i}(\pi v) = \phi_i(v)$, *for all* $i \in N$.

Sh2. $\Sigma_{i \in \mathcal{N}} \phi_i(v) = v(\mathcal{N})$.

Sh3. *For any two games* v *and* w *having the same set of players,* $\mathcal{N}$. $\boldsymbol{\phi}(v + w) = \boldsymbol{\phi}(v) + \boldsymbol{\phi}(w)$.

Sh1 is a symmetry condition which, recognizing that player i in game v and πi in game πv are the same player in the same game, asserts that the value of the game to the player must be identical in both cases.[12] Sh2 requires that the sum of the values over all the players must be equal to the amount which the coalition consisting of all players can guarantee itself.[13] The game $v + w$ is taken to be defined in the natural way. Thus the coalition $\mathcal{K}$ in the game $v + w$ can get $v(\mathcal{K}) + w(\mathcal{K})$, etc. The value of the

[11]As the characteristic function v contains all the information about the game (number of players, list of coalitions, etc.) it is reasonable to refer to a particular characteristic function as a game.

[12]This condition, Sh1, is like Nash's condition N5 in §2.1.1.

[13]This condition, Sh2, and others in different models which require that equilibria be Pareto optimal, are called *group rationality*.

combined game for a player should equal the sum of the values of the two games to him. That Sh3 is stated for two games having the same set of players is no real restriction. Let v and w have sets of players $\mathcal{N}_v$ and $\mathcal{N}_w$, respectively. Let $\mathcal{N} = \mathcal{N}_v \cup \mathcal{N}_w$. We can now define an augmented game based on v which has $\mathcal{N}$ as its set of players. Let $\mathcal{K}$ be any subset of $\mathcal{N}$ and let the augmented game be denoted $\bar{v}$. $\mathcal{K}$ may be partitioned into $\mathcal{K}_v$ and $\mathcal{K}_w$ with $\mathcal{K}_v = \mathcal{K} \cap \mathcal{N}_v$ and

$$\mathcal{K}_w = \mathcal{K} - \mathcal{K}_v,$$
$$\bar{v}(\mathcal{K}) = v(\mathcal{K}_v), \qquad \text{for all } \mathcal{K}_v \subseteq \mathcal{N}_v \text{ and all } \mathcal{K} \subseteq \mathcal{N}. \tag{11.12}$$

Thus the players in $\mathcal{N}_w$ who are not also in $\mathcal{N}_v$ have no effect on the augmented game $\bar{v}$. $\bar{v}$ is precisely v with some superfluous players added.[14] After doing the same with w to get an augmented game $\bar{w}$, the two games $\bar{v}$ and $\bar{w}$ may be combined as Sh3 indicates.

3.2. *On the interpretation of the value function*

The value function is found to assign to each player a value which reflects the contribution made by that player to all the coalitions of which he is a member. Roughly speaking, this is done by finding the marginal value of each coalition and splitting that marginal value equally among all players who are members. Then, for one player, the value of the game is the sum of all the shares of marginal contributions which he receives.

Before proceeding to the formal results, consider what is meant by the *marginal value of a coalition.* The amount a coalition $\mathcal{K}$ can get, $v(\mathcal{K})$, results from the combined opportunities of the members of $\mathcal{K}$ jointly, the combined opportunities of (smaller) subsets of $\mathcal{K}$ and from what the several members can do on their own. For example, let $v(\{1\}) = 2$, $v(\{2\}) = 5$ and $v(\{1, 2\}) = 10$. By themselves, the two players can get a total of 7; however, in a coalition, they can get 10. The marginal value of the coalition is 3. Let $c_{\mathcal{K}}(v)$ denote the marginal value of the coalition $\mathcal{K}$. This value may be defined recursively by setting

$$c_{\{i\}}(v) = v(\{i\}), \qquad \text{for all } i \in \mathcal{N}. \tag{11.13}$$

[14]Shapley would call the set $\mathcal{N}_v$ a *carrier* of $\bar{v}$. The carrier notion is made superfluous to the present exposition by restricting attention to games having only a finite number of players.

Then for coalitions $\mathcal{K}$ for which $k \geq 2$,

$$c_{\mathcal{K}}(v) = v(\mathcal{K}) - \sum_{\mathcal{L} \subset \mathcal{K}} c_{\mathcal{L}}(v), \qquad \text{for all } \mathcal{K} \subseteq \mathcal{N} \text{ with } k \geq 2. \tag{11.14}$$

In other words, the marginal value of the coalition $\mathcal{K}$ equals $v(\mathcal{K})$, the characteristic function value, minus the marginal values of all smaller coalitions contained in $\mathcal{K}$.[15] An equivalent way to compute $c_{\mathcal{K}}(v)$ is

$$c_{\mathcal{K}}(v) = \sum_{\mathcal{L} \subseteq \mathcal{K}} (-1)^{l-1} v(\mathcal{L}). \tag{11.15}$$

It turns out that

$$\phi_i(v) = \sum_{\mathcal{K} \subseteq \mathcal{N}, \mathcal{K} \ni i} c_{\mathcal{K}}(v)/k, \qquad \text{for all } i \in \mathcal{N}. \tag{11.16}$$

Equation (11.16) expresses the notion that the value of the game to player i equals the sum of his shares of the marginal values of all the coalitions of which he is a member. His share is a fraction, $1/k$, of $c_{\mathcal{K}}(v)$ for any coalition $\mathcal{K}$ to which he belongs.

The $c_{\mathcal{K}}(v)$ can be used to construct a set of games which sum to the game v. For each $c_{\mathcal{K}}(v)$, the corresponding member of the set is a game in which all coalitions except those which contain $\mathcal{K}$ get zero. If $\mathcal{K} \subseteq \mathcal{L}$, then $\mathcal{L}$ gets $c_{\mathcal{K}}(v)$. The game v is equal to the sum over all coalitions of the games just described. These games are used to prove that a unique $\boldsymbol{\phi}$ exists. The game based on $c_{\mathcal{K}}(v)$ is denoted $c_{\mathcal{K}}(v)v_{\mathcal{K}}$ where $v_{\mathcal{K}}$ is a game defined by

$$v_{\mathcal{K}}(\mathcal{L}) = \begin{cases} 1, & \text{if } \mathcal{K} \subseteq \mathcal{L}, \\ 0, & \text{if } \mathcal{K} \not\subseteq \mathcal{L}. \end{cases} \tag{11.17}$$

3.3. Existence and uniqueness of the value function

Proof that the value, $\boldsymbol{\phi}$, exists and is unique is carried out in several steps. The first is to establish the existence and uniqueness of $\boldsymbol{\phi}$ for a game of the family $cv_{\mathcal{K}}$. The second is to show that a game v is a linear combination of games, $c_{\mathcal{K}}v_{\mathcal{K}}$.

[15] Despite the superadditivity assumption, C11, it is possible for a coalition to have a negative marginal value.

Lemma 11.5. *For $c \geq 0$ and finite $k > 0$, $\phi_i(cv_{\mathcal{K}}) = c/k$ for $i \in \mathcal{K}$ and $\phi_i(cv_{\mathcal{K}}) = 0$ for $i \notin \mathcal{K}$.*

□ Consider $i \notin \mathcal{K}$. Then $cv_{\mathcal{K}}(\mathcal{K} + \{i\}) = c = cv_{\mathcal{K}}(\mathcal{K})$. From C11,

$$cv_{\mathcal{K}}(\mathcal{K} + \{i\}) \geq cv_{\mathcal{K}}(\mathcal{K}) + cv_{\mathcal{K}}(\{i\}), \tag{11.18}$$

hence $cv_{\mathcal{K}}(\{i\}) = 0$. By symmetry, Sh1, $\phi_i(cv_{\mathcal{K}}) = \phi_j(cv_{\mathcal{K}})$ for all i and j in $\mathcal{K}$; therefore, $\phi_i(cv_{\mathcal{K}}) = c/k$ for all $i \in \mathcal{K}$. □

Lemma 11.6. *Any game which satisfies C9–C11 is a linear combination of symmetric games $v_{\mathcal{K}}$:*

$$v = \sum_{\mathcal{K} \subseteq \mathcal{N}, \mathcal{K} \neq 0} c_{\mathcal{K}}(v) v_{\mathcal{K}}, \tag{11.19}$$

where the $c_{\mathcal{K}}(v)$ are given by eq. (11.15).

□ It must be shown that

$$v(\mathcal{K}) = \sum_{\mathcal{L} \subseteq \mathcal{N}, \mathcal{L} \neq 0} c_{\mathcal{L}}(v) v_{\mathcal{L}}(\mathcal{K}), \tag{11.20}$$

for all $\mathcal{K} \subseteq \mathcal{N}$. Using eq. (11.15), eq. (11.20) becomes

$$\begin{aligned} v(\mathcal{K}) &= \sum_{\mathcal{L} \subseteq \mathcal{K}} \sum_{\mathcal{M} \subseteq \mathcal{L}} (-1)^{l-m} v(\mathcal{M}) \\ &= \sum_{\mathcal{M} \subseteq \mathcal{K}} \left[\sum_{l=m}^{k} (-1)^{l-m} v(\mathcal{M})(k-m)!/((l-m)!(k-l)!) \right] \\ &= \sum_{\mathcal{M} \subseteq \mathcal{K}} \left[\sum_{l=m}^{k} (-1)^{l-m} C_{l-m}^{k-m} \right] v(\mathcal{M}). \end{aligned} \tag{11.21}$$

Except for $k = m$, the expression in square brackets in eq. (11.21) is zero, which reduces to $v(\mathcal{K}) = v(\mathcal{K})$. Thus the correctness of eq. (11.20) is verified. □

Corollary 11.7. *If v, w and $v - w$ are all games which satisfy C9–C11, then $\boldsymbol{\phi}(v - w) = \boldsymbol{\phi}(v) - \boldsymbol{\phi}(w)$.*

□ This is an immediate consequence of lemma 11.6. □

Note, by the way, that the $c_{\mathcal{K}}(v)$ defined by eq. (11.15) need not all be nonnegative, despite the superadditivity condition, C11.

Taking the result of lemma 11.5 and using it in eq. (11.19) gives

$$\phi_i(v) = \sum_{\mathcal{K} \subseteq \mathcal{N}, \mathcal{K} \ni i} c_{\mathcal{K}}(v)/k, \qquad i = 1, \ldots, n. \tag{11.22}$$

By substituting from (11.15), eq. (11.22) may be rewritten

$$\begin{aligned} \phi_i(v) &= \sum_{\mathcal{K} \subseteq \mathcal{N}, \mathcal{K} \ni i} v(\mathcal{K})(k-1)!(n-k)!/n! - \sum_{\mathcal{K} \subseteq \mathcal{N}, \mathcal{K} \not\ni i} v(\mathcal{K})k!(n-k-1)!/n! \\ &= \sum_{\mathcal{K} \subseteq \mathcal{N}} [v(\mathcal{K}) - v(\mathcal{K} - \{i\})](k-1)!(n-k)!/n! \\ &= \sum_{\mathcal{K} \subseteq \mathcal{N}} [v(\mathcal{K}) - v(\mathcal{K} - \{i\})]/(kC^n_k), \qquad i = 1, \ldots, n. \end{aligned} \tag{11.23}$$

In eq. (11.22), the value of the game to the ith player is given by adding up his share of the marginal value of each coalition of which he is a member. His share is the fraction $1/k$ when there are k members of the coalition. Equation (11.23), which is equivalent, adds up the marginal contribution of the ith player to each coalition of which he is a member. Where $i \in \mathcal{K}$, this marginal contribution is taken to be the difference between what that coalition can get ($v(\mathcal{K})$) and what it would get if i were not a member ($v(\mathcal{K} - \{i\})$), multiplied by a weight which is related to the number of k player coalitions in the game. The value function, $\boldsymbol{\phi}$, defined above, satisfies the conditions Sh1–Sh3 and is clearly the only function which does so. Thus the following theorem is proved:

Theorem 11.8. *For a game satisfying C9–C11, a unique value function exists which satisfies Sh1–Sh3. This function is given by eq. (11.22) or eq. (11.23).*

4. Harsanyi's generalization of Nash and Shapley

Harsanyi (1963) provides a model which generalizes both Shapley and Nash. It assumes nontransferable, noncomparable utility and allows for n players. When $n = 2$, it reduces to a Nash model and if utility is taken to be comparable and transferable, it becomes a Shapley model. Other models providing a value for general n-person games of the sort dealt with by Harsanyi may be found in Shapley (1969) and Owen (1972).

4.1. General description

In a way analogous to Shapley, Harsanyi (1963) calculates a marginal value for each coalition in a game. Because utility is not assumed transferable, there is naturally a set of marginal values associated with a coalition. Each member of the set is a vector of attainable values with dimension equal to the number of players in the coalition. This set of marginal values also depends on the strategies being chosen by the players outside the coalition in question. The members of a coalition have two problems associated with the choice of a marginal value vector (strictly speaking, with the choice of a joint strategy). The first is to make the pie available to the coalition as large as possible. The second is to determine its division. The model provides a set of weights for the players of the game which are used to determine the relative payoffs to the members of a coalition. The weights may also be used to calculate a scalar number to represent the value (or the marginal value) of each coalition. The weights solve the second problem of how to divide the pie attainable by a coalition.

With the method of division of payoffs inside each coalition fixed, each coalition is quite like a single player. Thus a coalition $\mathcal{K}$ and its complementary coalition $\tilde{\mathcal{K}}(=\mathcal{N}-\mathcal{K})$ may be thought to form a two-person game. If each coalition accepts the Nash bargaining model axioms as an arbitration scheme, then $\mathcal{K}$ and $\tilde{\mathcal{K}}$ may be regarded as playing a two-person cooperative game in which each must choose a threat strategy as discussed above in §2.2.

An n person game has $2^n - 2$ nonempty coalitions which are smaller than $\mathcal{N}$; therefore, there are $2^{n-1} - 1$ distinct "two-person" games consisting of nonempty coalition pairs, $\mathcal{K}$ and $\tilde{\mathcal{K}}$. Thus each coalition $\mathcal{K}$ chooses its strategy with a view to maximizing its final payoff in the game on the assumption that the actions of all other coalitions except $\tilde{\mathcal{K}}$ are committed; and on the assumption that the strategies which it and $\tilde{\mathcal{K}}$ choose determine a threat point from which final payoffs are found.

Two matters are left to be decided. One is the strategy choice of $\mathcal{N}$, the coalition of the whole, and the other is the determination of the weights which the coalitions use. They are decided together. Taking all the strategy choices of the smaller coalitions as given, a threat payoff vector, $\boldsymbol{P}^T_{\mathcal{N}}$, is determined. Measuring from this threat point, the chosen strategy of $\mathcal{N}$ is one which maximizes the Nash product (the product of gains from agreement) of the players. The weights are associated with this final payoff point.

Let $P^T_{i\mathscr{K}}$ denote the threat payoff of the ith player relative to the coalition $\mathscr{K}$; and let $P^*_{i\mathscr{K}}$ denote his equilibrium payoff relative to $\mathscr{K}$. These are defined for $i \in \mathscr{K}$. Then, from the standpoint of $\mathscr{N}$, $\boldsymbol{P}^T_{\mathscr{N}} = (P^T_{1\mathscr{N}}, \ldots, P^T_{1\mathscr{N}})$ is given and $s_{\mathscr{N}}$, the strategy of $\mathscr{N}$, is chosen so that $\boldsymbol{P}^*_{\mathscr{N}} = \boldsymbol{P}(\boldsymbol{s}_{\mathscr{N}})$ satisfies

$$\prod_{i \in \mathscr{N}} (P^*_{i\mathscr{N}} - P^T_{i\mathscr{N}}) = \max_{\boldsymbol{P} \geq \boldsymbol{P}^T_{\mathscr{N}}, \boldsymbol{P} \in \mathscr{H}} \prod_{i \in \mathscr{N}} (P_i - P^T_{i\mathscr{N}}), \tag{11.24}$$

where $\mathscr{H}$ is the set of attainable payoffs. Associated with $\boldsymbol{P}^*_{\mathscr{N}}$ are the weights

$$\rho_i = [1/(P^*_{i\mathscr{N}} - P^T_{i\mathscr{N}})] / \sum_{j \in \mathscr{N}} [1/(P^*_{j\mathscr{N}} - P^T_{j\mathscr{N}})], \qquad i \in \mathscr{N}. \tag{11.25}$$

Thus the equilibrium is characterized by a system of weights associated with the maximized Nash product, along with $2^{n-1} - 1$ two-player Nash cooperative games. The weights are used to aggregate the payoffs of the members of each coalition. For each nonempty pair of coalitions $\mathscr{K}$ and $\tilde{\mathscr{K}}$ the actions of all other coalitions are taken to be known and given and the two coalitions are regarded as playing a two-person cooperative game. The expressions which describe equilibrium payoffs look somewhat like Shapley value expressions.

In §4.2 a version of Harsanyi's model is presented. §4.3 contains a discussion of the relationship between Harsanyi's equilibrium and the Shapley value; and §4.4 discusses some possible generalizations. Throughout chs. 11 and 12, players are always assumed to have complete information. That is, each player knows the payoff functions of all players and the strategy sets of all coalitions. In Harsanyi (1967, 1968a, b) and Harsanyi and Selten (1972) cooperative games are examined in which players lack complete information.

4.2. Harsanyi's equilibrium for an n-person game without transferable utility

The model in this section is that of a finite game. §4.2.1 contains a presentation and discussion of the assumptions of the model. §4.2.2 gives the definition and motivation of the equilibrium; while §4.2.3 contains the existence theorem and its proof.

4.2.1. Assumptions for the model

Each player is assumed to have a finite number, m_i, of pure strategies. For each combination of pure strategies, the payoff to any player is finite. Let $\mathcal{S}_i$ denote the unit simplex of dimension $m_i - 1$, which is the (mixed) strategy set of the ith player. In a cooperative game, the members of a coalition cannot only choose strategies jointly, they may jointly randomize. The members of $\mathcal{K}$, $i_1, \ldots, i_k$, have at their disposal $\Pi_{i\in\mathcal{K}}\, m_i = m_{\mathcal{K}}$ pure strategies, and their mixed strategy set $\mathcal{S}_{\mathcal{K}}$ is the unit simplex of dimension $m_{\mathcal{K}} - 1$. $\mathcal{S}_{\mathcal{K}}$ is a much larger strategy set than the product of the k players' individual strategy sets, $\Pi_{i\in\mathcal{K}}\,\mathcal{S}_i$. As a hint of how much larger $\mathcal{S}_{\mathcal{K}}$ is than $\Pi_{i\in\mathcal{K}}\,\mathcal{S}_i$, consider an example in which there are three players in a coalition, each of whom has four pure strategies. Denote a pure strategy for $\mathcal{K} = \{1, 2, 3\}$ by a triple of numbers such as (2, 3, 3). They could choose to put probability 0.25 on each of (v, v, v) for $v = 1, 2, 3, 4$ and zero on all the other sixty possibilities. Such a joint strategy is impossible without coordination. That is, there is no $(s_1, \ldots, s_4)$ with $s_i \in \mathcal{S}_i$ $(i = 1, \ldots, 4)$ which yields these probabilities; however, the choice is easily within the grasp of the coalition.

The payoff function of a player is denoted P_i; however, the argument of the function may be variously represented. For example, $P_i(s_1, \ldots, s_n)$, $P_i(s_{\mathcal{K}}, s_{\bar{\mathcal{K}}})$ and $P_i(s_{\mathcal{N}})$ all have clear meanings. They are not equivalent, in that each assumes a different partition of the players into coalitions; hence, the implied strategy sets (i.e. the domain of P_i) vary from one to the other. As a consequence, $P_i(s)$ is an ambiguous notation (because the domain is left unclear) and is not used. Where an upper case letter is used to denote a coalition (e.g. $\mathcal{K}$, $\mathcal{L}$, $\mathcal{M}$), the corresponding lower case letter (e.g., k, l, m) denotes the number of members. Say $\mathcal{K} = \{i_1, \ldots, i_k\}$. Then $\mathbf{P}_{\mathcal{K}}(s_{\mathcal{K}}, s_{\bar{\mathcal{K}}}) = (P_{i_1}(s_{\mathcal{K}}, s_{\bar{\mathcal{K}}})), \ldots, P_{i_k}(s_{\mathcal{K}}, s_{\bar{\mathcal{K}}}))$. Stating the assumptions formally:

C12. *The strategy set of the ith player is the $m_i - 1$ dimensional (closed) unit simplex. The joint strategy set of the coalition $\mathcal{K}$ is the $\Pi_{i\in\mathcal{K}}\, m_i - 1 = m_{\mathcal{K}} - 1$ dimensional unit simplex $(i = 1, \ldots, n$ and $\mathcal{K} \subseteq \mathcal{N})$.*

C13. *Let $\mathcal{K}_1, \ldots, \mathcal{K}_r$ be an arbitrary partition of $\mathcal{N}$ into r coalitions. The payoff function $P_i(s_{\mathcal{K}_1}, \ldots, s_{\mathcal{K}_r})$ is bounded and linear in each of the $s_{\mathcal{K}_i}$ $(i = 1, \ldots, r)$.*

The attainable set of payoffs, $\mathcal{H}$ or $\mathcal{H}_{\mathcal{N}}$, is

$$\mathscr{H}_{\mathcal{N}} = \{\mathbf{P}(s_{\mathcal{N}}) | s_{\mathcal{N}} \in \mathscr{S}_{\mathcal{N}}\}. \tag{11.26}$$

For each coalition $\mathscr{K}$ there is an attainable set which depends on $s_{\bar{\mathscr{K}}}$,

$$\mathscr{H}_{\mathscr{K}}(s_{\bar{\mathscr{K}}}) = \{\mathbf{P}_{\mathscr{K}}(s_{\mathscr{K}}, s_{\bar{\mathscr{K}}}) | s_{\mathscr{K}} \in \mathscr{S}_{\mathscr{K}}\}, \quad \text{for } s_{\bar{\mathscr{K}}} \in \mathscr{S}_{\bar{\mathscr{K}}}. \tag{11.27}$$

Note that $\bigcup_{s_{\bar{\mathscr{K}}} \in \mathscr{S}_{\bar{\mathscr{K}}}} \mathscr{H}_{\mathscr{K}}(s_{\bar{\mathscr{K}}})$ may be smaller than the projection of $\mathscr{H}_{\mathcal{N}}$ onto coordinates corresponding to the members of $\mathscr{K}$, because $\mathscr{S}_{\mathcal{N}}$ is larger than $\mathscr{S}_{\mathscr{K}} \times \mathscr{S}_{\bar{\mathscr{K}}}$.

There is one additional assumption, of an ad hoc character, which is added:

C14. *Any player can reduce any payoff of his own by any amount with no effect on the payoffs of the others.*

C14 is a *free disposal* assumption which has the effect of enlarging $\mathscr{H}_{\mathscr{K}}$ to include anything *below* it. Calling this extended set $\mathscr{H}'_{\mathscr{K}}$, it is defined by

$$\mathscr{H}'_{\mathscr{K}}(s_{\bar{\mathscr{K}}}) = \{\mathbf{P}_{\mathscr{K}} | \mathbf{P}_{\mathscr{K}} \leq \mathbf{P}'_{\mathscr{K}} \text{ for at least one } \mathbf{P}'_{\mathscr{K}} \in \mathscr{H}_{\mathscr{K}}(s_{\bar{\mathscr{K}}})\}, \quad \mathscr{K} \subseteq \mathcal{N}. \tag{11.28}$$

Figure 11.6 illustrates the relationship of $\mathscr{H}_{\mathscr{K}}$ and $\mathscr{H}'_{\mathscr{K}}$. $\mathscr{H}_{\mathscr{K}}$ is heavily shaded and $\mathscr{H}'_{\mathscr{K}}$ is $\mathscr{H}_{\mathscr{K}}$ plus the area which is lightly shaded. Both sets are convex. The role played by C14 is made apparent in the proof of the main result in §4.2.3.

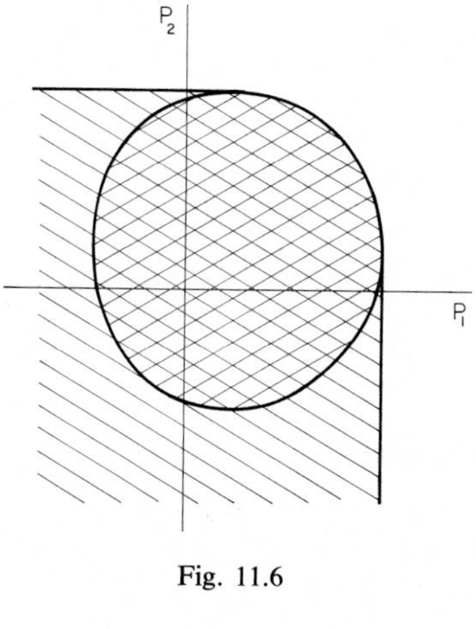

Fig. 11.6

4.2.2. The characteristics of the equilibrium

The equilibrium is characterized by $2^n - 1$ strategies and $n2^{n-1}$ associated payoffs. There is one strategy $s_{\mathcal{K}}$ for each nonempty $\mathcal{K} \subseteq \mathcal{N}$; for each $\mathcal{K} \subset \mathcal{N}$ there are payoff vectors $(\boldsymbol{P}_{\mathcal{K}}(s_{\mathcal{K}}, s_{\bar{\mathcal{K}}}), \boldsymbol{P}_{\bar{\mathcal{K}}}(s_{\mathcal{K}}, s_{\bar{\mathcal{K}}}))$, and for $\mathcal{N}$ there is $\boldsymbol{P}(s_{\mathcal{N}})$. The coalition $\mathcal{K} \subset \mathcal{N}$ chooses $s_{\mathcal{K}}$ in a way intended to maximize $\boldsymbol{P}_{\mathcal{K}}(s_{\mathcal{N}})$, the payoff to the members of $\mathcal{K}$ which is associated with $\mathcal{N}$. None of these $s_{\mathcal{K}}$ for $\mathcal{K} \subset \mathcal{N}$ are actually played. The strategy actually played is $s_{\mathcal{N}}$. This is analogous to the threat strategies and payoffs of the Nash (1953) model in which threat strategies are not used nor threat payoffs received. The threat strategies determine threat payoffs which, in turn, determine the solution strategy vector and payoffs.

It is helpful to define a vector to represent the marginal value of a coalition to its members. This is analogous to the marginal value of §3.2 for Shapley's model and to the excess of solution payoff over threat payoff in the Nash model. For each $\mathcal{K} \subseteq \mathcal{N}$ assume an arbitrary strategy $s_{\mathcal{K}}$. Marginal values are defined in relation to these arbitrary strategies.[16] The marginal value of the ith player to the coalition $\mathcal{K}$, when $i \in \mathcal{K}$, is

$$c_{i\mathcal{K}} = \sum_{\mathcal{L} \ni i, \mathcal{L} \subseteq \mathcal{K}} (-1)^{k-l} P_{i\mathcal{L}}, \qquad i \in \mathcal{K}, \quad \mathcal{K} \subseteq \mathcal{N}. \tag{11.29}$$

$\boldsymbol{c}_{\mathcal{K}} = (c_{i_1,\mathcal{K}}, \ldots, c_{i_k,\mathcal{K}})$, the vector of marginal values of $\mathcal{K}$ to its members, is a function of all the strategies, $s_{\mathcal{L}}$, of coalitions contained in $\mathcal{K}$ and their complements; that is of $s_{\mathcal{L}}$ and $s_{\bar{\mathcal{L}}}$ for each $\mathcal{L} \subseteq \mathcal{K}$ such that $i \in \mathcal{L}$. An implication of eq. (11.29) is

$$P_{i\mathcal{K}} = \sum_{\mathcal{L} \ni i, \mathcal{L} \subseteq \mathcal{K}} c_{i\mathcal{L}}, \qquad i \in \mathcal{K}, \quad \mathcal{K} \subseteq \mathcal{N}. \tag{11.30}$$

The payoff associated with player i in coalition $\mathcal{K}$ ($\boldsymbol{P}_{i\mathcal{K}}$) is the sum of the marginal values imputed to him for all $\mathcal{L}$ contained in $\mathcal{K}$ of which he is a member.

Let $s^*_{\mathcal{K}}$ ($\mathcal{K} \subseteq \mathcal{N}$) denote a collection of strategies, $P^*_{i\mathcal{K}} = P_i(s^*_{\mathcal{K}}, s^*_{\bar{\mathcal{K}}})$, $c^*_{i\mathcal{K}}$ be defined according to eq. (11.29) using the $P^*_{i\mathcal{K}}$ and $\boldsymbol{\rho}^*$ be a vector of n weights such that $\boldsymbol{\rho}^* > 0$ and $\Sigma_{i \in \mathcal{N}} \rho^*_i = 1$. These strategies, payoffs, marginal values and weights characterize a Harsanyi equilibrium if

$$\sum_{i \in \mathcal{N}} \rho^*_{i\mathcal{N}} P^*_{i\mathcal{N}} = \max_{\boldsymbol{P} \in \mathcal{H}_{\mathcal{N}}} \sum_{i \in \mathcal{N}} \rho^*_i P_i, \tag{11.31}$$

$$\rho^*_i c^*_{i\mathcal{N}} = \rho^*_j c^*_{j\mathcal{N}}, \qquad \text{for all } i, j \in \mathcal{N}, \tag{11.32}$$

[16] Arguments are dropped in the interest of concise notation: $P_{i\mathcal{K}} = P_i(s_{\mathcal{K}}, s_{\bar{\mathcal{K}}})$, etc.

$$\sum_{i\in\mathcal{K}} \rho_i^* P_{i\mathcal{K}}^* - \sum_{j\in\tilde{\mathcal{K}}} \rho_j^* P_{j\tilde{\mathcal{K}}}^* = \max_{s_{\mathcal{K}}\in\mathcal{S}_{\mathcal{K}}} \min_{s_{\tilde{\mathcal{K}}}\in\mathcal{S}_{\tilde{\mathcal{K}}}} \left[\sum_{i\in\mathcal{K}} \rho_i^* P_i(s_{\mathcal{K}}, s_{\tilde{\mathcal{K}}}) - \sum_{j\in\tilde{\mathcal{K}}} \rho_j^* P_j(s_{\mathcal{K}}, s_{\tilde{\mathcal{K}}})\right], \tag{11.33}$$

subject to

$$\rho_i^* c_{i\mathcal{K}}^* = \rho_{i'}^* c_{i'\mathcal{K}}^*, \qquad \text{for all } i, i' \in \mathcal{K}, \tag{11.34}$$

$$\rho_j^* c_{j\tilde{\mathcal{K}}}^* = \rho_{j'}^* c_{j'\tilde{\mathcal{K}}}^*, \qquad \text{for all } j, j' \in \tilde{\mathcal{K}}. \tag{11.35}$$

The threat payoff of the ith player in relation to coalition $\mathcal{K}$ (of which he is a member) is $P_{i\mathcal{K}}^T = P_{i\mathcal{K}}^* - c_{i\mathcal{K}}^*$ or

$$P_{i\mathcal{K}}^T = \sum_{\mathcal{L}\ni i, \mathcal{L}\subset\mathcal{K}} c_{i\mathcal{L}}^*, \qquad \text{for any } \mathcal{K} \subseteq \mathcal{N} \text{ where } k \geq 2. \tag{11.36}$$

Equations (11.31) and (11.32) are the conditions requiring that $\boldsymbol{P}_{\mathcal{N}}^*$ be chosen to maximize the Nash product.

$$\prod_{i\in\mathcal{N}} (P_{i\mathcal{N}}^* - P_{i\mathcal{N}}^T) = \max_{\boldsymbol{P}\in\mathcal{K}_{\mathcal{N}}, \boldsymbol{P}\geq\boldsymbol{P}_{\mathcal{N}}^T} \prod_{i\in\mathcal{N}} (\boldsymbol{P}_i - P_{i\mathcal{N}}^T). \tag{11.37}$$

Equations (11.33)–(11.35) describe the maximization conditions for the "two-person" game between coalitions $\mathcal{K}$ and $\tilde{\mathcal{K}}$. Equation (11.34) is a set of conditions imposed on the members of $\mathcal{K}$ requiring that the gains from the coalition to its members be in prescribed proportions. Equation (11.35) is a parallel set of conditions for $\tilde{\mathcal{K}}$. The imposition of these conditions causes a perfect coincidence between the interests of the various members of $\mathcal{K}$, and similarly for $\tilde{\mathcal{K}}$. There is no externally imposed requirement on how the payoffs to members of $\mathcal{K}$ are related to members of $\tilde{\mathcal{K}}$.

In order to see that eq. (11.33) does express the interest of the members of $\mathcal{K}$ and $\tilde{\mathcal{K}}$, let

$$P_{0\mathcal{K}} = \sum_{i\in\mathcal{K}} \rho_i P_{i\mathcal{K}}, \qquad \text{for all } \mathcal{K} \subseteq \mathcal{N}, \tag{11.38}$$

$$c_{0\mathcal{K}} = \sum_{i\in\mathcal{K}} \rho_i c_{i\mathcal{K}}, \qquad \text{for all } \mathcal{K} \subseteq \mathcal{N}. \tag{11.39}$$

$P_{0\mathcal{K}}$ is a scalar measure of the payoff to $\mathcal{K}$, using $\boldsymbol{\rho}$ as weights. $c_{0\mathcal{K}}$ is a similar measure of the gain to $\mathcal{K}$. From eqs. (11.38), (11.39), (11.29), (11.30), (11.34) and (11.35),

$$P_{0\mathcal{K}} = \sum_{\mathcal{L}\subseteq\mathcal{K}} (-1)^{k-l} c_{0\mathcal{L}}, \qquad \text{for all } \mathcal{K} \subseteq \mathcal{N}. \tag{11.40}$$

Recall that the members of $\mathscr{K}$ wish to choose $s_{\mathscr{K}}$ to make their final payoffs (the $P_{i\mathcal{N}}$, $i \in \mathscr{K}$) as large as possible. Consider the final payoff to player $i \in \mathscr{K}$. From eq. (11.40) and noting that $\rho_i c_{i\mathscr{K}} = c_{0\mathscr{K}}/k$,

$$P_{i\mathcal{N}} = \sum_{\mathscr{L} \ni i, \mathscr{L} \in \mathcal{N}} [P_{0\mathscr{L}} - P_{0\bar{\mathscr{L}}}]/(\rho_i l C_l^n)$$

$$= P_{0\mathcal{N}}/(\rho_i n) + [P_{0\mathscr{K}} - P_{0\bar{\mathscr{K}}}]/(\rho_i k C_k^n) + \gamma_i, \tag{11.41}$$

where $C_l^n = n!/(l!(n-l)!)$, $P_{0\bar{\mathcal{N}}} = 0$, and γ_i is all the terms involving coalitions other than $\mathscr{K}$, $\bar{\mathscr{K}}$ and $\mathcal{N}$. The payoffs to members of $\mathscr{K}$ rise and fall in a lock step given by eq. (11.34); therefore, an action which maximizes the final payoff of one particular member of $\mathscr{K}$ also maximizes the payoff of each of the others. The weights $\boldsymbol{\rho}$ are given; and further, $P_{0\mathcal{N}}$ is chosen so that it has a maximum value relative to the weights. Therefore, the only way that $\mathscr{K}$ can increase its payoff is by increasing $P_{0\mathscr{K}} - P_{0\bar{\mathscr{K}}}$. For $j \in \bar{\mathscr{K}}$ the parallel relation to eq. (11.41) is

$$P_{j\mathcal{N}} = P_{0\mathcal{N}}/(\rho_i n) + [P_{0\mathscr{K}} - P_{0\bar{\mathscr{K}}}]/(\rho_i k C_{n-k}^n) + \gamma_j. \tag{11.42}$$

Clearly, by the same reasoning, the interests of $\bar{\mathscr{K}}$ are best served by minimizing $P_{0\mathscr{K}} - P_{0\bar{\mathscr{K}}}$. With the character of the equilibrium established, attention may now be turned to proving its existence.

4.2.3. Existence of the Harsanyi equilibrium

In keeping with the usage that a solution is an equilibrium which is either unique, or in some important sense, uniquely distinguished, a solution to the present model cannot be proved to exist.[17] An equilibrium can. The equilibrium is described above in eqs. (11.31) to (11.35) without reference to the free disposal assumption, C14. In $\mathscr{H}_{\mathscr{K}}(s_{\bar{\mathscr{K}}})$ there is a point at which eq. (11.34) is satisfied for maximized values of the $c_{i\mathscr{K}}$. In $\mathscr{H}'_{\mathscr{K}}(s_{\bar{\mathscr{K}}})$, the point which maximizes the $c_{i\mathscr{K}}$, subject to eq. (11.34), is found by first finding a point in $\mathscr{H}_{\mathscr{K}}(s_{\bar{\mathscr{K}}})$ at which

$$\min_{i \in \mathscr{K}} (P_i(s_{\mathscr{K}}, s_{\bar{\mathscr{K}}}) - P_{i\mathscr{K}}^T) \tag{11.43}$$

is maximized with respect to $s_{\mathscr{K}} \in \mathscr{S}_{\mathscr{K}}$. At this point, if there are any $\rho_i c_{i\mathscr{K}}$ which exceed the minimum, they are reduced by free disposal. The

[17] An example of a uniquely distinguished equilibrium is the noncooperative equilibrium chosen by the tracing procedure in ch. 7, §4.3.

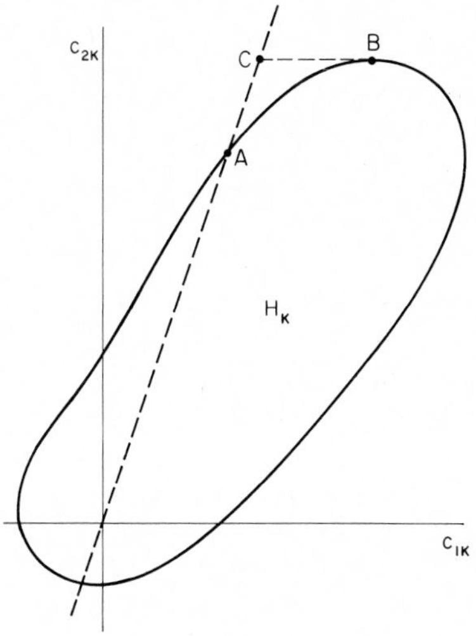

Fig. 11.7

situation is illustrated in fig. 11.7. The broken line through the origin is the set of points satisfying eq. (11.34). The point A is the optimal point in $\mathcal{H}_{\mathcal{K}}$ which satisfies the condition; however, it is at $B \in \mathcal{H}_{\mathcal{K}}$ that

$$\max_{s_{\mathcal{K}} \in \mathcal{S}_{\mathcal{K}}} \min_{i \in \mathcal{K}} c_{i\mathcal{K}} \tag{11.44}$$

is found. At B, player 1 gets too much ($\rho_1 c_{1\mathcal{K}} > \rho_2 c_{2\mathcal{K}}$); hence, he exercises free disposal, moving the chosen point to C. Note that each player's payoff function is concave in $s_{\mathcal{K}}$ and the minimum of a family of concave functions is also concave; therefore, the function defined by

$$\min_{i \in \mathcal{K}} (P_i(s_{\mathcal{K}}, s_{\bar{\mathcal{K}}}) - P^T_{i\mathcal{K}}), \tag{11.45}$$

is concave in $s_{\mathcal{K}}$. To take account of C14, eq. (11.33) must be rewritten. Let $\mathcal{S}'_{\mathcal{K}}$ denote an *extended strategy set* which consists of $\mathcal{S}_{\mathcal{K}}$ and the exercise of free disposal.[18]

$$\sum_{i \in \mathcal{K}} \rho_i^* P^*_{i\mathcal{K}} - \sum_{j \in \bar{\mathcal{K}}} \rho_j^* P^*_{j\bar{\mathcal{K}}}$$

$$= \max_{s_{\mathcal{K}} \in \mathcal{S}'_{\mathcal{K}}} \min_{s_{\bar{\mathcal{K}}} \in \mathcal{S}_{\bar{\mathcal{K}}}} \left[\sum_{i \in \mathcal{K}} \rho_i^* P_i(s_{\mathcal{K}}, s_{\bar{\mathcal{K}}}) - \sum_{j \in \bar{\mathcal{K}}} \rho_j^* P_j(s_{\mathcal{K}}, s_{\bar{\mathcal{K}}}) \right]. \tag{11.46}$$

[18]It is possible to give a formal representation to $\mathcal{S}'_{\mathcal{K}}$; however, doing so seems unnecessarily pedantic.

Theorem 11.9. *For a cooperative game model satisfying C9, C12–C14, there is a set of weights $\boldsymbol{\rho}^* > 0$, $\Sigma_{i\in\mathcal{N}} \rho_i^* = 1$ and a collection of strategies $s^*_{\mathcal{K}}$ ($\mathcal{K} \subseteq \mathcal{N}$) which satisfy eqs. (11.31), (11.32), (11.34), (11.35) and (11.46). These strategies and weights, and the associated payoffs, characterize the Harsanyi equilibrium.*

□ The method of proof is to construct a noncooperative game for which an equilibrium corresponds to a Harsanyi equilibrium, then to show that this noncooperative equilibrium exists. The noncooperative game has $2^n - 1$ players, one for each nonempty coalition. For $\mathcal{K} \subset \mathcal{N}$ the objective is to maximize with respect to $s_{\mathcal{K}} \in \mathcal{S}'_{\mathcal{K}}$, and subject to eq. (11.34),

$$\sum_{i\in\mathcal{K}} \rho_i P_i(s_{\mathcal{K}}, s_{\bar{\mathcal{K}}}) - \sum_{j\in\bar{\mathcal{K}}} \rho_j P_j(s_{\mathcal{K}}, s_{\bar{\mathcal{K}}}). \tag{11.47}$$

The weights $\boldsymbol{\rho}$ which appear in eq. (11.47) are the strategy choice of the player representing $\mathcal{N}$ (or *player* $\mathcal{N}$). Note that maximizing eq. (11.47) with respect to $s_{\mathcal{K}} \in \mathcal{S}'_{\mathcal{K}}$, subject to eq. (11.34) is equivalent to maximizing

$$\sum_{i\in\mathcal{K}} \rho_i [\min_{j\in\mathcal{K}} \rho_j (P_j(s_{\mathcal{K}}, s_{\bar{\mathcal{K}}}) - P^T_{j\mathcal{K}})] + \sum_{i\in\mathcal{K}} \rho_i P^T_{i\mathcal{K}} - \sum_{j\in\bar{\mathcal{K}}} \rho_j P_j(s_{\mathcal{K}}, s_{\bar{\mathcal{K}}}) \tag{11.48}$$

over $s_{\mathcal{K}} \in \mathcal{S}_{\mathcal{K}}$ and subject to eq. (11.34). This latter form of the objective function of player $\mathcal{K}$ ($\subset \mathcal{N}$) is one for which his strategy set is clearly compact and convex and his payoff function, eq. (11.48), is bounded, continuous in all arguments and concave in $s_{\mathcal{K}}$.[19] The arguments of player $\mathcal{K}$'s objective function are $\boldsymbol{\rho}$ as well as $s_{\mathcal{L}}$ and $s_{\bar{\mathcal{L}}}$ for all $\mathcal{L} \subseteq \mathcal{K}$. Thus the best reply function for $\mathcal{K}$ is upper semicontinuous with convex image sets for all $\mathcal{K} \subset \mathcal{N}$.

Player $\mathcal{N}$ chooses $\boldsymbol{\rho}$. It remains to show that his best reply function is upper-semicontinuous with convex image sets. This is done by directly specifying his best reply function without reference to a hypothetical objective function. The latter would only be concocted to justify a convenient best reply function anyway. The best reply function of $\mathcal{N}$ is a best reply to a collection of coalition strategy choices, $s_{\mathcal{K}}$ ($\mathcal{K} \subset \mathcal{N}$). Such a collection determines a unique threat point relative to $\mathcal{N}$, $\boldsymbol{P}^T_{\mathcal{N}}$. Except if $\boldsymbol{P}^T_{\mathcal{N}}$ is a Pareto optimal element of $\mathcal{H}_{\mathcal{N}}$, there is a unique $\boldsymbol{P}^*_{\mathcal{N}}$ associated with $\boldsymbol{P}^T_{\mathcal{N}}$ which maximizes the Nash product (eq. (11.24)). Using $\boldsymbol{P}^*_{\mathcal{N}}$ and

[19] It is because the objective function of $\mathcal{K}$ may be put into this form that it is unnecessary to specify formally the free disposal assumption in terms of an explicit extended strategy set.

$P^T_{\mathcal{N}}$, the best reply is calculated using eq. (11.25):

$$\rho_i = [1/P^*_{i\mathcal{N}} - P^T_{i\mathcal{N}})]/\sum_{j\in\mathcal{N}} [1/(P^*_{j\mathcal{N}} - P^T_{j\mathcal{N}})]. \tag{11.25}$$

If $P^T_{\mathcal{N}}$ is Pareto optimal in $\mathcal{H}_{\mathcal{N}}$, then we look at all the conceivable candidates for threat point which would have $P^T_{\mathcal{N}}$ as the associated Nash product maximizer. For each threat point candidate there is a weight vector given by eq. (11.25) with $P^T_{\mathcal{N}}$ taking the place of $P^*_{\mathcal{N}}$ and the threat point candidate taking the place of $P^T_{\mathcal{N}}$. For all threat point candidates which lead to $P^T_{\mathcal{N}}$, let $r(P^T_{\mathcal{N}})$ be the set of associated weight vectors. $r(P^T_{\mathcal{N}})$ is the set of best replies to the strategies which give rise to $P^T_{\mathcal{N}}$.

It is easily verified that the best reply mapping for $\mathcal{N}$ is upper-semicontinuous with convex image sets. Indeed if $P^T_{\mathcal{N}}$ is at a Pareto optimal point where the Pareto optimal surface is smooth (i.e. is differentiable), then $r(P^T_{\mathcal{N}})$ consists of exactly one point. Where $r(P^T_{\mathcal{N}})$ has more than one point, it is convex because the set of P which would be threat points leading to $P^T_{\mathcal{N}}$ is a convex cone. Where $P^T_{\mathcal{N}}$ is not Pareto optimal, the best reply is clearly continuous; because $P^T_{\mathcal{N}}$ is a continuous function of the coalition strategies and eq. (11.25) is continuous with respect to $P^T_{\mathcal{N}}$. Otherwise, consider a convergent sequence of coalition strategies, each of which is associated with a non-Pareto optimal $P^T_{\mathcal{N}}$, but which converge to give a Pareto optimal $P^T_{\mathcal{N}}$. Indexing the members of the sequence by $l = 1, 2, \ldots$, each has an associated ρ^l. While the sequence ρ^l, $l = 1, 2, \ldots$, need not converge, it has convergent subsequences. The limit of any convergent subsequence is in $r(P^T_{\mathcal{N}})$ where $P^T_{\mathcal{N}}$ is the limit of the threat point sequence. Thus, the best reply mapping has a fixed point. A fixed point of the best reply mapping is a set of weights and strategies which, along with their associated payoffs, form a Harsanyi equilibrium. □

A final point to note concerning the equilibrium and Pareto optimality is that the equilibrium payoffs $P^*_{\mathcal{N}}$ cannot be weakly dominated unless the associated threat point $P^T_{\mathcal{N}}$ is a Pareto optimal point which can be weakly dominated.

4.3. *The relationship of the Harsanyi equilibrium and the Shapley value*

Not only is it obvious that when $n = 2$, the game reduces to the Nash general two-person cooperative game, the role of the two person Nash theory is prominent and shows itself in the conditions which define the equilibrium. This is most striking in the network of two-person games which are imbedded into the model and in the requirement that the equilibrium should maximize an n-person Nash product.

The relationship to the Shapley value is less immediate; however, it comes through in two ways. First, the calculation of marginal values, in eq. (11.29), is a generalization of the similar calculation in eq. (11.15) for Shapley's model. The second connection is strongly hinted at in eq. (11.41). An alternative way to write eq. (11.23) is

$$\phi_i(v) = \sum_{\mathscr{L} \ni i, \mathscr{L} \subseteq \mathscr{N}} (v(\mathscr{L}) - v(\hat{\tilde{\mathscr{L}}}))/(lC_l^n). \tag{11.49}$$

Equations (11.49) and (11.41) are essentially in the same form. Indeed, if all the $v(\mathscr{L})$ are set equal to $P_{0\mathscr{L}}$ for all $\mathscr{L} \subseteq \mathscr{N}$ and $\phi_i(v)$ is taken to be $\rho_i P_{i\mathscr{N}}$; then they are the same.

4.4. *A brief comment on generalizing the finite game model*

It would be desirable to have a form of the Harsanyi model in which mixed strategies were either ruled out or dominated by pure strategies, as such a model would be better suited to oligopoly applications. The sticking point in such a formulation appears to be the quasi-concavity requirements of eq. (11.47) which is the objective function of the coalition $\mathscr{K}$. Other than to merely assert that eq. (11.47) is quasi-concave in $s_{\mathscr{K}}$ and quasi-convex in $s_{\bar{\mathscr{K}}}$, I do not see how to build these requirements into the assumptions.[20] Of course, they should be made at a fundamental level so that their effect on the basic structure of the model is clear.

5. *Concluding comments*

Four related models are reviewed in this chapter; Nash two-person fixed threat (called the bargaining model), Nash two-person variable threat, Shapley n-person with transferable utility, and Harsanyi n-person. Though restricted to two players, the Nash models do not assume transferable utility; hence, Shapley's model does not contain those of Nash as a subset. Shapley could be said to gain the ability to deal with n players by accepting transferable utility. Harsanyi provides a generaliza-

[20]If $-f(\boldsymbol{x})$ is quasi-concave, then $f(\boldsymbol{x})$ is quasi-convex.

tion of both in his model of n-person cooperative games without transferable utility.

With all the models, it is instructive to look for the part played by the characteristic function. The Nash bargaining model is presented in a form which is very nearly characteristic function form, and the characteristic functions for the three coalitions are simple to construct. Each player can guarantee himself the payoff he gets at the no trade point, and the coalition $\{1, 2\}$ can guarantee itself any point in the attainable set of the game. Indeed, it should be obvious that in any game of pure trade, with no externalities characteristic functions may be unambiguously defined. Any coalition can guarantee to itself whatever payoffs can be associated with redistributions of their own endowments among themselves. Insofar as people outside a coalition cannot force actions on those inside, whatever the outsiders do to divide their own resources has no effect on the payoffs to those inside.

The Nash variable threat model is quite another matter. What player 1 can *guarantee* himself depends on what player 2 chooses to do. In discussing the model, no mention is made of what a player can guarantee himself irrespective of the actions of the other. In contrast, the Shapley model is formulated in characteristic function form. Each coalition has payoffs it can guarantee itself. It is not said whether, in some circumstances, a coalition might do better. Such possibilities play no role.

The Harsanyi model is like the Nash variable threat model in its underlying formulation. It, too, is given in normal form; and characteristic functions really do not appear. For a coalition $\mathcal{K}$, there is $\mathcal{H}_{\mathcal{K}}(s_{\tilde{\mathcal{K}}})$, which is a set of attainable payoffs for the coalition when the complementary coalition $\tilde{\mathcal{K}}$ uses $s_{\tilde{\mathcal{K}}}$. $\mathcal{H}_{\mathcal{K}}(s_{\tilde{\mathcal{K}}})$ is not a set of payoffs which $\mathcal{K}$ can guarantee itself; because it cannot guarantee the behavior of $\tilde{\mathcal{K}}$. $\mathcal{K}$ can guarantee itself $\bigcap_{s_{\tilde{\mathcal{K}}} \in \mathcal{S}_{\tilde{\mathcal{K}}}} \mathcal{H}_{\mathcal{K}}(s_{\tilde{\mathcal{K}}}) = \mathcal{V}_{\mathcal{K}}$; because $\mathcal{V}_{\mathcal{K}}$ consists of those payoffs which are attainable no matter what strategy the other coalition chooses. These sets played no role in §4. They make a possible definition of characteristic function; however, forcing a coalition to base its actions on what it can guarantee itself (in the sense of $\mathcal{V}_{\mathcal{K}}$) may seem excessively conservative. After all, some of the $s_{\tilde{\mathcal{K}}}$ may be so unfavorable to $\tilde{\mathcal{K}}$ that is is highly unlikely that they would ever be chosen. The equilibrium, which has associated with it a strategy for each coalition, does only use a system of strategies which there is some reason to choose. That is, $s_{\mathcal{K}}$ is selected because it is a best reply for the members of $\mathcal{K}$ against the whole collection of strategies actually chosen by the other coalitions. It is well to keep this in mind as core theory is examined in the following chapter, for the approach there is quite different to that taken in the present chapter.

CHAPTER 12

COOPERATIVE GAMES – THE CORE

1. Edgeworth, the core and the von Neumann–Morgenstern "solution"

Just as the noncooperative equilibrium of Nash (1951) is a generalization of the oligopoly equilibrium of Cournot (1838), the *core*, proposed by Gillies (1953), is a generalization of the *contract curve* of Edgeworth (1881).[1] Edgeworth described a market in which there are two consumers and two commodities. Each consumer has an endowment of one or both commodities, and the only choices the consumers have is (a) to consume their respective endowments or (b) to trade with one another and then consume their respective holdings after trade. Such a market is represented in fig. 12.1 in the familiar *Edgeworth box*. One point in the box simultaneously represents the position of both consumers; and the dimensions of the box are fixed by the sum of the two endowments. The origin for trader A is the lower left corner and for B, the upper right-hand corner. For B, increasing amounts are downward and leftward. Indifference curves are drawn for each. Any point in the box is a feasible, efficient allocation of the total endowment among the two traders; it allocates nonnegative amounts of each good to each trader; and the total allocated equals the total of the two endowments.

The curve labelled *abcd* is the Pareto optimal curve. It consists of those allocations which cannot be dominated. If a point is on the curve *abcd*, there is no other point in the box which gives greater utility to one consumer without giving less to the other. The initial endowment point is at *E*. The shaded area is also of special significance. It consists of those allocations for which the utility of each consumer is at least as great as he would receive if he consumed his initial endowment.

[1]In Edgeworth (1881) the terminology is different from that which I will use, and different from what is now common usage. What I call the *contract curve* Edgeworth called the *final settlements*, and what I call the *Pareto optimal curve* Edgeworth called the *contract curve.*

Fig. 12.1

That part of the Pareto optimal curve which lies in the shaded area, *bc*, is the contract curve. In the terminology of game theory, it is the core. The points in the core, or on the contract curve, are all the possible trades which one or both of the two players would not surely rule out as possible points of agreement. To start, A would not accept a trade which afforded him lower utility than he gets at *E*. Similarly for B. Nothing can force them to accept one of these inferior trades; hence, all points outside the shaded area are ruled out as equilibrium points by one consumer or the other. Inside the shaded area, consider a point which is not on *bc*, such as *D*. Both consumers together would rule out *D* as a trade; because they could as well agree on a point inside the heavily shaded area and both be better off than at *D*. The only points in the shaded area which cannot be ruled out in this way are the points on the segment *bc*. The nature of the core in this example and the criteria used to define it illustrate its characteristics. The term contract curve is reserved to mean the core of an economic model of trade; while core is used to mean essentially the same thing for more general models.

Before going on to study the core more closely, I want to mention two matters. The first is the behavior of the core in the Edgeworth trading model as the market is *enlarged*, and the second is the von Neumann–Morgenstern "solution." Edgeworth considered the way in which the core changes when the market is made larger by replicating the traders. He considered having *k* traders exactly like A and *k* exactly like B. *Exactly like* means having both the same preferences and the same endowment. Such a way of enlarging the market is ingenious because it remains possible to represent the market in a two-dimensional diagram, and, for any k, the core points are a subset of the points on the segment bc. As $k \to \infty$, the only core points which remain are the points on bc which are competitive equilibria. The reader interested in a fuller treatment is referred to Debreu and Scarf (1972) where additional references are also given.

The von Neumann–Morgenstern "solution" is related to the core, but is a less satisfactory concept. It is described in ch. VI of von Neumann and Morgenstern (1944) and in §§8.6–9.1 of Luce and Raiffa (1957). It is not a solution in the sense that word is used in the present volume, but rather an equilibrium. Imagine an n-person cooperative game in characteristic function form. A vector of payoffs, $\boldsymbol{P}$, is *dominated* or *can be improved upon* if there is some subset of players which can guarantee to its members larger payoffs than they get from $\boldsymbol{P}$. The von Neumann–Morgenstern solution consists of a set of payoff vectors which are attainable and which are dominated by any other member of the set. Any payoff vector outside the solution is dominated by one which is in it. Thus, $\boldsymbol{P}'$ may be in the solution yet be dominated by something not in the solution.

The solution is nonunique in two senses. First, of course, it may contain more than one payoff vector. Second, more than one set may form a solution. It is clear, however, that the core, which is a unique set, is a subset of every solution. That is because the core consists of those attainable payoff vectors which are undominated by any other attainable payoff vector. The unsatisfactory element of the von Neumann–Morgenstern solution is that some $\boldsymbol{P}'$ may be in a solution, yet be dominated by an attainable payoff vector $\boldsymbol{P}''$. The reader may, of course, judge for himself whether he finds this characteristic damning.

§2 and §3 are based on papers by Scarf. §2 follows Scarf (1967) in showing that the core is not empty for a class of games in characteristic function form. In §3 the results of §2 are extended to a class of models in normal form, following Scarf (1971). These sections contain the main results of the chapter. Generalizations of some of Scarf's results may be found in Billera (1970) and Shapley (1972). In §4 results due to Farrell (1970), Shitovitz (1973) and Aumann (1959), (1960), (1961) are briefly reviewed without proofs. Farrell and Shitovitz both deal with a trading game in which a few traders are very large and the rest very small. The point of their work is to provide a general equilibrium model of an economy in which an oligopoly is imbedded. This contrasts with the usual treatments of oligopoly in which a single oligopolistic market is modelled in isolation from the rest of the economy. Both of their papers, though more clearly that of Farrell, have Edgeworth (1881) as a point of departure. Aumann is concerned with a class of cooperative supergames. §5 contains comments on the comparison between the core models and the value models. The last section, §6, contains concluding comments on cooperative games and applications of them to oligopoly.

2. *Existence of a nonempty core for games in characteristic function form*

In this section a class of games is specified in characteristic function form. It is proved that for a member of this class, the core is not empty. The method of proof is particularly interesting for the way in which it contrasts with many of the existence proofs found in earlier chapters. Proofs of existence of noncooperative equilibrium, as well as the existence proof for the Harsanyi equilibrium, make use of a fixed point argument. Such an argument establishes that an equilibrium exists; but does not either construct an equilibrium point or suggest a method of finding one. By contrast, the proof which Scarf uses, reproduced below, proceeds by specifying a computational algorithm which, when used, necessarily leads to the discovery of an equilibrium point. In the present application, it leads to a point in the core. Thus the algorithm is capable of doing double duty. It is a vehicle for the proof that the core is not empty; and it can actually be used in connection with a specific game to find a core point.

The algorithm has very great interest for applications outside the one with which we are directly concerned. In its original form, it was used to prove the existence of, and to find, an equilibrium point for a two-person noncooperative game of the sort studied by Nash (1951). For this work, see Lemke and Howson (1964) and Lemke (1965). In the course of extending and adapting the algorithm of Lemke to other uses, Scarf and others found that it can be used to prove existence of competitive equilibrium, as well as to compute an equilibrium point; and that the algorithm can be used to prove the Brouwer fixed point theorem. The history of these developments, in addition to the results themselves, can be found in Scarf (1973).

In §2.1 the game model is described. §2.2 contains a description of the algorithm and some intermediate results on the way it works. §2.3 presents the statement and proof of the main result.

2.1. The model

The model is specified in characteristic function form. $\mathcal{V}_i$ is the set of payoff levels which the ith player can guarantee for himself, no matter what the set of remaining players, $\mathcal{N}-\{i\}$ decides to do. Letting $\mathcal{K}$ denote a coalition whose members are indexed by $i_1, \ldots, i_k$, $\mathcal{V}_{\mathcal{K}}$ is the set of payoff vectors, $\boldsymbol{P}_{\mathcal{K}}=(P_{i_1}, \ldots, P_{i_k})$ which the coalition $\mathcal{K}$ can guarantee for itself irrespective of the behavior of the complementary coalition $\tilde{\mathcal{K}}$.[2] Finally, let

$$\underline{P}_i=\max_{P_i \in \mathcal{V}_i} P_i, \qquad i=1, \ldots, n. \tag{12.1}$$

$\underline{P}_i$ is the largest payoff which the ith player can guarantee himself on his own. $\underline{P}_i$ is the only point in $\mathcal{V}_i$ which is of interest. In addition to C9 (finite number of players) and C14 (free disposal), the remaining assumptions are that each coalition K can attain at least $\underline{\boldsymbol{P}}_{\mathcal{K}}=(\underline{P}_{i_1}, \ldots, \underline{P}_{i_k})$; and the set of attainable payoff vectors for $\mathcal{K}$ which is at least as large as $\underline{\boldsymbol{P}}_{\mathcal{K}}$ is closed and bounded.

C15. *The set* $\underline{\mathcal{V}}_{\mathcal{K}}=\{\boldsymbol{P}_{\mathcal{K}} | \boldsymbol{P}_{\mathcal{K}} \in \mathcal{V}_{\mathcal{K}},\ \boldsymbol{P}_{\mathcal{K}} \geq \underline{\boldsymbol{P}}_{\mathcal{K}}\}$ *is nonempty and compact, for all* $\mathcal{K} \subseteq \mathcal{N}$.

The basic assumptions of the model are C9, C14, C15, and that the game is *balanced*, although other assumptions are made in order to obtain intermediate results. Any such extra assumptions are eventually dropped; therefore, the theorem proved in §2.3 is that a balanced game satisfying C9, C14 and C15 has a nonempty core.

To give a formal definition of the core, first the concept *improve upon* must be defined. Let $\boldsymbol{P}' \in \mathcal{V}_{\mathcal{N}}$ and let $\boldsymbol{P}'_{\mathcal{K}}=(P'_{i_1}, \ldots, P'_{i_k})$ be the projection of $\boldsymbol{P}'$ onto the coordinate subspace of the members of $\mathcal{K}$. $\boldsymbol{P}'$ can be *improved upon* by $\mathcal{K}$ if there is $\boldsymbol{P}''_{\mathcal{K}} \in \mathcal{V}_{\mathcal{K}}$ for which $\boldsymbol{P}''_{\mathcal{K}} \gg \boldsymbol{P}'_{\mathcal{K}}$. Thus a payoff vector $\boldsymbol{P}'$ can be improved upon if there is some coalition which can simultaneously guarantee each of its members a larger payoff than he gets under $\boldsymbol{P}'$. A payoff vector $\boldsymbol{P}' \in \mathcal{V}_{\mathcal{N}}$ is in the core if there is no coalition $\mathcal{K}$ which can improve upon it.[3]

In the basic model, there remain several things to specify. First are the concepts of *balanced collection of coalitions* and *balanced game.* Then a

[2]In §3 the $\mathcal{V}_{\mathcal{K}}$ are derived from a game in normal form.

[3]Traditionally *block* has been the name used for the concept *improve upon.* I am following the suggestion of Shapley (1973) and using the latter instead because it is more suggestive.

special assumption is introduced which allows a convenient matrix representation of the game.

A notion which plays a key role in the proofs to follow is that of a *balanced collection of coalitions* and the related *balanced game.* Let $\mathcal{W}$ be an arbitrary collection of coalitions for an n-person game. $\mathcal{W}$ is a *balanced collection of coalitions* if it is possible to find a set of weights, $\delta_{\mathcal{K}} > 0$, $\mathcal{K} \subseteq \mathcal{W}$, such that

$$\sum_{\mathcal{K} \in \mathcal{W}, \mathcal{K} \ni i} \delta_{\mathcal{K}} = 1, \qquad i = 1, \ldots, n. \tag{12.2}$$

In other words, for an arbitrary collection of coalitions, a positive weight is assigned to each member. Then, the weights are summed for only those coalitions to which a specific player i belongs. If the weights sum to unity for each i, the collection is balanced. An n-person game is balanced if, for every balanced collection $\mathcal{W}$, a vector $\boldsymbol{P} \in \mathcal{V}_{\mathcal{N}}$ if $\boldsymbol{P}_{\mathcal{K}} \in \mathcal{V}_{\mathcal{K}}$ for all $\mathcal{K} \in \mathcal{W}$.

An example may help in understanding both balanced collection and balanced game. Imagine a three-person game. A suitable arbitrary collection of coalitions is {1, 2}, {1}, {2, 3}, {1, 3}. Let the four weights be $\frac{1}{4}, \frac{1}{2}, \frac{3}{4}$ and $\frac{1}{4}$. The sum of weights for player 1 is $\frac{1}{4}+\frac{1}{2}+\frac{1}{4}=1$; for player 2, $\frac{1}{4}+\frac{3}{4}=1$; and for player 3, $\frac{3}{4}+\frac{1}{4}=1$. Thus the collection is balanced; because it is possible to find weights which satisfy eq. (11.2).[4] To illustrate the condition that a game is balanced, let $(6, 3) \in \mathcal{V}_{\{1,2\}}$, $(5) \in \mathcal{V}_{\{1\}}$, $(2, 8) \in \mathcal{V}_{\{2,3\}}$ and $(10, 6) \in \mathcal{V}_{\{1,3\}}$. Now construct the element of $\mathcal{V}_{\mathcal{N}}$ having as its ith coordinate the smallest payoff which the ith player receives in any of the preceding four vectors in which he is represented. For player 1, 5 is chosen, being the minimum of 6, 5 and 10. For player 2, 2 which is the minimum of 3 and 2; and for player 3, 6, which is the minimum of 8 and 6. This must be in $\mathcal{V}_{\mathcal{N}}$ if the game is balanced because $(5, 2) \in \mathcal{V}_{\{1,2\}}$, $(5) \in \mathcal{V}_{\{1\}}$, $(2, 6) \in \mathcal{V}_{\{2,3\}}$ and $(5, 6) \in \mathcal{V}_{\{1,3\}}$.

A game is called *finite cornered* if the characteristic function for each coalition can be represented by a finite number of payoff vectors in this way:

C16. *For finite $r_{\mathcal{K}}$ there are $\mathbf{P}^1_{\mathcal{K}}, \boldsymbol{P}^2_{\mathcal{K}}, \ldots, \boldsymbol{P}^{r_{\mathcal{K}}}_{\mathcal{K}} \in \underline{\mathcal{V}}_{\mathcal{K}}$ such that*

$$\underline{\mathcal{V}}_{\mathcal{K}} = \{\boldsymbol{P}_{\mathcal{K}} \mid \underline{\boldsymbol{P}}_{\mathcal{K}} \leq \boldsymbol{P}_{\mathcal{K}} \leq \boldsymbol{P}^r_{\mathcal{K}} \text{ for at least one } r (=1, \ldots, r_{\mathcal{K}})\},$$
$$\text{for all } \mathcal{K} \subset \mathcal{N},\ k \geq 2. \tag{12.3}$$

[4]Many sets of weights would do in this case. For example $\delta_1 = \delta$, $\delta_2 = 1 - 2\delta$, $\delta_3 = 1 - \delta$, $\delta_4 = \delta$ for any $\delta \in (0, 0.5)$ gives weights which satisfy eq. (12.2).

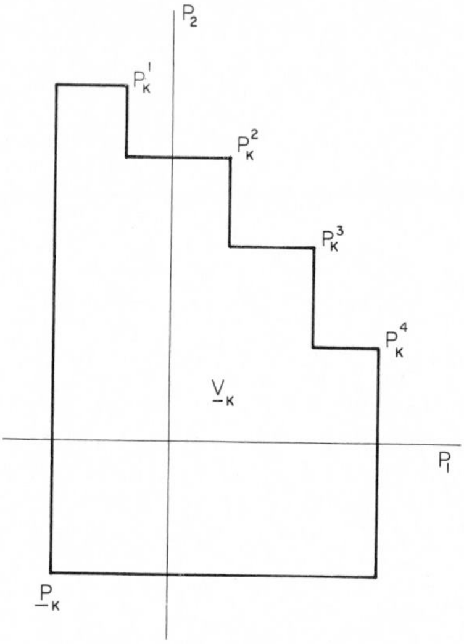

Fig. 12.2

Under C16, $\underline{\mathcal{V}}_{\mathcal{K}}$ is described by $\underline{\boldsymbol{P}}_{\mathcal{K}}$ and a finite list of maximal payoff vectors. Any payoff vector which is (component wise) no smaller than $\underline{\boldsymbol{P}}_{\mathcal{K}}$ and no larger than one of the maximal vectors is in $\underline{\mathcal{V}}_{\mathcal{K}}$. Such a set is illustrated in fig. 12.2. C16 is used to establish that a balanced game satisfying C9 and C14–C16 has a nonempty core. Then that result is used to show that the core is not empty even when C16 does not hold.

It is now possible to give a complete representation of the game using two finite matrices. The first of these, $\boldsymbol{B}$, is a matrix in which each column corresponds to one of the payoff vectors used to define $\underline{\mathcal{V}}_{\mathcal{K}}$ in eq. (12.3). As shown below, the first n columns

$$\boldsymbol{B} = \begin{array}{c} \begin{array}{cccccccc} (1) & (2) & & (n\} & 1,\{1,2\} & & r_{\{1,2\}},\{1,2\} & \end{array} \\ \left[\begin{array}{cccccccc} \underline{P}_1 & M_{12} & \dots & M_{1n} & P^1_{1,\{1,2\}} & \dots & P^{\{1,2\}}_{1,\{1,2\}} & \dots \\ M_{21} & \underline{P}_2 & \dots & M_{2n} & P^1_{2,\{1,2\}} & \dots & P^{\{1,2\}}_{2,\{1,2\}} & \dots \\ \cdot & \cdot & & \cdot & \cdot & & \cdot & \\ \cdot & \cdot & & \cdot & \cdot & & \cdot & \\ \cdot & \cdot & & \cdot & \cdot & & \cdot & \\ M_{n1} & M_{n2} & \dots & \underline{P}_n & M_{n,n+1} & \dots & M_{n,n+r\{1,2\}} & \dots \end{array}\right] \end{array}$$

$$
= \begin{bmatrix} b_{11} & b_{12} & \dots & b_{1n} & b_{1,n+1} & \dots & b_{1,n+r\{1,2\}} \\ b_{21} & b_{22} & \dots & b_{2n} & b_{2,n+1} & \dots & b_{2,n+\{1,2\}} \\ \cdot & \cdot & & \cdot & \cdot & & \cdot \\ \cdot & \cdot & & \cdot & \cdot & & \cdot \\ \cdot & \cdot & & \cdot & \cdot & & \cdot \\ b_{n1} & b_{n2} & & b_{nn} & b_{n,n+1} & \dots & b_{n,n+r\{1,2\}} \end{bmatrix}, \quad (12.4)
$$

correspond, respectively, to $\underline{P}_1, \dots, \underline{P}_n$. $\underline{P}_i$ is the (unique) maximal element of $\mathscr{V}_i$. The entries M_{ij} are very large numbers which are described more fully below. Each coalition is represented by $r_{\mathscr{K}}$ columns, one for each of the maximal members of $\underline{\mathscr{V}}_{\mathscr{K}}$. The M_{ij} are dummy entries which play no role other than to allow each vector $\boldsymbol{P}^r_{\mathscr{K}}$ to appear as a vector of dimension n. Their values are chosen so that they do not "get in the way." This is accomplished by making all the M_{ij} different from one another and strictly larger than any other values in the matrix.

The second matrix, $\boldsymbol{I}$, is the *incidence matrix*. It is given below. For each column in $\boldsymbol{B}$ there is a

$$
\boldsymbol{I} = \begin{bmatrix} 1 & 0 & \dots & 0 & 1 & \dots & 1 & \dots \\ 0 & 1 & \dots & 0 & 1 & \dots & 1 & \dots \\ \cdot & \cdot & & \cdot & \cdot & & \cdot & \\ \cdot & \cdot & & \cdot & \cdot & & \cdot & \\ \cdot & \cdot & & \cdot & \cdot & & \cdot & \\ 0 & 0 & & 1 & 0 & \dots & 0 & \dots \end{bmatrix}, \quad (12.5)
$$

corresponding column of $\boldsymbol{I}$. That column has a "1" in the ith row if i is a member of the coalition represented in that column; and has a "0" otherwise. Thus the first n columns of $\boldsymbol{I}$ form an identity matrix of order n. The next $r_{\{1,2\}}$ columns have "1" in the first two rows and "0" in the remaining rows, etc. The matrices $\boldsymbol{B}$ and $\boldsymbol{I}$ contain all the information needed to find a point in the core of the game which they represent.

2.2. The computational algorithm

2.2.1. The feasible basis and the ordinal basis

The matrices $\boldsymbol{I}$ and $\boldsymbol{B}$ are in *standard form* if the ith column, for $i = 1, \dots, n$, corresponds to the coalition $\{i\}$ and if the arbitrary entries in the first n columns $b_{ij}(i \neq j,\ j \leq n)$ are chosen so that $b_{ij} > b_{ik}(k > n)$. In eqs. (12.4) and (12.5) $\boldsymbol{B}$ and $\boldsymbol{I}$ are specified so that they meet the former condition. Meeting the latter condition, that, for each row, the arbitrary entries in the first n columns are larger than the arbitrary entries in the remaining columns, poses no problem.

Associated with each of the matrices is a *basis* which has an interesting interpretation and which plays an important part in the algorithm. Let $\boldsymbol{e}$ be a vector which has n components, all of which are equal to 1; and let $\boldsymbol{x} \geq 0$ be a vector with as many components as $\boldsymbol{I}$ has columns. The collection of columns of $\boldsymbol{I}$ numbered $j_1, \ldots, j_n$ forms a *feasible basis* for $\boldsymbol{I}$ if there exists $\boldsymbol{x}$ such that

$$x_j \begin{cases} >0, & \text{for } j = j_1, \ldots, j_n, \\ =0, & \text{otherwise}, \end{cases} \tag{12.6}$$

$$\boldsymbol{Ix} = \boldsymbol{e}. \tag{12.7}$$

If the columns $j_1, \ldots, j_n$ are a feasible basis for $\boldsymbol{I}$, then the collection of coalitions corresponding to those columns is balanced and $x_{j_1}, \ldots, x_{jn}$ are the associated weights. From a feasible basis there is a way to find a point in $\underline{\mathcal{V}}_{\mathcal{N}}$. Let $\boldsymbol{P}$ be defined by

$$P_i = \min\,(b_{i,j_1}, \ldots, b_{i,j_n}), \qquad i = 1, \ldots, n. \tag{12.8}$$

Then $\boldsymbol{P} \in \underline{\mathcal{V}}_{\mathcal{N}}$ because the game is balanced.[5]

Now let $j_1, \ldots, j_n$ be n columns of $\boldsymbol{B}$, let $\boldsymbol{P}$ be defined by eq. (12.8), and denote the kth column of $\boldsymbol{B}$ by $\boldsymbol{b}_k$. Then the columns $j_1, \ldots, j_n$ are an *ordinal basis* for $\boldsymbol{B}$ if

$$\boldsymbol{P} \nless \boldsymbol{b}_k, \qquad \text{for all } k. \tag{12.9}$$

That is, there must be no column of $\boldsymbol{B}$ which is strictly larger in each component than $\boldsymbol{P}$. Recall that the given column k is associated with a particular coalition. If, for one member, i, of that coalition, $P_i \geq b_{ik}$, then that coalition cannot improve upon $\boldsymbol{P}$. Thus if $j_1, \ldots, j_n$ forms an ordinal basis, the payoff vector $\boldsymbol{P}$, defined by eq. (12.8), cannot be improved upon. Therefore, if $j_1, \ldots, j_n$ were both a feasible basis and an ordinal basis, $\boldsymbol{P}$ would be in $\underline{\mathcal{V}}_{\mathcal{N}}$ and it could not be improved upon by any coalition; because, for any column of $\boldsymbol{B}$ there would be at least one member of the corresponding coalition whose payoff would not be increased over what he gets from $\boldsymbol{P}$ by adopting the payoff vector in that

[5]This construction is used in the example in §2.1.

column. Furthermore, if a particular coalition $\mathcal{K}$ cannot improve upon $\boldsymbol{P}$ by the use of one of the columns with its label, then it cannot improve upon $\boldsymbol{P}$ by any other member of $\mathcal{V}_{\mathcal{K}}$. All other members of $\underline{\mathcal{V}}_{\mathcal{K}}$ are dominated by at least one of the columns already considered.

To repeat, if a given collection of n columns is an ordinal basis (of $\boldsymbol{B}$), the associated payoff vector $\boldsymbol{P}$ (defined by eq. (12.8)) cannot be improved upon by any member of $\underline{\mathcal{V}}_{\mathcal{N}}$. If it is a feasible basis, then $\boldsymbol{P}$ is attainable ($\boldsymbol{P} \in \underline{\mathcal{V}}_{\mathcal{N}}$). If both conditions are met, $\boldsymbol{P}$ is in the core.

2.2.2. The pivot steps of the algorithm

The algorithm operates by finding as a starting point a collection of n columns which form an ordinal basis and another collection of n columns which form a feasible basis. The two collections have $n-1$ columns in common. From this starting point either of two steps may be imagined. The first is to change from one feasible basis to another by adding the column which is in the ordinal basis and not matched in the feasible basis. Doing this requires following rules which force one of the original columns to be dropped. If the column which is dropped is the original mismatched column, the core point is found. If not, there are still a pair of bases which have $n-1$ columns in common. One feature of the algorithm is that when a pivot step occurs to carry one feasible basis into another, there is only one possible column to drop. This means that if a pivot step is to be taken which changes one feasible basis for another and the column to be dropped is specified, then there is only one possible pivot step. The preceding is the content of lemma 12.1.

Now consider a pivot step on the ordinal basis. A natural step is to remove that column which is not matched in the feasible basis. In so doing, a new column is added. Either it, too, is a mismatch or it corresponds to the mismatched member of the feasible basis. In the latter event, the core point has been found; in the former, the new ordinal basis and the old feasible basis have $n-1$ columns in common. When a pivot step is taken on the ordinal basis and a particular column is chosen to be removed, there is only one column which can possibly be added. This is the content of lemma 2.

To summarize, the process starts with a pair of bases which have $n-1$ columns in common. The algorithm proceeds by pivoting on the ordinal basis, then on the feasible, then on the ordinal, etc. For an ordinal pivot step, the mismatched column is removed. For a feasible pivot step, the column is added which corresponds to the mismatch. Making the pivot steps in this way guarantees that after each step, the two bases have at least $n-1$ columns in common. With these rules and the results of lemmas 12.1 and 12.2, at an intermediate step in the algorithm, say just after an ordinal pivot, there are only two possible choices. One is to make an ordinal pivot, which will precisely reverse the step which has just been taken by removing the column which was just added and replacing it with the one most recently removed. The other is to make a feasible pivot. After making a feasible pivot, there are still only two possible moves, one of which is the backward move which reverses the feasible pivot step. If the bases remain mismatched, it is always possible to make either of two steps, one of which reverses the most recent step. The process stops only when a core point is found.

To establish some of the results, nondegeneracy assumptions are needed.

C17. *All the x_j associated with the n columns of a feasible basis are greater than zero and unique.*

C18. *No two elements in the same row of $\boldsymbol{B}$ are equal.*

C17 may be met by making small perturbations in the positive elements of $\boldsymbol{I}$, so that any collection of n columns is linearly independent if no row contains only zeros. Similarly, C18 may be met by making tiny perturbations in the appropriate elements of $\boldsymbol{B}$ to eliminate any instances of equality.

Lemma 12.1. *For a game satisfying C9 and C14–C18, let $j_1, \ldots, j_n$ be the columns of a feasible basis and let j^* be an arbitrary column not in the basis. $\boldsymbol{x}$ is the vector of weights associated with the basis; hence*

$$\boldsymbol{Ix} = \boldsymbol{e}, \tag{12.10}$$

$$x_j > 0, \qquad j = j_1, \ldots, j_n, \tag{12.11}$$

$$x_j = 0 \text{ otherwise.} \tag{12.12}$$

Then if the convex set

$$\{\boldsymbol{x} \,|\, \boldsymbol{x} \geq \boldsymbol{0}, \boldsymbol{Ix} = \boldsymbol{e}\} \tag{12.13}$$

is bounded, there is a unique feasible basis consisting of column j^* and $n-1$ of the original columns.

□ Introducing j^* and keeping the constraints $\boldsymbol{Ix} = \boldsymbol{e}$ satisfied is precisely a linear programming pivot step. Given the nondegeneracy assumption C17, there is only one column which could possibly drop out.[6] Briefly, this may be seen as follows: let $\boldsymbol{x}^1 = (x^1_{j_1}, \ldots, x^1_{j_n})$ be the original nonzero weights and let $\boldsymbol{I}^1$ be the columns of $\boldsymbol{I}$ to which they correspond. Let x_{j^*} be the weight associated with the column and denote that column by i_{j^*}. Then if a positive weight is assigned to j^* and the weights attached to the columns $j_1, \ldots, j_n$ are adjusted by y_{j_i}, $i = 1, \ldots, n$, to keep eq. (12.10) satisfied, then, letting $y = (y_{j_1}, \ldots, y_{j_n})$,

$$\boldsymbol{I}^1[\boldsymbol{x}^1 + \boldsymbol{y}] + x_{j^*}\boldsymbol{i}_{j^*} = e. \tag{12.14}$$

Recalling that

$$\boldsymbol{I}^1\boldsymbol{x}^1 = \boldsymbol{e}, \tag{12.15}$$

implies

$$\boldsymbol{I}^1\boldsymbol{y} = -x_{j^*}\boldsymbol{i}_{j^*}, \tag{12.16}$$

or

$$\boldsymbol{y} = -x_{j^*}[\boldsymbol{I}^1]^{-1}\boldsymbol{i}_{j^*}. \tag{12.17}$$

Thus there is a one to one relationship between x_{j^*} and $\boldsymbol{y}$. The boundedness condition implies that some of the y_{j_i} must be negative; hence, as x_{j^*} is increased, eventually $x_{j_i} - y_{j_i} = 0$ for one i. At this point the pivot step is complete. The analogy to linear programming is clear. □

Lemma 12.2. *Under the assumptions of lemma 1, let $j_1, \ldots, j_n$ be an ordinal basis of **B** and let j_1 be an arbitrary one of these columns. Assume, further, that $j_2, \ldots, j_n$ are not all among the first n columns of **B**. If **B** is in standard form, there is a unique column $j^* \neq j_1$ such that $j^*, j_2, \ldots, j_n$ is an ordinal basis.*

□ The proof proceeds in three parts. In the first an ordinal pivot step is defined. Second, it is shown that such a step is unique and leads to a new ordinal basis. Third, it is shown that no other step leads to a new ordinal basis.

To define the ordinal pivot step, note first that among the columns $j_2, \ldots, j_n$, exactly one of them contains two entries which are the smallest in their respective rows. One of these entries is a row minimizer for the

[6] See a text on linear programming such as Gass (1975).

original basis; the other is not. Let i^* be the row in which the row minimizer for the original basis is found, let the column in question be j_m and let the new row minimizer be b_{l,j_m}. Find all the columns in $\boldsymbol{B}$ for which $b_{ik} > P_i$, $i \neq i^*, l$ and $b_{lk} > b_{l,j_m}$. $\boldsymbol{P}$ is the original vector of row minimizers, defined by eq. (12.8). Among these find the one which maximizes $b_{i^*,k}$. This column is introduced. Call it j^*.

To see that the pivot step is unique and leads to a new ordinal basis, denote the new row minimizing elements $\boldsymbol{P}'$. For $i \neq i^*$, l, $P'_i = P_i$, $P'_l = b_{l,j_m}$, and $P'_{i^*} = b_{i^*j^*}$. The columns omitted from the original basis divide into those chosen initially in the search for j^* and those not so chosen. For the latter, each has $b_{ik} < P'_i$ for at least one $i \neq i^*$. If one of these would have been introduced with $i \neq i^*$, l, one of the other basis members would have been removed instead of j_1. Among the former, if any other than j^* had been introduced, $\boldsymbol{b}_{j^*} \gg \boldsymbol{P}'$, which violates the definition of an ordinal basis (see eq. (12.9)).

It only remains to show that a new column (j^*) could not be chosen with its row minimizers in row l. Were that done, $P'_i = P_i$ for all $i \neq l$. Therefore, $b_{l,j^*} \geq b_{l,j_1}$ which implies that column j_1 could not improve upon $\boldsymbol{P}'$. Also, if j^* can be introduced, $b_{l,j^*} > P_l$; however, recalling that $j_1, \ldots, j_n$ is an ordinal basis, $b_{i,j^*} \leq P_i$ for at least one i. This leaves only $i = l$. Therefore, $b_{l,j^*} \leq P_l = b_{l,j_1}$ which implies $b_{l,j^*} = b_{l,j_1}$; so j^* could only be j_1. Therefore, a new column, $j^* \neq j_1$, cannot be chosen with its row minimizers in row l. The choice of j^* is unique, with its row minimizers in i^*. □

2.3. Existence of a point in the core

It is now possible to prove a theorem stating that a game satisfying the assumptions of lemma 12.1 has a nonempty core. From lemmas 12.1 and 12.2, it is clear that starting from a pair of bases, one ordinal and one feasible, which have $n-1$ columns in common, there are at most two pivot steps which may be taken which would preserve the $n-1$ column overlap. It must be shown that this feature can be made to lead to a pair of bases which use the same n columns.

Theorem 12.3. *A balanced game satisfying C9 and C14–C18 has a nonempty core.*

□ Let there be a feasible basis consisting of columns $j_1, \ldots, j_n$ and an ordinal basis consisting of columns $j_2, \ldots, j_{n+1}$. Conditions under which the theorem is proved are these: (a) There is a unique $j_1, \ldots, j_n$ and $j_2, \ldots, j_{n+1}$ such that only one pivot is possible. This pair of bases provides a starting point for the algorithm. (b) Once the pivot step is taken from the unique starting point, from lemmas 12.1 and 12.2, there are exactly two moves. One is to remove the nonoverlapping column from the ordinal basis, and the other is to add to the feasible basis the column which corresponds to the nonoverlapping column in the ordinal basis. That is, if j_m is the nonoverlapping column in the ordinal basis, either remove it from the ordinal basis, or add it to the feasible. One of these moves is *backward* to the previous position, so the other is the only move to take. (c) Because they are only a finite number of conceivable bases, the process, which starts from the unique beginning of (a) and continues as outlined in (b), must terminate. It cannot terminate at the (unique) beginning point because that would imply two possible pivots from there, contrary to fact. It cannot continue forward and come back to some intermediate point, because that would imply three possible pivots from that point, which is impossible. Finally, the process cannot terminate with an overlap of $n-1$ because, if (a) can be shown, there is only one pair of basis with an $n-1$ overlap from which there is only one pivot. The only remaining possibility is termination at a pair of bases which use the same n columns. The $\boldsymbol{P}$ associated with these bases is in the core of the game.

Thus, given lemmas 12.1 and 12.2, all that remains to be shown is (a), that there is a unique pair of bases having $n-1$ columns in common and from which only one pivot is possible. This starting point consists of a feasible basis of $1, 2, \ldots, n$ and an ordinal basis of $2, 3, \ldots, n, j^*$. A pivot is clearly possible by introducing j^* into the feasible basis, but a pivot on the ordinal basis by removing j^* is not possible. To see this, note that $\boldsymbol{b}_n$ has two row minimizers. These are for rows 1 and n. For all $\boldsymbol{b}_j$ outside the basis, $b_{ij} > P_i$, $i = 2, \ldots, n-1$. Furthermore $b_{nj} > P_n$ $(= b_{nn})$; therefore, any column brought in must minimize for $i = 1$, as it is impossible to keep $\boldsymbol{b}_n$ and find a new minimizer for $i = n$. For $i = 1$, the only admissible column is j^*.

So, there is a unique starting point from which the algorithm gives a series of steps which must terminate in finding a pair of bases which use the same column. The associated $\boldsymbol{P}$, defined by eq. (12.8), is in the core. □

It is now possible to remove the nondegeneracy assumptions, C17 and C18, and prove that a game satisfying the remaining assumptions has a point in the core. After that is done, the *finite cornered* assumption, C16 is, similarly removed.

Theorem 12.4. *A balanced game satisfying C9 and C14–C16 has a nonempty core.*

□ Let the matrices $\boldsymbol{I}^l$ and $\boldsymbol{B}^l$, $l = 1, 2, \ldots$, be members of a sequence which satisfy C9 and C14–C18 and which converge to $\boldsymbol{I}$ and $\boldsymbol{B}$ which satisfy C9 and C14–C16. Also let $\boldsymbol{B}^l$ converge subject to the condition that the ordinal ranking of entries in the matrix is preserved. Let $\boldsymbol{x}^l$ be the weights for game l and $\boldsymbol{P}^l$ the core point found by the algorithm. A cluster point of $\{\boldsymbol{x}^l\}$ is a solution for $\boldsymbol{Ix} = \boldsymbol{e}$. A subsequence for which the $\boldsymbol{x}^l$ converge has a convergent subsequence for which the $\boldsymbol{P}^l$ also converge. These limits give the weights and a core point for a game satisfying the assumptions of the theorem. □

Theorem 12.5. *A balanced game satisfying C9, C14 and C15 has a nonempty core.*

□ For a game which is not finite cornered, it is possible to use an approximation to it which is finite cornered. A sequence of approximations can be made which are dense in the limit. These games have associated core vectors. A cluster point of the sequence of core vectors is in the core of the original game. □

3. Existence of a nonempty core for games in normal form

Many games are much more naturally represented in normal form than in characteristic function form. In Scarf (1971) the results of §2 are used to prove that the core of a class of games in normal form is not empty. After specifying the model of the game in normal form, a characteristic function form is then derived from the normal form. It is then shown that the game, as represented in characteristic function form, satisfies C9, C14, C15 and is balanced.

The game in normal form is assumed to obey C1 (compact convex strategy sets in a finite space for each player), C9, C14 and C19.

C19. *$P_i(s)$ is bounded and quasiconcave in s, $i = 1, \ldots, n$.*

For any coalition $\mathscr{K}$, the joint strategy set $\mathscr{S}_{\mathscr{K}}$ is taken to be the product of the strategy sets of the members. This contrasts with the model in ch. 11 §4, based on Harsanyi's work, in which $\mathscr{S}_{\mathscr{K}}$ is larger than the product of the $\mathcal{S}_i (i \in \mathcal{K})$. The quasi-concavity assumption C19 is stronger than the counterpart assumption made in ch. 7 for noncooperative games, where quasi-concavity is required of P_i only with respect to s_i.

The characteristic function $\mathcal{V}_{\mathcal{K}}$ for the coalition $\mathcal{K}$ consists of the payoff vectors which the members can assure themselves, no matter what choices are made by the members of $\tilde{\mathcal{K}}$. So, $\boldsymbol{P}^0_{\mathcal{K}} \in \mathcal{V}_{\mathcal{K}}$ if there is $s^0_{\mathcal{K}}$ such that

$$P_i(s^0_{\mathcal{K}}, s_{\tilde{\mathcal{K}}}) \geq P^0_i, \qquad \text{for all } s_{\tilde{\mathcal{K}}} \in \mathcal{S}_{\tilde{\mathcal{K}}}, \qquad i \in \mathcal{K}. \tag{12.18}$$

In other words, for $\boldsymbol{P}^0_{\mathcal{K}}$ to be in $\mathcal{V}_{\mathcal{K}}$ there must be a particular strategy in $\mathcal{S}_{\mathcal{K}}$ which guarantees at least P^0_i to each player, $i \in \mathcal{K}$, no matter what actions are taken by the complementary coalition.

From C19 it is immediate that C15 is satisfied. The sets $\mathcal{V}_{\mathcal{K}}$ are, themselves, nonempty and compact; and $\underline{\boldsymbol{P}}_{\mathcal{K}} \in \mathcal{V}_{\mathcal{K}}$, which means $\mathcal{V}_{\mathcal{K}}$ cannot be empty. Of course, $\underline{P}_i$ is defined by eq. (12.18) with $\mathcal{K} = \{i\}$ and $\tilde{\mathcal{K}} = \mathcal{N} - \{i\}$. To show there is a point in the core it is only necessary to show that the game is balanced.

Theorem 12.6. *A game satisfying C9, C14 and C19 has a nonempty core.*

□ If the game is balanced, the core is nonempty. Let $\mathcal{T}$ be a balanced collection of coalitions and let $\boldsymbol{P}^0$ be a payoff vector such that $\boldsymbol{P}^0_{\mathcal{K}} \in \mathcal{V}_{\mathcal{K}}$ for all $\mathcal{K} \in \mathcal{T}$. If $\boldsymbol{P}^0 \in \mathcal{V}_{\mathcal{N}}$, the game is balanced.

For each coalition $\mathcal{K}$ there is $s^0_{\mathcal{K}}$ such that

$$P_i(s^0_{\mathcal{K}}, s_{\tilde{\mathcal{K}}}) \geq P^0_i, \qquad i \in K \text{ for all } s_{\tilde{\mathcal{K}}} \in \mathcal{S}_{\tilde{\mathcal{K}}}, \tag{12.19}$$

$$s^0_{\mathcal{K}} = (s_{i_1,\mathcal{K}}, \ldots, s_{i_k,\mathcal{K}}), \tag{12.20}$$

where $i_1, \ldots, i_k$ are the members of $\mathcal{K}$. Because $\mathcal{T}$ is a balanced collection, there is an associated set of weights, $\{\delta_k\}$ such that

$$\sum_{\mathcal{K} \in \mathcal{T}, \mathcal{K} \ni i} \delta_{\mathcal{K}} = 1. \tag{12.21}$$

These weights and the strategies $s^0_{\mathcal{K}}$ described in eqs. (12.21) and (12.20) are used to construct a joint strategy which yields a point in $\mathcal{V}_{\mathcal{N}}$. This strategy s^* is

$$s^*_i = \sum_{\mathcal{K} \in \mathcal{T}, \mathcal{K} \ni i} \delta_{\mathcal{K}} s_{i\mathcal{K}}, \qquad i = 1, \ldots, n. \tag{12.22}$$

It is shown that $P_i(s^*) \geq P^0_i$ for all i. This is done for $i = 1$, and the same argument may be repeated for any player i. Thus it is sufficient to demonstrate the result for one player. Quasi-concavity (which has played no role thus far) is crucial to the result.

It is possible to find a collection of strategy vectors $s^{\mathcal{K}} = (s_1^{\mathcal{K}}, \ldots, s_n^{\mathcal{K}})$, $\mathcal{K} \in \mathcal{T}$, such that

$$s^* = \sum_{\mathcal{K} \in \mathcal{T}, \mathcal{K} \ni 1} \delta_{\mathcal{K}} s^{\mathcal{K}}. \tag{12.23}$$

Such a collection is defined by letting

$$s_i^{\mathcal{K}} = s_{i\mathcal{K}}, \qquad \text{if } i \in \mathcal{K}, \tag{12.24}$$

$$s_i^{\mathcal{K}} = \Big[\sum_{\mathcal{L} \in \mathcal{T}, i \in \mathcal{L}, 1 \notin \mathcal{L}} \delta_{\mathcal{L}} s_{i\mathcal{L}}\Big] \Big/ \Big[\sum_{\mathcal{L} \in \mathcal{T}, i \in \mathcal{L}, 1 \notin \mathcal{L}} \delta_{\mathcal{L}}\Big], \qquad \text{if } i \notin \mathcal{K}. \tag{12.25}$$

Note that $s_i^{\mathcal{K}}$ is identical over all $\mathcal{K}$ which do not contain 1. Also, for all $\mathcal{K}$ of which 1 is a member, $s_i^{\mathcal{K}} = s_{i\mathcal{K}}$, which implies that $s_{\mathcal{K}}^{\mathcal{K}} = s_{\mathcal{K}}^0$ for all $\mathcal{K}$ to which 1 belongs. Therefore,

$$P_1(s_{\mathcal{K}}^{\mathcal{K}}, s_{\bar{\mathcal{K}}}^{\mathcal{K}}) = P_1(s^{\mathcal{K}}) \geq P_1^0, \qquad \text{for all } \mathcal{K} \in \mathcal{T}, \quad 1 \in \mathcal{K}. \tag{12.26}$$

By quasi-concavity, a convex combination of the $s^{\mathcal{K}}$ appearing in eq. (12.26) gives a payoff at least as large as P_1^0. It remains to show that s^* is a convex combination of the $s^{\mathcal{K}}$ (for $\mathcal{K} \in \mathcal{T}$, $1 \in \mathcal{K}$):

$$s_i^* = \sum_{\mathcal{K} \in \mathcal{T}, 1 \in \mathcal{K}} \delta_{\mathcal{K}} s_i^{\mathcal{K}}, \qquad i = 1, \ldots, n. \tag{12.27}$$

Starting from eq. (12.27),

$$\sum_{\mathcal{K} \in \mathcal{T}, 1 \in \mathcal{K}} \delta_{\mathcal{K}} s_i^{\mathcal{K}} = \sum_{\mathcal{K} \in \mathcal{T}, \{1,i\} \subseteq \mathcal{K}} \delta_{\mathcal{K}} s_{i\mathcal{K}} +$$

$$\sum_{\mathcal{K} \in \mathcal{T}, 1 \in \mathcal{K}, i \notin \mathcal{K}} \delta_{\mathcal{K}} \Big[\sum_{\mathcal{L} \in \mathcal{T}, 1 \notin \mathcal{L}, i \in \mathcal{L}} \delta_{\mathcal{L}} s_{i\mathcal{L}}\Big] \Big/ \Big[\sum_{\mathcal{L} \in \mathcal{T}, 1 \notin \mathcal{L}, i \in \mathcal{L}} \delta_{\mathcal{L}}\Big]$$

$$= \sum_{\mathcal{K} \in \mathcal{T}, \{1,i\} \subseteq \mathcal{K}} \delta_{\mathcal{K}} s_{i\mathcal{K}} + \gamma \sum_{\mathcal{K} \in \mathcal{T}, i \in \mathcal{K}, 1 \notin \mathcal{K}} \delta_{\mathcal{K}} s_{i\mathcal{K}}, \tag{12.28}$$

where

$$\gamma = \Big[\sum_{\mathcal{K} \in \mathcal{T}, 1 \in \mathcal{K}, i \notin \mathcal{K}} \delta_{\mathcal{K}}\Big] \Big/ \Big[\sum_{\mathcal{K} \in \mathcal{T}, i \in \mathcal{K}, 1 \notin \mathcal{K}} \delta_{\mathcal{K}}\Big]. \tag{12.29}$$

If $\gamma = 1$, then both eqs. (12.22) and (12.27) are satisfied, establishing the required condition

$$P_1(s^*) \geq P_1^0, \tag{12.30}$$

for player 1. By a parallel argument for any other player, it can be shown that

$$P_i(s^*) \geq P_i^0. \tag{12.31}$$

To see the final step, that $\gamma = 1$,

$$\begin{aligned}\sum_{\mathcal{K}\in\mathcal{T},1\in\mathcal{K}} \delta_{\mathcal{K}} &= \sum_{\mathcal{K}\in\mathcal{T},\{1,i\}\subseteq\mathcal{K}} \delta_{\mathcal{K}} + \sum_{\mathcal{K}\in\mathcal{T},1\in\mathcal{K},i\notin\mathcal{K}} \delta_{\mathcal{K}}\\ &= 1\\ &= \sum_{\mathcal{K}\in\mathcal{T},\{1,i\}\subseteq\mathcal{K}} \delta_{\mathcal{K}} + \sum_{\mathcal{K}\in\mathcal{T},i\in\mathcal{K},1\notin\mathcal{K}} \delta_{\mathcal{K}}\\ &= \sum_{\mathcal{K}\in\mathcal{T},i\in\mathcal{K}} \delta_{\mathcal{K}}. \end{aligned} \tag{12.32}$$

Therefore $\gamma = 1$ and the game is balanced, which implies it has a point in the core. □

4. *A brief account of other core models*

There are two developments which are mentioned in this section. The first is due to Farrell (1970) and Shitovitz (1973); and the second to Aumann (1959, 1960, 1961).

4.1. The Edgeworth box with unequal numbers of trader types

Both Farrell and Shitovitz are concerned with models of pure trade which incorporate oligopolistic elements by having a large number of traders who are *small* and a small number who are *large*. Farrell does this within the framework of a model having two commodities and two types of traders. The second type is fixed at n_2 while n_1, the number of traders of the first type, is allowed to grow indefinitely. This is a direct generalization of the model of Edgeworth (1881) in which the number of traders of each type is the same and grow together. The Shitovitz model is much more general, taking a measure theoretic approach under which most traders are small in the sense that a point in a set is small compared with the size of the set itself (i.e. has measure zero). Then other traders are large (have positive measure). There may be several large traders; they may be divided into types; and there may be more than one of each type.

The exposition to follow is carried out in terms of the Farrell model because it illustrates the basic ideas very well and requires less advanced mathematics.

Farrell's result is that as n_1 goes to infinity, if n_2 is constant and ≥ 2, the only core points left in the limit are the competitive equilibria. Shitovitz' result for the more general model is that as long as there are at least two of each type of large trader, the core consists only of the competitive equilibria. The mechanism which appears to remove points from the core as the number of traders in the market increases is relatively easy to illustrate.

Let fig. 12.3 represent an economy in which there are two traders of type A and two of type B. The box may be thought of as drawn to the scale of one trader of each type. The segment *ab* would be the core if there were only one trader of each type. It is easy to see that the point *a*, and some other points on *ab* near to *a* cannot be in the core when there are two traders of each type; because two A traders and one B trader can make an agreement which leaves the two A traders at the point A_i and the B trader at B_1. All three are better off than they would be at *a*. Such a trade leaves the other B trader at *E*, the initial endowment point. But the important thing to see is that an agreement on the part of all four to choose *a* can be improved upon by a coalition of two A's and one B. Two important facts in relation to this model are (1) a competitive equilibrium,

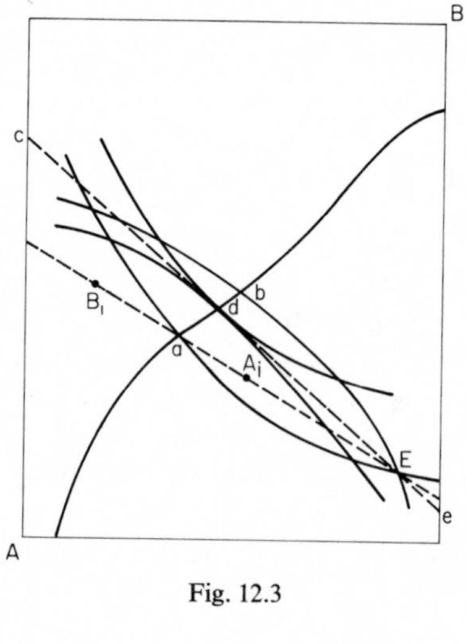

Fig. 12.3

such as *d*, is always in the core, no matter how many traders there are; and (2) any point on the segment *ab* which is not a competitive equilibrium is, for some finite $n = n_1 = n_2$ (and larger n) no longer in the core. For details see Debreu and Scarf (1972).

The remarkable result of Farrell is that if n_2 is held at some fixed, finite level (larger than 1) and n_1 is allowed to increase; it remains true that in the limit the only core points are competitive equilibria. To anyone raised on traditional economic theory this result is likely to seem highly counterintuitive. An economy in which all industries are competitive on the buyers' side and all but one are competitive on the sellers' side, with the remaining market a duopoly, is not expected to settle at a competitive equilibrium. The *market power* of the two firms, even if they do not collude with one another, is expected to result in prices which are above marginal cost for their product. Indeed consult chs. 2 or 3 for the conditions describing the Cournot equilibrium. Price exceeds marginal cost, which violates the competitive conditions.

This raises a question of whether one model or the other is wrong. In a logical sense, neither is wrong. The difference is because the traditional oligopoly models, of which those in chs. 2 and 3 are typical examples, assume the buyers cannot form coalitions. In the Farrell and Shitovitz models any subset of players can form a coalition. As a result, as long as there are at least two players of each type, no player has something unique which he can withhold. One way to think of the lack of special power of one of the large traders is to note that if all of the small traders band together, they are as *large* as any large trader.

There remains an issue of whether one sort of model is superior to the other because it better describes the workings of real economies. Here it is all too easy to claim victory for a model which is very familar because it is the model one was brought up on in economics; or to do so for a model in which one has a vested interest because of having done research with it. Neither of these attitudes is warranted (nor is either to be found in the papers of Farrell or Shitovitz). Determination of the truth of this question requires a large and sophisticated body of empirical work which is yet to be done. Furthermore, when answers are known, they may turn out to be that there are certain circumstances in which one model is better and certain circumstances in which the other is better.

The possibilities for the core model to be a good descriptor would seem to depend on whether it is possible for very large numbers of economic agents to organize relatively cheaply into coalitions.

4.2. Cooperative supergames

Within a short space it is not possible to do justice to Aumann's papers; however, an idea of their content can be conveyed. Imagine a game of the sort found in Nash (1951) in which each of n players has a finite number of pure strategies. Aumann defines a supergame as the temporal repetition of one such game. That is a game which, in the terminology of ch. 8, would be a stationary supergame without time dependence. Unlike the models of chs. 8–10, there is no discounting of future payoffs; so the objective function is taken as the limit, as the number of periods goes to infinity, of the payoff per period. For general n, there are no existence of equilibrium results; however, there are some results on the nature of equilibria (should they chance to exist). Important among these is that the set of strong equilibrium points for the supergame corresponds to the *β-core* of the constituent game. Strong equilibrium points are defined in ch. 7 §5. A strategy vector corresponds to a strong equilibrium point if there is no coalition which can improve upon the payoff it gets at that point, with the strategy choice of the complementary coalition remaining fixed. Thus a strong equilibrium is a Nash noncooperative equilibrium; but the converse is not true in general.

The core concepts, α-core and β-core, are first given those names in Aumann and Peleg (1960). They are based on the notion of α-effectiveness and β-effectiveness, which are defined as follows:[7] "A coalition $\mathcal{K}$ is said to be α-effective for the payoff vector $\boldsymbol{P}$ if there is a strategy for $\mathcal{K}$, such that for each strategy used by $\tilde{\mathcal{K}}$, each member i of $\mathcal{K}$ receives at least P_i. A coalition $\mathcal{K}$ is said to be β-effective for the payoff vector $\boldsymbol{P}$ if, for each strategy used by $\tilde{\mathcal{K}}$, there is a strategy for $\mathcal{K}$ such that each member i of $\mathcal{K}$ receives at least P_i." A payoff vector $\boldsymbol{P}$ is in the α-core if it is attainable by $\mathcal{N}$ and if no coalition is α-effective for it. Similarly, a payoff vector $\boldsymbol{P}$ is in the β-core if it is attainable by $\mathcal{N}$ and there is no coalition which is β-effective for it.

α-effectiveness is the improvement criterion used by Scarf and leads to the more common core concept. If a coalition $\mathcal{K}$ is α-effective for $\boldsymbol{P}$, then the members of $\mathcal{K}$ can guarantee themselves at least $(P_{i_1}, \ldots, P_{i_k})$ no matter what $\tilde{\mathcal{K}}$ chooses to do. If $\mathcal{K}$ is β-effective for $\boldsymbol{P}$, then $\tilde{\mathcal{K}}$ cannot prevent $\mathcal{K}$ from obtaining $(P_{i_1}, \ldots, P_{i_k})$. Thus the β-core is contained in the α-core and is, in many models, smaller. It is, of course, possible for the latter to be empty when the former is not.

[7]See Aumann and Peleg (1960). The notation is changed to conform to notation in the present volume.

5. Value equilibria versus core equilibria

Can it be said that one type of equilibrium is *better* than the other? Very likely there are those, who for reasonable cause, do prefer one to the other. Perhaps there are also circumstances in which one and not the other is a more natural cooperative equilibrium. On balance I see no reason to wish to promote one in particular and forget the other.

There are some points of contrast which it is well to see. To my view, the core has a simplicity which is appealing and a conservatism which is not. Consider again the way the α-core is defined. A payoff vector is in the α-core if no coalition can improve upon it, which means that no coalition can absolutely guarantee to itself higher payoffs (member by member) than are received under the proposed payoff vector. It is hard to dismiss a core point as being among the outcomes to which the players of a game might agree. Indeed, the converse could easily be argued – if a point is not in the core, it should not be a cooperative equilibrium point, precisely because some coalition can improve upon it. It is appealing to make as part of the conditions to be met by any cooperative equilibrium that it should be in the core.[8] This is not always done; however, something along those lines is done when, for example, a cooperative equilibrium is required to be Pareto optimal and to give more to each player than the player could get on his own.[9]

The conservative element of the core appears in the definition of what a coalition can guarantee itself. The guarantee is ironclad. Payoffs can be guaranteed by a coalition only if they can be assured no matter what actions the complementary coalition takes. There may be actions which it would not be reasonable for the complementary coalition to take; hence, which it would be reasonable for the coalition to disregard in figuring what it ought to expect to be able to do for itself. To define some sort of core this way requires criteria by which certain strategies of the complementary coalition would be ruled out.

Generally acceptable criteria may be very difficult to find. One simple criterion is that a coalition may never be expected to use a strategy which is dominated. That is, $s_{\mathcal{K}}$ is dominated for the coalition $\mathcal{K}$ if there is $s'_{\mathcal{K}}$ for which $\boldsymbol{P}_{\mathcal{K}}(s'_{\mathcal{K}}, s_{\tilde{\mathcal{K}}}) \geq \boldsymbol{P}_{\mathcal{K}}(s_{\mathcal{K}}, s_{\tilde{\mathcal{K}}})$ for all $s_{\tilde{\mathcal{K}}} \in \mathcal{S}_{\tilde{\mathcal{K}}}$. No matter what $\tilde{\mathcal{K}}$ chooses, $\mathcal{K}$ does better with $s'_{\mathcal{K}}$ than with $s_{\mathcal{K}}$. A reason to refuse to rule out the use of $s_{\mathcal{K}}$ is that $\tilde{\mathcal{K}}$ might do worse under $s_{\mathcal{K}}$ than $s'_{\mathcal{K}}$ (at least for some $s_{\tilde{\mathcal{K}}}$),

[8]Luce and Raiffa (1957, p. 194) make this argument.
[9]These less stringent conditions are called *group rationality* and *individual rationality*.

which makes $s_{\mathcal{K}}$ a useful threat strategy if it can be made credible. Credibility can be easily attained if the rules of the game allow players and coalitions to make commitments about what they choose under various hypothetical circumstances. Such commitments are contracts between the coalition and the *court* or *umpire*. So the conservatism of the core may be both difficult to accept and difficult to rule out.

A value equilibrium, such as Harsanyi's, does not consider the core directly. A network of strategies is arrived at, one per coalition, from which final payoffs are determined. The *threats* are the specific actions named by each coalition – again, one action per coalition – not the general possibility that any coalition might choose any action in its strategy set. The strategy choices arise out of a set of criteria which define the characteristics of an equilibrium. It is convenient to think of these as being an arbitration scheme. The features of the arbitration scheme might not all be acceptable. One's reservations about such features is the price paid to have an equilibrium in which strategies play no role except when there are clear circumstances in which they might be used.

Another point of contrast between the two equilibria concerns uniqueness. With the core, intuition suggests that if the core is not empty, it is likely to be a *large* set. Indeed, it is likely to be a set of payoff vectors which is convex and of the same dimension as the number of players in the game (a set of positive measure relative to $\mathcal{R}^n$). While I know of no results, intuition suggests that Harsanyi's equilibrium may be unique for a limited, though interesting, class of models. Furthermore, in most models for which the equilibrium exists, the number of equilibria is probably finite, or at least, countable. If this conjecture is correct, then the number of equilibria is of a smaller order of magnitude than the (conjectured) size of the core. Finally, perhaps it is possible to adapt Harsanyi's tracing procedure (see ch. 7 §4.3) to select a unique solution from a set of equilibria. Another question of interest is whether Harsanyi equilibria are necessarily in the core.

6. Concluding comments

The value type of cooperative equilibria, reviewed in ch. 11, and the core theory of the present chapter are both far more satisfactory equilibrium concepts for oligopoly than the sort of thing which has pervaded the oligopoly literature. Typically, what is found is the assertion that cooperating firms maximize total industry profits. Often, the whole discussion is carried out with reference to a *symmetric* industry in which all firms are identical. In the latter instance, any reasonable equilibrium treats all firms identically; hence, if firms are cooperating, they maximize industry profits and arrange to have equal profits. But, if firms are not symmetric, then maximization of total profits poses some problems. Unless the firms can make side payments (i.e. distribute the given total any way they see fit), the maximization of joint profits implies a particular distribution of the total. That distribution need not be acceptable to all firms. Indeed, it may be that one firm gets huge profits and all others get little or nothing. Then, even if side payments can be made, what side payments are made? That is, what is the final settlement? These questions tend not to be answered.

The core and value equilibria are much more satisfactory than the vague statements about joint maximization. Work still needs to be done to implement these concepts in oligopoly models. That is, to determine classes of models for which they exist and to attempt to learn more about the nature of such equilibria in oligopoly models.

APPENDIX A

NOTATION AND SYMBOLS

Throughout the book, type style is used in a consistent way to distinguish various sorts of entities. Italic type is used for scalars: a, b, c, A, B, C, α, β, γ, etc. Vectors and matrices are in bold face italic: $\boldsymbol{a}$, $\boldsymbol{b}$, $\boldsymbol{c}$, $\boldsymbol{A}$, $\boldsymbol{B}$, $\boldsymbol{C}$, $\boldsymbol{\alpha}$, $\boldsymbol{\beta}$, $\boldsymbol{\gamma}$, etc. Sets are in script or bold face roman: a, b, c, $\mathcal{A}$, $\mathcal{B}$, $\mathcal{C}$, $\boldsymbol{\alpha}$, $\boldsymbol{\beta}$, $\boldsymbol{\gamma}$, etc. Upper and lower case are not used in any special way, nor are Roman versus Greek letters. *Correspondences* (i.e. point-to-set mappings) are in script or bold roman because the image of a point is a set.

Vector inequalities are denoted by $\geq$, $>$, and $\gg$. $\boldsymbol{a} \geq \boldsymbol{b}$ means that each element of $\boldsymbol{a}$ is at least as large as the corresponding element of $\boldsymbol{b}$. $\boldsymbol{a} > \boldsymbol{b}$ means $\boldsymbol{a} \geq \boldsymbol{b}$ and $\boldsymbol{a} \neq \boldsymbol{b}$ – that is, at least one element of $\boldsymbol{a}$ is larger than its counterpart in $\boldsymbol{b}$ and none is smaller. $\boldsymbol{a} \gg \boldsymbol{b}$ means that each element of $\boldsymbol{a}$ is larger than the corresponding element of $\boldsymbol{b}$.

The symbols $\subseteq$ and $\subset$ are used to denote set inclusion. $\mathcal{A} \subseteq \mathcal{B}$ means each element of $\mathcal{A}$ is contained in $\mathcal{B}$. $\mathcal{A} \subset \mathcal{B}$ means $\mathcal{A} \subseteq \mathcal{B}$, but the two sets are not identical. There are some members of $\mathcal{B}$ which are not in $\mathcal{A}$. Sets are often defined with the help of curly brackets. Thus $\{x \mid x \in \mathcal{R}, x > 4\}$ is the set of all x which are real numbers ($x \in \mathcal{R}$) and greater than 4.

The symbol $\rightarrow$ means *goes to*. Thus $x \rightarrow \infty$ means x goes to infinity. A slash through a symbol denotes the negative. Thus, $\boldsymbol{a} \notin \mathcal{A}$ means $\boldsymbol{a}$ is not a member of $\mathcal{A}$. $\boldsymbol{a} \not> \mathbf{A}$ means $\boldsymbol{a}$ is not larger than $\mathbf{A}$, etc.

Partial derivatives are usually denoted by superscripts. Thus f^j denotes the first partial derivative of f with respect to its jth argument. f^{jk} means the second partial derivative with respect to the jth and kth arguments.

Mathematical terms and symbols which are not defined above are either defined in the text where they are first encountered or they may be found in Bartle (1964) or other books at the same level.

There follows below a list of symbols, together with a brief definition. They are organized by the section in which they first occur. The list is far from exhaustive, and tends to include the more *important* symbols. That is those which are used extensively and which appear in more than one place with the same meaning.

Chapter 2 §2

Q	market demand
p	market price
$F(p)$	market demand function
$f(Q)$	$F^{-1}(Q)$
q_i	output level of ith firm
$C_i(q_i)$	total cost function of ith firm
n	number of firms
$\boldsymbol{q}$	$(q_1, \ldots, q_n)$
$\pi_i(\boldsymbol{q})$	profit function for the ith firm
$\boldsymbol{\pi}(\boldsymbol{q})$	$(\pi_1(\boldsymbol{q}), \ldots, \pi_n(\boldsymbol{q}))$
$\mathscr{R}^n$	n-dimensional Euclidean space
$\mathscr{H}$	set of payoff vectors which are Pareto optimal with respect to the firms

Chapter 2 §3

$\boldsymbol{q}^c$	Cournot equilibrium output levels

Chapter 2 §5

$w_i(\bar{q}_i)$	Cournot reaction function for the ith firm
$\boldsymbol{w}(\boldsymbol{q})$	$(w_1(\bar{q}_1), \ldots, w_n(\bar{q}_n))$

Chapter 3 §3

p_i	price of the ith firm
$\boldsymbol{p}$	$(p_1, \ldots, p_n)$
$F_i(\boldsymbol{p})$	demand function for the ith firm
$\mathscr{R}^n_+$	nonnegative orthant of $\mathscr{R}^n$ (vectors whose components are all nonnegative)
$\mathring{\mathscr{A}}_i$	price vectors for which $F_i(\boldsymbol{p}) > 0$
$\mathscr{A}_i$	closure of $\mathring{\mathscr{A}}_i$
a_i	$\mathscr{A}_i - \mathring{\mathscr{A}}_i$
$\mathscr{A}$	$\bigcap_{i=1}^n \mathscr{A}_i$
$\boldsymbol{p}^+$	maximal element of $\mathscr{A}$
$\pi_i(\boldsymbol{p})$	profit function for the ith firm

Chapter 3 §5

$\boldsymbol{p}^c$	Cournot equilibrium price vector

Chapter 4 §1

$\psi_i(\boldsymbol{p}_{t-1})$	reaction function for the ith firm

Chapter 4 §3

α_i	discount parameter of the ith firm

Chapter 5 §1

$\boldsymbol{\psi}(\boldsymbol{p})$	$(\psi_1(\boldsymbol{p}), \ldots, \psi_n(\boldsymbol{p}))$

Chapter 7 §1

s_i	strategy of the ith player
s	$(s_1, \ldots, s_n)$
$\mathscr{S}_i$	strategy set of the ith player
$\mathscr{S}$	$\mathscr{S}_1 \times \cdots \times \mathscr{S}_n$
$P_i(s)$	payoff to the ith player
$\boldsymbol{P}(s)$	$(P_1(s), \ldots, P_n(s))$
$\mathscr{N}$	set of players
$r_i(\bar{s}_i)$	best reply function
$\boldsymbol{r}(s)$	$(r_1(\bar{s}_1), \ldots, r_n(\bar{s}_n))$
$\mathscr{K}, \mathscr{L}, \mathscr{M}$	coalitions
$\mathscr{V}_{\mathscr{K}}$	the set of payoff vectors attainable by $\mathscr{K}$

Chapter 7 §3

$\mathscr{r}_i(\bar{s}_i)$	best reply correspondence
$\mathscr{r}(s)$	$\mathscr{r}_1(\bar{s}_1) \times \cdots \times \mathscr{r}_n(\bar{s}_n)$

Chapter 8 §2

$\mathscr{H}$	set of attainable payoffs
$\mathscr{H}^*$	set of Pareto optimal payoffs
$\boldsymbol{\sigma}_i$	supergame strategy for the ith player
$\boldsymbol{\sigma}$	$(\boldsymbol{\sigma}_1, \ldots, \boldsymbol{\sigma}_n)$

Chapter 9 §1

$F_i(\boldsymbol{\sigma})$	supergame payoff to the *i*th player

Chapter 10 §1

$\boldsymbol{\Omega}$	set of states
$q(k'\|k, s_k)$	probability that k' is the next state given that k is the present state and s_k is the present action
$\boldsymbol{\theta}_i$	stationary policy for the *i*th player
$\boldsymbol{\theta}$	$(\boldsymbol{\theta}_1, \ldots, \boldsymbol{\theta}_n)$
$\boldsymbol{\Theta}_i$	set of stationary policies for the *i*th player
$\boldsymbol{\Theta}$	$\boldsymbol{\Theta}_1 \times \cdots \times \boldsymbol{\Theta}_n$
$\boldsymbol{q}(k, s_k)$	$(q(1\|k, s_k), q(2\|k, s_k), \ldots)$
$\boldsymbol{P}_{i\boldsymbol{\theta}}$	$(P_{i1}(s_1), P_{i2}(s_2), \ldots)$
$\boldsymbol{q}_{\boldsymbol{\theta}}$	$(\boldsymbol{q}(1, s_1), \boldsymbol{q}(2, s_2), \ldots)$
$\boldsymbol{F}_{i\boldsymbol{\theta}}$	$\sum_{t=1}^{\infty} \alpha_i^{t-1} \boldsymbol{q}_{\boldsymbol{\theta}}^{t-1} \boldsymbol{P}_{i\boldsymbol{\theta}}$

Chapter 11 §2

$\mathcal{H}$	set of attainable payoffs which dominate $\boldsymbol{P}^T$

Chapter 11 §3

$v(\mathcal{K})$	characteristic function for a transferable utility game
$\boldsymbol{\phi}(v)$	*Shapley value* function
$\bar{\mathcal{K}}$	$\mathcal{N}-\mathcal{K}$, the coalition complementary to $\mathcal{K}$
$\mathcal{H}_{\mathcal{K}}(s_{\bar{\mathcal{K}}})$	set of attainable payoffs for $\mathcal{K}$, conditional upon $s_{\bar{\mathcal{K}}}$

ASSUMPTIONS

Part I

A1. *The demand function, $f(Q)$, is defined and continuous for all $Q \geq 0$. There is $\bar{Q} > 0$ such that $f(Q) = 0$ for $Q \geq \bar{Q}$ and $f(Q) > 0$ for $Q < \bar{Q}$. Furthermore, $f(0) = \bar{p} < \infty$, and, for $0 < Q < \bar{Q}$, f has a continuous second derivative and $f'(Q) < 0$.*

A2. *The cost function of the ith firm is denoted $C_i(q_i)$ where q_i is the output level of the ith firm. C_i is defined and continuous for all output levels $0 \leq q_i$. $C_i(0) \geq 0$. C_i has a continuous second derivative for $0 < q_i$ and $C'_i(q_i) > 0$.*

A3. *The demand of the ith firm is given by $q_i = F_i(p_1, \ldots, p_n) = F_i(\mathbf{p}) \geq 0$. F_i is defined, continuous and bounded for all $\mathbf{p} \geq \mathbf{0}$. For $p_j^0 > p_j^1$ and $\bar{\mathbf{p}}_j^0 \geq 0$, $F_i(p_j^1, \bar{\mathbf{p}}_j^0) \leq F_i(p_j^0, \bar{\mathbf{p}}_j^0)$, $(i \neq j)$ and $F_j(p_j^1, \bar{\mathbf{p}}_j^0) \geq F_j(p_j^0, \bar{\mathbf{p}}_j^0)$.*

A4. *$F_i(\mathbf{p})$ is twice continuously differentiable for all $\mathbf{p} \gg \mathbf{0}$, except for $\mathbf{p} \in \cup_j \alpha_j$. If $\mathbf{p} \in \alpha_j$ for $j = i_1, \ldots, i_k$ and $\mathbf{p} \notin \alpha_j$ for $j = i_{k+1}, \ldots, i_n$, then continuous second partial derivatives with respect to $p_{i_{k+1}}, \ldots, p_{i_n}$ exist at $\mathbf{p}$. All derivatives are bounded, and, if $F_i(\mathbf{p}) = 0$, $\mathbf{p} \notin \mathscr{A}$, all derivatives of F_j $(j = 1, \ldots, n)$ with respect to p_i are zero. For $\mathbf{p} \in \mathring{\mathscr{A}}_i \cap \mathring{\mathscr{A}}_j$ $(j \neq i)$, $F_i^j(\mathbf{p}) > 0$, and, for $\mathbf{p} \in \mathring{\mathscr{A}}_i \cap \mathring{\mathscr{A}}_{i_1} \cap \cdots \cap \mathring{\mathscr{A}}_{i_k}$, $F_i^i(\mathbf{p}) + \Sigma_{j=1}^k F_i^{i_j}(\mathbf{p}) < 0$.*

A5. *$\mathscr{A}$ is bounded.*

A6. *For all $\mathbf{q}$ such that $q_i > 0$ and $Q < \bar{Q}$, $\partial^2 \pi_i(\mathbf{q}) / \partial q_i^2 < 0$, $i = 1, \ldots, n$.*

A7. *For all $\mathbf{q}$ such that $\mathbf{q} \gg \mathbf{0}$ and $Q < \bar{Q}$, $\partial^2 \pi_i(\mathbf{q}) / \partial q_i^2 + \Sigma_{j \neq i} |\partial^2 \pi_i(\mathbf{q}) / \partial q_i \partial q_j| < 0$, $i = 1, \ldots, n$.*

A8. *For all* $\mathbf{p} \in \mathring{\mathscr{A}}_i^*$, $\partial^2\pi_i/\partial p_i^2 < 0$.

A9. *For all* $\mathbf{p}$ *in the interior of* $\mathscr{A}^*$, $\partial^2\pi_i/\partial p_i^2 + \Sigma_{j\neq i}\,|\partial^2\pi_i/\partial p_i\partial p_j| < 0$.

A10. $p_1^0 > C_1'(F_1(p_1^0, \mathbf{0})) = C_1'(0)$.

A11. *For the interior of the domain on which* $\boldsymbol{\psi}$ *is defined,* $\boldsymbol{\psi}$ *has continuous second partial derivatives. Furthermore,* $\psi_j^k(\mathbf{p}) > 0$ *and* $\Sigma_{k=1}^n \psi_j^k(\mathbf{p}) \leq \lambda < 1$. *Finally, for any* $\mathbf{p}$ *in the domain of* $\boldsymbol{\psi}$, $\boldsymbol{\psi}(\mathbf{p})$ *is in the domain.*

A12. *For any* $\mathbf{p}_t \in \mathring{\mathscr{A}}_1^*$, $\mathbf{p}_{t+1}, \ldots, \mathbf{p}_{t+s-1} \in \mathscr{D}$ *and* $s \geq 1$, $G_{1s}^1(\mathbf{p}_t) \leq -\epsilon$; *for* $\epsilon > 0$.

A13, *There is a set* $\mathscr{D}_1^* = \{\mathbf{p}|\mathbf{p}_* \leq \mathbf{p} \leq \mathbf{p}^*\}$, *where* $\mathbf{p}_* \ll \mathbf{p}^*$ *and* $\mathbf{p}_*$, $\mathbf{p}^* \in \mathscr{B}_1$, *such that*

$$\pi_1^1(p_{*1}, \bar{\boldsymbol{\psi}}_1(\mathbf{p}_*)) = \pi_1^1(\mathbf{p}_*) \geq 0, \tag{5.20}$$

$$\pi_1^1(p_1^*, \bar{\boldsymbol{\psi}}_1(\mathbf{p}^*)) + \sum_{l=1}^{\infty} \alpha_1^l \sum_{j=2}^{n} \pi_1^j(p_1^*, \bar{\boldsymbol{\psi}}_1(\mathbf{p}^*))M_{1,j,l}(\mathbf{p}^*)$$

$$= \pi_1^1(\mathbf{p}^*) + \sum_{l=1}^{\infty} \alpha_1^l \sum_{j=2}^{n} \pi_1^j(\mathbf{p}^*)M_{1,j,l}(\mathbf{p}^*) \leq 0, \tag{5.21}$$

and, for any $\mathbf{p}^1, \mathbf{p}^2 \in \mathscr{D}_1^*$ *with* $\mathbf{p}^1 \geq \mathbf{p}^2$,

$$\pi_1^j(\mathbf{p}^1) \geq \pi_1^j(\mathbf{p}^2), \qquad j = 2, \ldots, n, \tag{5.22}$$

$$\psi_j^k(\mathbf{p}^1) \geq \psi_j^k(\mathbf{p}^2), \qquad j = 2, \ldots, n \text{ and } k = 1, \ldots, n. \tag{5.23}$$

A14. *For all* $\mathbf{p} \in \mathscr{D}$, $\Sigma_{j=2}^n |G_{1s}^j(\mathbf{p})|/G_{1s}^1(\mathbf{p})$ *is bounded in absolute value independently of s.*

A15. *The bounds of A14 are sufficient to insure that the Lipschitz condition obeyed by the* ϕ_{1s} *is the same as is obeyed by* $\bar{\boldsymbol{\psi}}_1$, *namely,* $\lambda < 1$. *The first partial derivatives of* π_1 *are bounded.*

A16. *For all* $\mathbf{p} \in \mathscr{D}$ *and* $s \geq 1$, $G_{is}^j(\mathbf{p}) \geq 0$, $j \neq i$, $i = 1, \ldots, n$.

A17. $\mathscr{D}^* = \cap_{i=1}^n \mathscr{D}_i^*$ *has a nonempty interior which contains* $\mathbf{p}^c$.

E1. *The demand function,* $q_t = F(\mathbf{p}_t)$, *is defined, nonnegative and con-*

tinuous on the interval $[0, \infty)$. For a positive, finite p^+, $F(p) = 0$ when $p \geq p^+$. For $p \in (0, p^+)$, $F(p)$ is bounded and twice differentiable with $F' < 0$.

E2. *$C(q_t, K_t^*)$ is defined and continuous for $q_t \in [0, F(0)]$ and $K_t^* \in [0, \infty)$, and has continuous second partial derivatives on the interior of its domain. $C(0, 0) = 0$, $C^1(q, K^*) > 0$, $C^2(Q, K^*) < 0$, $C^{11}(q, K^*) \geq 0$ and $C^{12}(q, K^*) < 0$. There is a minimal $\underline{K}^* < \infty$ such that $C^1(0, \underline{K}^*) \leq p^+$.*

E3. *$\pi(p, K^*)$ is concave with respect to $(p, K^*) \in \mathcal{D}$. The profit function $\pi(p, K^*)$ is the rate of flow of profit per unit of time, not taking into account the cost of acquiring capital.*

E4. *The discounted profits of the firm are a quasi-concave function of the firm's own strategy, $(\mathbf{t}_i, \mathbf{p}_i, \mathbf{K}_i)$.*

Part II

Chapters 7, 8

G1. *n, the number of players, is finite.*

G2. *$\mathcal{S}_i$, the strategy set of the ith player, is a compact convex subset of $\mathcal{R}^m$, $i = 1, \ldots, n$.*

G3. *$P_i(s)$, the payoff function of the ith player, is a scalar valued function defined for all $s \in \mathcal{S}$ which is continuous and bounded everywhere, $i = 1, \ldots, n$.*

G4. *$P_i(s)$ is strictly quasi-concave with respect to s_i, $i = 1, \ldots, n$.*

G5. *$P_i(s)$ is continuous and bounded on $\mathcal{T}_i \subseteq \mathcal{S}$. For any $\bar{s}_i \in \bar{\mathcal{S}}_i$ there is at least one $s_i \in \mathcal{S}_i$ such that $(s_i, \bar{s}_i) \in \mathcal{T}_i$. $\mathcal{T}_i$ is compact and convex, $(i = 1, \ldots, n)$.*

G6. *$P_i(s)$ is quasi-concave with respect to s_i, $i = 1, \ldots, n$.*

G7. *$\mathcal{S}_i$, the strategy set of the ith player, is a closed, convex and bounded subset of a Banach space $\mathcal{E}$ (for $i = 1, \ldots, n$).*

G8. *There is* $s^+ \in \mathscr{E}$ *such that for any* s, $s' \in \mathscr{S}_i$, $|s_j - s'_j| \leq s_j^+$ *and* $\|s^+\| < \infty$, $i = 1, \ldots, n$.

G9. *If* $\mathbf{P}' \ll \mathbf{P}''$ *and* $\mathbf{P}'$, $\mathbf{PP}'' \in \mathscr{H}$, *then for* $\mathbf{P}' \leq \mathbf{P} \leq \mathbf{P}''$, $\mathbf{P} \in \mathscr{H}$.

G10. $\mathscr{H}^*$ *is a concave surface. That is, if* $\mathbf{P}'$, $\mathbf{P}'' \in \mathscr{H}^*$ *and* $\mathbf{P}^\lambda = \lambda \mathbf{P}' + (1 - \lambda)\mathbf{P}''$, *there is* $\mathbf{P}^0 \in \mathscr{H}^*$ *such that* $\mathbf{P}^0 \geq \mathbf{P}^\lambda$.

G11. *The correspondence* $\mathscr{P}^{-1}$ *from* $\mathscr{H}^*$ *into* $\mathscr{S}$ *is lower semicontinuous with convex image sets.*

G12. *The constituent game payoff functions,* P_{it}, *are bounded on* $\mathscr{S}_t$ *independently of t.* $(i = 1, \ldots, n)$.

G13. *The discount parameters are declining over time and have a bounded sum.* $\alpha_{i1} = 1$, $\alpha_{it} \geq \alpha_{i,t+1} > 0$, $t = 1, 2, \ldots$ *and* $\Sigma_{t=1}^{\infty} \alpha_{it} < \infty$.

Chapters 9, 10

S1. *There are a finite number of players, n.*

S2. $\mathscr{S}_{it}$, *the strategy set of the ith player in the tth time period, is a compact, convex subset of* $\mathscr{R}^m$, $i = 1, \ldots, n$, $t = 1, 2, \ldots$ *The strategy sets are bounded independently of t. Thus* $\|\mathscr{S}_{it}\| \leq \bar{m} < \infty$, $i = 1, \ldots, n$, $t = 1, 2, \ldots$.

S3. $P_{it}(s_t, s_{t-1})$ *is continuous on* $\mathscr{S}_t \times \mathscr{S}_{t-1}$. *For any* $(s_t, s_{t-1}) \in \mathscr{S}_t \times \mathscr{S}_{t-1}$, $|P_{it}(s_t, s_{t-1})| \leq \bar{P} < \infty$, $i = 1, \ldots, n$, $t = 1, 2, \ldots$

S4. *The payoff functions* $P_{it}(s_{it}, \bar{s}_{it}, s_{i,t-1}, \bar{s}_{i,t-1})$ *are concave with respect to* $(s_{it}, s_{i,t-1}) \in \mathscr{S}_{it} \times \mathscr{S}_{i,t-1}$, $i = 1, \ldots, n$, $t = 1, 2, \ldots$.

S5. *The discount parameters of the player are declining over time and bounded in sum.* $1 = \alpha_{i1} > \alpha_{it} > \alpha_{i,t+1}$, $t = 2, 3, \ldots$ $\Sigma_{t=1}^{\infty} \alpha_{it} < \infty$, $i = 1, \ldots, n$.

S6. $\boldsymbol{\phi}$ *is a continuous function which satisfies the following Lipschitz condition:*

$$\|\boldsymbol{\phi}(s_{t+1}, s_{t-1}) - \boldsymbol{\phi}(s'_{t+1}, s'_{t-1})\| \leq k_1 \|s_{t+1} - s'_{t+1}\| + k_2 \|s_{t-1}, s'_{t-1}\|,$$

where $k_1 + k_2 \leq k < 1$. (9.6)

S7. *$P^*_{it}(s_{t+1}, s_t, s_{t-1})$ is a quasi-concave function of s_{it}.*

S8. *Each correspondence ϕ_t has a selection f_t which obeys the following Lipschitz conditions:*

$$\|f_t(s'_{t+1}, s'_{t-1}) - f_t(s'_{t+1}, s''_{t-1})\| \le k_{t-}\|s'_{t-1} - s''_{t-1}\|, \tag{9.33}$$

$$\|f_t(s'_{t+1}, s'_{t-1}) - f_t(s''_{t+1}, s'_{t-1})\| \le k_{t+}\|s'_{t+1} - s''_{t+1}\|, \tag{9.34}$$

with $k_{t+} + k_{t-} < 1$ for any $s'_{t+1}, s''_{t+1} \in \mathcal{S}_{t+1}$, any $s'_{t-1}, s''_{t-1} \in \mathcal{S}_{t-1}$ and $t = 2, 3, \ldots$. For $t = 1$,

$$\|f_1(s'_2, s_0) - f_1(s''_2, s_0)\| < \|s'_2 - s''_2\|. \tag{9.35}$$

S9. *For any $(\rho_1, \ldots, \rho_n) > 0$ with $\Sigma_{i=1}^n \rho_i = 1$, and any $s_{t-1} \in \mathcal{S}_t$, there is $s_t \in \mathcal{T}(s_{t-1})$ such that*

$$[P_i(s_t, s_{t-1}) - P_i(s^c_t, s_{t-1})]/\sum_{j=1}^{n} [P_j(s_t, s_{t-1}) - P_j(s^c_t, s_{t-1})] = \rho_i, \qquad i = 1, \ldots, n, \tag{9.47}$$

and $\mathbf{P}(s_t, s_{t-1}) \in \mathcal{H}(s_{t-1})$.

S10. *$\mathcal{S}_{ik} \subset \mathcal{R}^m$ is a compact, convex set for $i = 1, \ldots, n$ and $k \in \mathbf{\Omega}$.*

S11. *The payoff functions, $P_{ik}(s_k)$ are continuous on $\mathcal{S}_k$. They are bounded by $|P_{ik}(s_k)| \le \bar{P} < \infty$. $i = 1, \ldots, n$, $k \in \mathbf{\Omega}$.*

S12. *$P_{ik}(s_{ik}, \bar{s}_{ik})$ is a concave function of s_{ik}. $i = 1, \ldots, n$, $k \in \mathbf{\Omega}$.*

S13. *$q(k'|k, s_k) \ge 0$, $\Sigma_{k' \in \Omega}\, q(k'|k, s_k) = 1$ for $k \in \mathbf{\Omega}$ and $s_k \in \mathcal{S}_k$. For fixed k, $\lambda \in [0, 1]$ and s_k, $s'_k \in \mathcal{S}_k$, $q(k'|k, \lambda s_k + (1-\lambda)s'_k) = \lambda q(k'|k, s_k) + (1-\lambda)q(k'|k, s'_k)$.*

Chapters 11, 12

C1. *The ith player has a strategy set $\mathcal{S}_i$ which is a compact and convex subset of $\mathcal{R}^m$, $i = 1, 2$.*

C2. *There is a distinguished strategy vector s^T, called the threat strategy, such that $P_j(s^T_i, \bar{s}_i) = P_j(s^T)$ for all $\bar{s}_i \in \bar{\mathcal{S}}_i$ and for $i, j = 1, 2$,*

C3. *$\mathcal{H}$ is compact and convex.*

C4. *$\mathcal{S}_i$, the strategy set of the ith player, is the unit simplex in $\mathcal{R}^{m_i-1}$, i = 1, 2.*

C5. *The payoff function of the ith player, $P_i(s_1, s_2)$, is bounded and linear in s_1 and s_2, i = 1, 2.*

C6. *The payoff function of the ith player, $P_i(s)$, is continuous and bounded, i = 1, 2.*

C7. *For any $\mathbf{P} \in \mathcal{H}$, $\mathbf{P}^* \in \mathcal{H}^*$, $\mathbf{P}^* > \mathbf{P}$ and $\lambda \in (0, 1)$, $\lambda \mathbf{P} + (1-\lambda)\mathbf{P}^* \in \mathcal{H}$.*

C8. *For all **P** for which ρ is defined,*

$$P_2(s) - \rho(\mathbf{P}(s))P_1(s),$$

is concave in s_2 and convex in s_1.

C9. *The number of players, n, is finite.*

C10. *$v(\mathbf{0}) = 0$.*

C11. *$v(\mathcal{K}) \geq v(\mathcal{K} \cap \mathcal{L}) + v(\mathcal{K} - \mathcal{L})$ for all $\mathcal{K}, \mathcal{L} \subseteq \mathcal{N}$.*

C12. *The strategy set of the ith player is the $m_i - 1$ dimensional (closed) unit simplex. The joint strategy set of the coalition $\mathcal{K}$ is the $\Pi_{i \in \mathcal{K}}\, m_i - 1 = m_{\mathcal{K}} - 1$ dimensional unit simplex (i = 1, . . . , n and $\mathcal{K} \subseteq \mathcal{N}$).*

C13. *Let $\mathcal{K}_1, \ldots, \mathcal{K}_r$ be an arbitrary partition of $\mathcal{N}$ into r coalitions. The payoff function $P_i(s_{\mathcal{K}_1}, \ldots, s_{\mathcal{K}_r})$ is bounded and linear in each of the $s_{\mathcal{K}_j}$ (i = 1, . . . , r).*

C14. *Any player can reduce any payoff of his own by any amount with no effect on the payoffs of the others.*

C15. *The set $\underline{\mathcal{V}}_{\mathcal{K}} = \{\mathbf{P}_{\mathcal{K}} | \mathbf{P}_{\mathcal{K}} \in \mathcal{V}_{\mathcal{K}}, \mathbf{P}_{\mathcal{K}} \geq \underline{\mathbf{P}}_{\mathcal{K}}\}$ is nonempty and compact, for all $\mathcal{K} \subseteq \mathcal{N}$.*

C16. *For finite $r_{\mathcal{K}}$ there are $\mathbf{P}^1_{\mathcal{K}}, \mathbf{P}^2_{\mathcal{K}}, \ldots, \mathbf{P}^{r_{\mathcal{K}}}_{\mathcal{K}} \in \underline{\mathcal{V}}_{\mathcal{K}}$ such that*

$$\underline{\mathcal{V}}_{\mathcal{K}} = \{\mathbf{P}_{\mathcal{K}} | \underline{\mathbf{P}}_{\mathcal{K}} \leq \mathbf{P}_{\mathcal{K}} \leq \mathbf{P}^r_{\mathcal{K}} \text{ for at least one } r\,(=1,\ldots,r_{\mathcal{K}})\}$$
$$\text{for all } \mathcal{K} \subset \mathcal{N},\ k \geq 2 \tag{12.3}$$

C17. *All the x_j associated with the n columns of a feasible basis are greater than zero and unique.*

C18. *No two elements in the same row of* **B** *are equal.*

C19. *$P_i(s)$ is bounded and quasi-concave in s, $i = 1, \ldots, n$.*

Nash axioms

N1. *The solution should be unique.*

N2. *Let two games be given by $\mathcal{H}_1$ and $\mathcal{H}_2$. If $\mathcal{H}_1$ can be transformed into $\mathcal{H}_2$ by a positive linear transformation of P_1 and another of P_2, then the solutions to the two games are related by the same pair of transformations.*

N3. *The solution should be an undominated point $\mathbf{P}^*$ of $\mathcal{H}$, i.e. there should be no element* **P** *of $\mathcal{H}$ such that $\mathbf{P} > \mathbf{P}^*$.*

N4. *Let $\mathcal{H}_1$ and $\mathcal{H}_2$ be two games having the same threat point. If $\mathcal{H}_1 \subseteq \mathcal{H}_2$ and the solution of $\mathcal{H}_2$ is an element of $\mathcal{H}_1$, then both games have the same solution.*

N5. *If the set $\mathcal{H}$ is symmetric about a 45° line through the threat point, then the solution lies on that line.*

Shapley axioms

Sh1. *For each $\boldsymbol{\pi}$, $\phi_{\pi i}(\boldsymbol{\pi} v) = \phi_i(v)$, for all $i \in N$.*

Sh2. $\Sigma_{i \in \mathcal{N}}\, \phi_i(v) = v(\mathcal{N})$.

Sh3. *For any two games v and w having the same set of players, $\mathcal{N}$. $\boldsymbol{\phi}(v + w) = \boldsymbol{\phi}(v) + \boldsymbol{\phi}(w)$.*

REFERENCES

Arrow, Kenneth J. and Marc Nerlove, 1962, Optimal advertising policy under dynamic conditions, Economica N.S. 29, 129–142.

Aumann, Robert J., 1959, Acceptable points in general cooperative n-person games, in: Tucker and Luce (1959, 287–324).

Aumann, Robert J., 1960, Acceptable points in games of perfect information, Pacific Journal of Mathematics 10, 381–417.

Aumann, Robert J., 1961, The core of a cooperative game without side payments, Transactions of the American Mathematical Society 98, 539–552.

Aumann, Robert J., 1967, A survey of cooperative games without side payments, in: Shubik (1967, 3–27).

Aumann, Robert J., 1974, Subjectivity and correlation in randomized strategies, Journal of Mathematical Economics 1, 67–96.

Aumann, Robert J. and Bezalel Peleg, 1960, von Neumann–Morgenstern solutions to cooperative games without side payments, Bulletin of the American Mathematical Society 66, 173–179.

Bartle, Robert G., 1964, The elements of real analysis (John Wiley, New York).

Berge, Claude, 1963, Topological spaces (Macmillan, New York).

Bertrand, Joseph, 1883, Review of Théorie mathématique de la richesse sociale and Recherches sur les principes mathématiques de la théorie des richesses, Journal des Savants, 499–508.

Billera, Louis J., 1970, Some theorems on the core of an n person game, SIAM Journal of Applied Mathematics 18, 567–579.

Blackwell, David, 1965, Discounted dynamic programming, Annals of Mathematical Statistics 36, 226–235.

Bliss, C. J., 1975, Capital theory and the distribution of income (North-Holland, Amsterdam).

Bohnenblust, H. F. and S. Karlin, 1950, on a theorem of Ville, in: Kuhn and Tucker (1950).

Borel, Émile, 1953a, The theory of play and integral equations with skew symmetric kernels, Econometrica 21, 97–100.

Borel, Émile, 1953b, On games that involve chance and the skill of the players, Econometrica 21, 101–115.

Borel, Émile, 1953c, On systems of linear forms of skew symmetric determinant and the general theory of play, Econometrica 21, 116–117.

Bowley, Arthur L., 1924, The mathematical groundwork of economics (Oxford University Press, Oxford).

Braithwaite, Dorothea, 1928, The economic effects of advertising, Economic Journal 38, 16–37.

Brems, Hans, 1951, Product equilibrium under monopolistic competition (Harvard University Press, Cambridge).

Chamberlin, Edward H., 1956, The theory of monopolistic competition, 7th ed. (Harvard University Press, Cambridge).
Choquet, Gustave, 1966, Topology, (Academic Press, New York).
Cournot, Augustin, 1838, Recherches sur les principes mathématiques de la théorie des richesses (Hachette, Paris).
Cournot, Augustin, 1960, Researches into the mathematical principles of the theory of wealth, English edition of Cournot (1838), translated by Nathaniel T. Bacon (Kelley, New York).
Cyert, Richard M. and Morris de Groot, 1970, Multiperiod decision models with alternating choice as a solution to the duopoly problem, Quarterly Journal of Economics 84, 410–429.
Dantzig, George, 1963, Linear programming and extensions (Princeton, Princeton).
Debreu, Gerard, 1952, A social equilibrium existence theorem, Proceedings of the National Academy of Science 38, 886–893.
Debreu, Gerard and Herbert Scarf, 1972, The limit of the core of an economy, in: McGuire and Radner (1972, 283–295).
Denardo, Eric V., 1971, Markov renewal programs with small interest rates, Annals of Mathematical Statistics 42, 477–496.
Dieudonné, Jean, 1960, Foundations of modern analysis (Academic Press, New York).
Downs, Anthony, 1957, An economic theory of democracy (Harper and Row, New York).
Dunford, Nelson and Jacob T. Schwartz, 1957, Linear operators part I: General theory (Interscience, New York).
Edgeworth, Francis Y., 1881, Mathematical psychics (Kegan Paul, London).
Edgeworth, Francis Y., 1925, The pure theory of monopoly, in: Edgeworth, Francis Y., Papers relating to political economy, vol. 1 (Macmillan, London) 111–142.
Fan, Ky, 1952, Fixed-point and minimax theorems in locally convex topological linear spaces, Proceedings of the National Academy of Sciences 38, 121–126.
Farrell, Michael J., 1970, Edgeworth bounds for oligopoly prices, Economica 37, 342–361.
Fellner, William J., 1949, Competition among the few (Knopf, New York).
Fenchel, Werner, 1953, Convex cones, sets and functions, mimeographed notes.
Fisher, Franklin M., 1961, The stability of the Cournot oligopoly solution: The effects of speeds of adjustment and increasing marginal costs, Review of Economic Studies 28, 125–135.
Frank, Charles R. and Richard E. Quandt, 1963, On the existence of Cournot equilibrium, International Economic Review 4, 92–96.
Fréchet, Maurice, 1953, Commentary on the three notes of Émile Borel, Econometrica 21, 118–124.
Friedman, James W., 1963, Individual behavior in oligopolistic markets: An experimental study, Yale Economic Essays 3, 359–417.
Friedman, James W., 1968, Reaction functions and the theory of duopoly, Review of Economic Studies 35, 257–272.
Friedman, James W., 1971a, A non-cooperative equilibrium for supergames, Review of Economic Studies 38, 1–12.
Friedman, James W., 1971b, A noncooperative view of oligopoly, International Economic Review 12, 106–122.
Friedman, James W., 1972, On the structure of oligopoly models with differentiated products, in: Sauermann, Heinz, Contributions to experimental economics, vol. 3 (J. C. B. Mohr, Tubingen) 28–63.
Friedman, James W., 1973a, A non-cooperative equilibrium for supergames: A correction, Review of Economic Studies 40, 435.
Friedman, James W., 1973b, On reaction function equilibria, International Economic Review 14, 721–734.
Friedman, James W., 1974, Non-cooperative equilibria in time-dependent supergames, Econometrica 42, 221–237.

Friedman, James W., 1976a, Reaction functions as Nash equilibria, Review of Economic Studies 43, 83–90.

Friedman, James W., 1976b, Cournot, Bowley, Stackelberg and Fellner, and the evolution of the reaction function, in: Balassa, Bela, and Richard R. Nelson, Economic progress, private values and public policy: Essays in honor of William Fellner, (North-Holland, Amsterdam).

Gabszewicz, Jean Jaskold and Jean-Philippe Vial, 1972, Oligopoly "à la Cournot" in a general equilibrium analysis, Journal of Economic Theory 4, 381–400.

Gass, Saul I., 1975, Linear Programming 4th ed., (McGraw-Hill, New York).

Gillies, Donald B., 1953, Some theorems on n person games, Ph.D. thesis (Department of Mathematics, Princeton University).

Graves, L. M., 1946, The theory of functions of real variables (McGraw-Hill, New York).

Hadar, Josef, 1966, Stability of oligopoly with product differentiation, Review of Economic Studies 33, 57–60.

Hahn, Frank H., 1962, The stability of the Cournot oligopoly solution, Review of Economic Studies 29, 329–331.

Harsanyi, John C., 1963, A simplified bargaining model for the n-person cooperative game, International Economic Review 4, 194–220.

Harsanyi, John C., 1967, Games with incomplete information played by "Bayesian" players, part I: The basic model, Management Science 14, 159–182.

Harsanyi, John C., 1968a, Games with incomplete information played by "Bayesian" players, part II: Bayesian equilibrium points, Management Science 14, 320–334.

Harsanyi, John C., 1968b, Games with incomplete information played by "Bayesian" players, part III: The basic probability distribution of the game, Management Science 14, 486–502.

Harsanyi, John C., 1974, The tracing procedure: A Bayesian approach to defining a solution for n-person noncooperative games. Part I, Working paper #15 (Institut für Mathematische Wirtschaftsforschung, Universität Bielefeld).

Harsanyi, John C. and Reinhard Selten, 1972, A generalized Nash solution for two-person bargaining games with incomplete information, Management Science 18, P-80–P-106.

Henderson, Alexander M., 1954, The theory of duopoly, edited after the author's death by F. Modigliani and H. A. Simon, Quarterly Journal of Economics 68, 565–584.

Herfindahl, Orris C., 1959, Copper costs and prices: 1870–1957 (Johns Hopkins Press, Baltimore).

Hoos, Sidney, 1959, The advertising and promotion of farm products: Some theoretical issues, Journal of Farm Economics 38, 1679–1691.

Hotelling, Harold, 1929, Stability in competition, Economic Journal 39, 41–57.

Howard, Ronald, 1960, Dynamic programming and Markov processes (M.I.T. Press, Cambridge).

Kakutani, Shizuo, 1941, A generalization of Brouwer's fixed point theorem, Duke Mathematical Journal 8, 457–459.

Kirman, Alan P. and Matthew J. Sobel, 1974, Dynamic oligopoly with inventories, Econometrica 42, 279–287.

Kuhn, Harold W. and Albert W. Tucker, 1950, Contributions to the theory of games, vol. I (Princeton, Princeton).

Kuhn, Harold W. and Albert W. Tucker, 1953, Contributions to the theory of games, Vol. II (Princeton, Princeton).

Lancaster, Kelvin, 1976, Socially optimal product differentiation, American Economic Review 66, forthcoming.

Launhardt, Wilhelm, 1885, Mathematische begründung der volkswirthschaftslehre (Leipsig).

Lemke, C. E., 1965, Bimatrix equilibrium points and mathematical programming, Management Science 11, 681–689.

Lemke, C. E. and J. T. Howson, 1964, Equilibrium points of bimatrix games, SIAM Journal of Applied Mathematics 12, 413–423.
Luce, R. Duncan and Howard Raiffa, 1957, Games and decisions (Wiley, New York).
McGuire, C. B., and Roy Radner, eds., 1972, Decision and organization (North-Holland, Amsterdam).
McKenzie, Lionel, 1960, Matrices with dominant diagonals and economic theory, in: Arrow, Kenneth J., Samuel Karlin and Patrick Suppes, eds., Mathematical methods in the social sciences, 1959 (Stanford University Press, Stanford) 47–62.
Malinvaud, Edmond, 1972, Lectures on microeconomic theory (North-Holland, Amsterdam and London).
Marschak, Thomas and Reinhard Selten, 1974, General equilibrium with price-making firms (Springer, New York).
Marshall, Alfred, 1922, Principles of economics, 8th ed. (Macmillan, London).
Marshall, Alfred, 1975, The early economic writings of Alfred Marshall, 1867–1890, edited and introduced by J. K. Whitaker (Macmillan, London).
Marshall, Alfred, 1961, Principles of economics, ninth (variorum) edition, with annotations by C. W. Guillebaud (Macmillan, London).
Mayberry, J. P., John F. Nash and Martin Shubik, 1953, A comparison of treatments of a duopoly situation, Econometrica 21, 141–154.
Michael, E., 1956, continuous selections I, Annals of Mathematics 63, 361–382.
Nash, John F., Jr., 1950, The bargaining problem, Econometrica 18, 155–162.
Nash, John F., Jr., 1951, Non-cooperative games, Annals of Mathematics 45, 286–295.
Nash, John F., Jr., 1953, Two-person cooperative games, Econometrica 21, 128–140.
Negishi, Takashi, 1961, Monopolistic competition and general equilibrium theory, Review of Economic Studies 28, 196–201.
Negishi, Takashi, and Koji Okuguchi, 1972, A model of duopoly with Stackelberg equilibrium, Zeitschrift fur Nationalokonomie 32, 153–162.
Nikaido, Hukukane, 1975, Monopolistic competition and effective demand (Princeton, Princeton).
Okuguchi, Koji, 1964, The stability of the Cournot solution: a further generalization, Review of Economic Studies 31, 143–146.
Okuguchi, Koji, 1969, On the stability of price adjusting oligopoly equilibrium under product differentiation, Southern Economic Journal 35, 244–246.
Okuguchi, Koji, 1973, Quasi-competitiveness and Cournot oligopoly, Review of Economic Studies 40, 145–148.
Owen, Guillermo, 1968, Game theory (Saunders, Philadelphia).
Owen, Guillermo, 1972, Values of games without side payments, International Journal of Game Theory 1, 95–109.
Parthasarathy, T., 1972, Selection theorems and their applications (Springer, Berlin).
Rapoport, Anatol, and Albert M. Chammah, 1965, Prisoner's dilemma, with the collaboration of Carol J. Orwant (University of Michigan Press, Ann Arbor).
Riker, William H. and Peter C. Ordeshook, 1973, An introduction to positive political theory (Prentice Hall, Englewood Cliffs).
Roemer, John, 1970, A Cournot duopoly problem, International Economic Review 11, 548–552.
Rogers, Philip D., 1969, Nonzero-sum stochastic games, # ORC 69–8 (Operations Research Center, University of California, Berkeley).
Ruffin, Roy, 1971, Cournot oligopoly and competitive behavior, Review of Economic Studies 38, 493–502.
Scarf, Herbert E., 1967, The core of an n person game, Econometrica 35, 50–69.
Scarf, Herbert E., 1971, On the existence of a cooperative solution for a general class of n-person games, Journal of Economic Theory 3, 169–181.
Scarf, Herbert E., 1973, The computation of economic equilibria (Yale, New Haven).

Schneider, Erich, 1962, Pricing and equilibrium: An introduction to static and dynamic analysis, English version by Esra Bennathan (Macmillan, New York).
Schumpeter, Joseph A., 1954, History of economic analysis, edited from manuscript by Elizabeth Boody Schumpeter (Oxford, New York).
Selten, Reinhard, 1965, Spieltheoretische behandlung eines oligopolmodells mit nachfragetragheit, Zeitschrift fur die gesamte Staatswissenschaft 121: Teil I, Bestimmung des dynamischen preisgleichgewichts, 301–324; Teil II, Eigenschaften des dynamischen preisgleichgewichts, 667–689.
Selten, Reinhard, 1968, An oligopoly model with demand inertia, Working Paper No. 250 (Center for Research in Management Science, University of California, Berkeley).
Selten, Reinhard, 1974, The chain store paradox, working paper (Institut für Mathematische Wirtschaftsforschung, Universität Bielefeld).
Shapley, Lloyd, 1953a, Stochastic games, Proceedings of the National Academy of Science 39, 1095–1100.
Shapley, Lloyd, 1953b, A value for *n*-person games, in: Kuhn and Tucker (1953, 307–317).
Shapley, Lloyd S., 1969, Utility comparison and the theory of games, in: Guilbaud, G. Th., La decision, aggregation et dynamique, Colloques Internationaux du Centre de la Recherche Scientifique No. 171, Editions C.N.R.S.
Shapley, Lloyd S., 1972, On balanced games without side payments, in: Hu, T. C. and Stephen M. Robinson, Mathematical programming (Academic Press, New York) 261–290.
Shapley, Lloyd, 1973, Let's block "block," Econometrica 41, 1201–1202.
Shapley, Lloyd and Martin Shubik, 1969, Price strategy oligopoly with product variation, Kyklos 22, 30–44.
Shitovitz, Benyamin, 1973, Oligopoly in markets with a continuum of traders, Econometrica 41, 467–501.
Shubik, Martin, 1959, Strategy and market structure: competition, oligopoly, and the theory of games (John Wiley, New York).
Shubik, Martin, 1967, Essays in mathematical economics in honor of Oskar Morgenstern (Princeton, Princeton).
Shubik, Martin, 1970, Price strategy oligopoly: Limiting behavior with product differentiation, Western Economic Journal 9, 226–232.
Smith, Adam, 1937, The wealth of nations, 5th ed., edited by Edwin Cannan (Modern Library, Random House, New York).
Sobel, Matthew J., 1971, Noncooperative stochastic games, Annals of Mathematical Statistics 42, 1930–1935.
Sobel, Matthew J., 1973, Continuous stochastic games, Journal of Applied Probability 10, 597–604.
Sraffa, Piero, 1926, The laws of return under cooperative conditions, Economic Journal 36, 535–550.
Stackelberg, Heinrich von, 1934, Marktform und Gleichgewicht (Julius Springer, Vienna).
Stigler, George and Kenneth E. Boulding, 1952, Readings in price theory (Richard D. Irwin, Homewood).
Telser, Lester G., 1972, Competition, collusion, and game theory (Aldine, Chicago).
Theocharis, R. D., 1960, On the stability of the Cournot solution on the oligopoly problem, Review of Economic Studies 27, 133–134.
Triffin, Robert, 1940, Monopolistic competition and general equilibrium theory (Harvard University Press, Cambridge).
Tucker, Albert W. and R. Duncan Luce, 1959, Contributions to the theory of games, vol. IV (Princeton, Princeton).
von Neumann, John, 1928, Zur theorie der Gesellschaftsspiele, Math. Annalen 100, 295–320.
von Neumann, John, 1959, On the theory of games of strategy, English version of von Neumann (1928), translated by Sonya Bargmann, in: Tucker and Luce (1959, 13–42).

von Neumann, John, and Oskar Morgenstern, 1944, Theory of games and economic behavior (Princeton University Press, Princeton).

Walras, Léon, 1954, Elements of pure economics: Or the theory of social wealth, English edition of the Edition Définitive (1926), translated by William Jaffé (Richard D. Irwin, Homewood).

Zabel, Edward, 1972, Multiperiod monopoly under uncertainty, Journal of Economic Theory 5, 524–536.

AUTHOR INDEX

SUBJECT INDEX

ADVANCED TEXTBOOKS IN ECONOMICS

Edited by **C. J. BLISS** and **M. D. INTRILIGATOR**

Vol. 1: **HADLEY, G.** and **M. C. KEMP,** *Variational Methods in Economics*, 1971.

Vol. 2: **MALINVAUD, E.,** *Lectures on Microeconomic Theory*, 1972, 2nd repr. 1974.

Vol. 3: **HEAL, G. M.,** *The Theory of Economic Planning*, 1973.

Vol. 4: **BLISS, C. J.,** *Capital Theory and the Distribution of Income*, 1975.

Vol. 5: **PHLIPS, L.,** *Applied Consumption Analysis*, 1974.

Vol. 6: **HILDENBRAND, W.** and **A. P. KIRMAN,** *Introduction to Equilibrium Analysis*, 1976.

Vol. 7: **MADANSKY, A.,** *Foundations of Economics*, 1976.

Vol. 8: **FRIEDMAN, J. W.,** *Oligopoly and the Theory of Games*, 1977.

Vol. 9: **INTRILIGATOR, M. D.,** *Econometric Models, Techniques and Applications*, in preparation (a co-production with Prentice-Hall).